Functionalized Polymers
and Their Applications

Functionalized Polymers and Their Applications

A. Akelah

Tanta University, Eygpt

and

A. Moet

Case Western Reserve University, USA

CHAPMAN AND HALL

LONDON • NEW YORK • TOKYO • MELBOURNE • MADRAS

UK   Chapman and Hall, 11 New Fetter Lane, London EC4P 4EE

USA   Chapman and Hall, 29 West 35th Street, New York NY10001

JAPAN   Chapman and Hall Japan, Thomson Publishing Japan
            Hirakawacho Nemoto Building, 7F, 1-7-11 Hirakawa-cho,
            Chiyoda-ku, Tokyo 102

AUSTRALIA   Chapman and Hall Australia, Thomas Nelson Australia,
            480 La Trobe Street, PO Box 4725, Melbourne 3000

INDIA   Chapman and Hall India, R. Seshadri, 32 Second Main
            Road, CIT East, Madras 600 035

First edition 1990

© 1990 A. Akelah and A. Moet

Typeset in 10/12 Times by
Thomson Press (India) Ltd, New Delhi
Printed in Great Britain by
St. Edmundsbury Press, Bury St. Edmunds, Suffolk.

ISBN 0 412 30290 X

British Library Cataloguing in Publication Data
Akelah, A.
   Functionalized polymers and their applications.
   1. Polymers
   I. Title   II. Moet A.
   547.7
ISBN 0–412–30290–X

Library of Congress Cataloging-in-Publication Data
                  Available

CONTENTS

viii   Contents

PREFACE

Rapid progress in the synthesis and utilization of functionalized polymeric materials has been noted in the recent past. Interest in the field is being enhanced by the possibility of creating systems that combine the unique properties of conventional active moieties and those of high molecular weight polymers. The successful utilization of these polymers based on their specific active functional groups is wide ranging and opens up a field of great growth potential. In view of their applications in the laboratory and industrial process, the study of reactive functionalized polymers has been mainly directed towards their production. Limited attention, however, has been paid to their physicochemical characterization.

This book presents a comprehensive review of the broad spectrum of research activities currently being undertaken in the field of functionalized polymers and their significant applications requirements in health, nutrition, environmental pollution control and economic developments. The book is structured in four parts.

Part I is a general review of the chemistry of functionalized polymers and is divided into sections on their synthesis and physical properties. The first section gives the background knowledge of the techniques necessary for designing reactive polymers through polymerization. In addition, the basic framework is considered that underlies the chemistry of the reactions by which functionalized polymers can be synthesized by chemical modifications on pendant groups attached either to synthetic organic polymers or to naturally occurring polymers such as polysaccharides or inorganic supports. This section also includes an explanation of the potential advantages and disadvantages of using various techniques to change the physical nature of the polymers. Pertinent physical properties are considered in the second section. Such properties, which depend on the conditions employed in preparation, in addition to the physical nature of the polymer, must be considered during the design of a new reactive polymer. Furthermore, design criteria for specific purposes are emphasized and the polymer microstructure is related to its reactivity.

The remainder of the subject is divided into three more parts. Part II describes the contribution of functionalized polymers in solving problems associated with conventional procedures of chemical reactions. This part is further subdivided into four sections which consider a broad range of chemical applications, including the various types of functionalized polymers and in particular those used as reagents, catalysts and carriers in separation and synthesis.

Part III provides the basic principles necessary for the practical aspects of functionalized polymers with respect to a variety of biological fields such as the pharmaceutical, agriculture and food industries. It is divided into two sections and deals with the applications of functionalized polymers in the broad area of biological applications, especially in controlled release formulations of drugs and agrochemicals. The discussion also covers the functionalized polymers that have considerable potential as useful materials in food additives and other related fields.

The last part focuses on the applications of reactive functionalized polymers in various technological processes, including polymer stabilization, the conversion and storage of energy, conductive polymers and other processes.

A. Akelah
A. Moet

Part One   Chemistry of Reactive Polymers: Preparation and Properties

In an attempt to examine the utility of reactive polymeric materials, this part is concerned with fundamental and background knowledge of the techniques necessary for their design and characterization. A brief description of the synthesis of functionalized polymers either by polymerization or by chemical modification techniques as well as an explanation of potential advantages and disadvantages of each technique are given.

Effective functionalization of a polymer depends on its physical form, solvation behaviour, porosity, chemical reactivity and stability. These factors are crucial in the design of a new reactive polymer and can be influenced by the conditions employed during preparation.

1
Preparation of functionalized polymers

Functional polymers are macromolecules to which chemically functional groups are attached; they have the potential advantages of small molecules with the same functional groups. Their usefulness is related both to the functional groups and to their polymeric nature whose characteristic properties depend mainly on the extraordinarily large size of the molecules.

The attachment of functional groups to a polymer is frequently the first step towards the preparation of a functional polymer for a specific use. However, the proper choice of the polymer is an important factor for successful application. In addition to the synthetic aliphatic and aromatic polymers, a wide range of natural polymers have also been functionalized and used as reactive materials. Inorganic polymers have also been modified with reactive functional groups and used in processes requiring severe service conditions. In principle, the active groups may be part of the polymer backbone or linked to a side chain as a pendant group either directly or via a spacer group. A required active functional group can be introduced onto a polymeric support chain (a) by incorporation during the synthesis of the support itself through polymerization and copolymerization of monomers containing the desired functional groups, (b) chemical modification of a suitably non-functionalized preformed support matrix and (c) by a combination of (a) and (b). Each of the two approaches has its own advantages and disadvantages, and one approach may be preferred for the preparation of a particular functional polymer when the other would be totally impractical. The choice between the two ways to the synthesis of functionalized polymers depends mainly on the required chemical and physical properties of the support for a specific application. Usually the requirements of the individual system must be thoroughly examined in order to take full advantage of each of the preparative techniques.

1.1 POLYMERIZATION

The goal of the polymerization technique is to obtain polymers with specific structures and properties which generally require specialized reaction conditions. Functionalized polymers prepared by the synthesis and polymerization of functional monomers are basically of two types: polycondensation or chain polymerization polymers. The resulting polymer, either a homopolymer or a copolymer, can be used in the desired application as it is or after further modification. The polymerization technique has a number of advantages.

1. The resulting polymer is truly homogeneous with more uniform functionalization.
2. The structure of the required functional group can be ascertained by analysis of the monomers prior to polymerization.
3. The degree of functionalization desired depends on the intended application of the support. In some cases such as polymeric reagents, the polymer should be prepared with high loading to maximize the concentration of reagent and to avoid the use of large amounts of support materials. In contrast, in other cases such as in solid phase peptide synthesis, loadings must be limited to minimize the change in solubility characteristics as the peptide chain is extended. Hence it is possible to control the loading and distribution of functional groups within the support and achieve the desired degree of functionalization.
4. The supports not contaminated by traces of other functional groups remaining from prior chemical transformations.

The main disadvantages of this technique include the following.

1. The introduction of a functional group during polymerization requires an appropriately substituted monomer. A wide variety of monomers can be obtained from commercial sources, but the majority are synthesized. However, the synthesis of monomers with desired functional groups is often difficult and they are generally obtained in low yields as a result of multistep synthesis.
2. Monomers of high purity must be prepared to obtain relatively high molecular weight polymer.
3. Some reactive monomers lack the required stability and display incompatibility during polymerization.
4. The polymerization of functional monomers to polymers with optimum molecular weight and of a desirable sequence distribution and compositional homogeneity of the copolymers can be achieved sometimes, though with difficulty.
5. In some cases great difficulty is also encountered in the copolymerization. An additional disadvantage of this approach is the necessity of evaluating copolymerization parameters in order to obtain high yields and good physical properties with a satisfactory physical form of the desired copolymer.

1.1.1 Condensation polymerization

Polycondensation is characteristic of compounds containing functional groups and proceeds between pairs of functional groups with the liberation of small molecules as byproducts. As a result, the composition of the polymer formed differs from that of the starting materials. The functional groups may be associated with a single molecule,

$$n(A—R—B) \longrightarrow A—[—Z—R]_{n-1}—B + (n-1)C$$

C is the byproduct; Z is the group bonding the residues of reacted molecules. Alternatively, the functional group may be present in ring form: the two groups are present within the ring in condensed form and the byproduct is eliminated during formation of the ring. The compositions of the monomers and the polymers are essentially the same as in addition polymerization, but the stepwise nature and the rate of reaction are more characteristic of condensation polymerization reactions:

$$n\,R \overbrace{} Z \longrightarrow -[Z—R—]_n$$

The functional groups may be associated with two different molecules in which all the reacting molecules have at least two reactive groups:

$$n(A—R—A) + n(B—R—B) \longrightarrow A—[R—Z—R—]_{n-1}—B + (n-1)C$$

Polycondensation through functional groups always proceeds in a stepwise manner, i.e. in stages, and is a reversible reaction with equilibrium properties. Moreover, each interaction is chemically identical and it is possible to interrupt or continue the reaction at any time without affecting the reactivity of the present polymer chain. The type of condensation reaction depends upon the functionality of the reactants, and the reactivity of functional groups at the ends of the polymer chains is similar to that of the corresponding functional groups in the monomer molecules. Since most polycondensations are slow reactions, vigorous conditions such as low pressures and high temperatures are needed. Polycondensation of bifunctional compounds gives linear polymers while polycondensation of polyfunctional compounds results in crosslinked polymers. Polycondensation polymers may contain the reactive functional groups as a part of the polymer backbone or as pendant substituent.

Although the mechanical properties of condensation polymers are often superior to those exhibited by addition vinyl polymerization, little attention has been directed toward the introduction of functional groups by polycondensation using an appropriately substituted monomer.

1.1.2 Addition polymerization

Polymerization through multiple bonds may be regarded as simply the joining together of unsaturated molecules without the formation of byproducts in which there is no difference in the relative positions of the atoms in the monomer and the structural unit of the polymer, i.e.

$$n(CH_2{=}CH{-}R) \longrightarrow \underset{\underset{R}{|}}{-\!\!\!\!-}CH_2{-}CH\underset{n}{-\!\!\!\!-}$$

In principle, chain polymerization consists of three primary stages.

(i) Initiation. Initiation is the formation of an active centre by transferring the active state from the initiator to the double bond of the vinyl monomer, which becomes capable of starting the polymerization of the unreactive monomer. Depending on the nature of the active centre, chain polymerization may be of three different types: free radical, ionic (cationic or anionic) or coordination polymerization. However, the most widely used in the preparation of functionalized polymers is the free radical technique. It is generally limited to monomers possessing olefinic links. The initiation occurs through an initiator which decomposes into free radicals,

$$I{-}I \longrightarrow 2I^{\bullet}$$

and these radicals react with the monomer to give an activated species capable of starting the polymerization of the unreactive monomer M, i.e.

$$I^{\bullet} + M \longrightarrow I{-}M^{\bullet}$$

The dissociation of the initiator requires a high activation energy and not every radical formed leads to the starting of a chain; a fraction of the radicals can disappear through recombination or through reaction with atmospheric oxygen or other inhibitors. The rate of the initiation reaction depends on (a) the rate of the initiator decomposition reaction which in turn depends on the temperature and the nature of the solvent and (b) the stability of the radical formed.

Free radical initiation is commonly induced chemically or by radiation techniques. Chemical initiators are energy-rich compounds such as peroxides or azo compounds. Redox initiation is used to increase the rate of decomposition of the peroxide initiator at low temperatures by reducing the activation energy of the decomposition reaction, which can be achieved by adding reducing agents as activators. Such redox systems decrease the possibility of side reactions which may change the properties of the resulting polymers. Alternatively, free radicals may also be generated by light, radiation or heat which result in the monomer itself becoming excited, i.e. the electrons are moved from the ground state to new orbitals corresponding to a higher energy and then form a biradical which changes into polymeric monoradicals:

$$CH_2{=}CH{-}R \xrightarrow{hv} CH_2{\overset{*}{=}}CH{-}R \longrightarrow {^{\bullet}}CH_2{-}{^{\bullet}}CH{-}R$$

This technique is often useful in avoiding contamination of the polymer from catalyst residues and has been successfully used for modifying polymers by anchoring organic reagents to polymer surfaces through radiation grafting.

(ii) Propagation. Chain propagation is the rapid addition of the activated species $(I—M^•)$ to a new monomer unit to form high molecular weight polymer. This results in a change of π-bonds to σ-bonds with the liberation of heat due to the difference in energies between them. In contrast with initiation, the rate of the propagation reaction

$$I—M^• + nM \longrightarrow I\text{-}[M]_n M^•$$

is less temperature dependent and the degree of polymerization is proportional to the ratio of the rate of the propagation reaction and the rate of the termination reaction.

(iii) Termination. Termination is the disappearance of the active centre. Radical polymerization must always involve disappearance of the unpaired electron because radicals have a strong tendency to react with each other either by disproportionation leading to two chains, i.e.

$$2I\text{-}[M]_n M^• \longrightarrow I\text{-}[M]_n M + I\text{-}[M]_n M^•$$

or by recombination of two growing chains, i.e.

$$2I\text{-}[M]_n M^• \longrightarrow I\text{-}[M]_n M—M\text{-}[M]_n I$$

Obviously, chains terminated by recombination are larger than those terminated by disproportionation. Deactivation of the growing chain radical may also occur through chain transfer, i.e. by abstraction of an atom from another molecule such as the initiator, monomer, solvent, completed polymer chain, modifier or impurities, and thus the chain radical becomes saturated. The molecule from which the atom has been abstracted will then become a free radical and start a new chain. Accordingly, the rate of polymerization does not decrease but the molecular weight decreases.

(a) Techniques of free radical polymerization

The choice of the method by which a monomer is converted to a high molecular weight polymer depends to a marked extent on the nature of the monomer, the end use of the polymer, the molecular weight, the rate of polymerization and control of side reactions. Various techniques have been utilized in free radical polymerizations and each method has its own advantages and disadvantages [1–3]; commercial polymerization processes are described elsewhere [4–7] and may involve one of the following methods.

(i) Bulk (Mass) Polymerization [8,9]. Bulk polymerization is the simplest technique and consists of carrying out the reactions on the pure monomers alone,

with or without initiator, in the absence of solvents. Either (a) the polymer is soluble in the monomer and thus there is an increase in viscosity as the polymerization progresses or (b) the polymer is insoluble in the monomer and the formed polymer is precipitated without increase in solution viscosity. The major advantages of the technique are the high purity and the high molecular weight of the formed polymers due to the decreased possibility of chain transfer.

(ii) Solution polymerization [10]. In solution polymerization the monomer is diluted by inert solvent (which may or may not be a solvent for the polymer) to assist in dissipation of the exothermic heat of reaction and to facilitate the contact between the monomer and the initiator. If the reaction solvent is compatible with the polymer, then the polymer is formed in solution and can be isolated by addition of a non-solvent which causes precipitation. However, if the solvent is not compatible with the polymer, then the polymer precipitates as it is formed. The major disadvantage of the method is the possibility that the solvent acts as a chain transfer agent and hence leads to low molecular weight polymer.

(iii) Suspension polymerization [11–14]. Suspension polymerization is a well-established procedure which has proved to be perhaps the most useful technique for synthesizing crosslinked polymeric support because of the extremely convenient physical form of the bead product, i.e. the regular spherical shape and surface area. Crosslinked polymer beads of both a swelling and a non-swelling type can be produced by this technique. The initiator is dissolved in the liquid monomer or comonomer mixture which is suspended in an immiscible solvent. Usually, to suspend the monomer in small droplets in the non-solvent, mechanical stirring of the reaction mixture and the use of a suspension stabilizer are required. The solvent acts as a heat exchanger to remove the heat of polymerization from each of the monomer droplets. Hydrophobic monomers, such as styrene, are suspended in water whereas for hydrophilic monomers, such as acrylamide, a reverse suspension polymerization is used in which the aqueous solution of the monomers is suspended in hydrocarbon solvent using calcium phosphate as suspension stabilizer. A free radical initiator soluble in the monomer phase is used because it remains dissolved in the monomer during the polymerization. As the polymerization starts by heating, the liquid monomer droplets become highly viscous and then continue to polymerize to form spherical solid polymer particles called beads or pearls. The suspension stabilizer, which does not interfere with the reaction mixture, prevents the coagulation of the highly viscous droplets into gel-like precipitates by keeping the monomer in the state of small droplets. Each of the suspended monomer droplets undergoes individual bulk polymerization and kinetically the system consists of a large number of bulk polymerization units. After the polymerization, the polymeric bead product is collected by filtration and washed free of stabilizer and other contaminants. The size of the polymer beads depends on the extent of dispersion

in solution, the amount of agitation, the temperature and the initiator used. Pore dimensions can be determined during preparation by regulating the amount of crosslinking and are also influenced by the solvent employed. A potential problem is associated with the presence of residual impurities from the suspension-stabilizing agents required in the polymerization procedure. Moreover, considerable difficulties are encountered when it is necessary to copolymerize a mixture of water-soluble and water-insoluble monomers.

(iv) Emulsion polymerization. Emulsion polymerization differs from suspension polymerization in the type of particles in which polymerization occurs and their smaller size and in the kind of initiator employed. The emulsion polymerization process has several distinct advantages. The physical state of the emulsion system makes it easy to control the thermal and viscosity problems of the process, which are much less significant than in the other techniques. Large decreases in the molecular weight of a polymer can be made without altering the polymerization rate by using chain transfer agents. However, large increases in molecular weight can only be made by decreasing the polymerization rate, by lowering the initiator concentration or lowering the reaction temperature. Emulsion polymerization is a unique process in that it affords a means of increasing the polymer molecular weight without decreasing the polymerization rate. The difficulty of removing all impurities such as soap residues from the polymer is the only disadvantage of this technique.

In conventional emulsion polymerization [15–20] a hydrophobic monomer is emulsified in water with an oil-in-water emulsifier and then the polymerization is initiated with a water-soluble initiator. Alternatively 'inverse emulsion polymerization' [21] can be carried out in which an aqueous solution of a hydrophilic monomer is emulsified in a hydrophobic oil phase with a water-in-oil emulsifier.

The main components of conventional emulsion polymerization are the monomer(s), dispersant, emulsifier and water-soluble initiator. The dispersant is the liquid, usually water, in which the various components are dispersed in an emulsion state by means of the emulsifier. The ratio of water to monomer(s) is generally in the range 70:30 to 40:60 (by weight). Deionized water may be used since the presence of foreign ions in uncontrolled concentrations can interfere with both the initiation process and the action of the emulsifier. The initiators used in emulsion polymerizations are water-soluble initiators which may decompose either thermally, e.g. ammonium persulphate, hydrogen peroxide, or by redox reactions, e.g. persulphate with ferrous ion. The action of the emulsifier (surfactant, soap) results from the fact that its molecules have both hydrophilic and hydrophobic segments. Anionic emulsifiers are the most commonly used in emulsion polymerization. The shape of the micelles depends on the surfactant concentration. At lower surfactant concentrations (1%–2%) the micelles are small and spherical, but at higher concentrations they are larger and rod-like in shape. The surfactant molecules are arranged in a micelle with their hydrocarbon ends pointing towards the interior of the micelle and their ionic ends outwards toward the water. Other components as antifreeze additives are used to allow polymerization at temperatures below 0 °C.

When a water-insoluble monomer is added a very small fraction dissolves and goes into solution. Small portions of the monomer enter the interior hydrocarbon part of the micelles and the largest portion of the monomer is dispersed as monomer droplets whose size depends on the intensity of agitation. Thus, in a typical emulsion polymerization, the system consists of three types of particles: relatively large monomer droplets, inactive micelles in which polymerization is not occurring, and active micelles which are the focus of the polymerization resulting from the incorporation of free radicals into the micelles. A further difference between micelles and monomer droplets is that the micelles have a much greater total surface area. The site of polymerization is the water phase where the initiating radicals are produced rather than the monomer droplets since the initiators employed are insoluble in the organic monomer. This distinguishes emulsion polymerization from suspension polymerization in which oil-soluble initiators are used and reaction occurs in the monomer droplets. Polymerization takes place almost exclusively in the interior of the micelles. The micelles act as a meeting place for the organic monomer and the water-soluble initiator. As polymerization proceeds, the micelles grow by the addition of monomer from the aqueous solution; the concentration of monomer is replenished by dissolution of monomer from the monomer droplets.

1.1.3   Copolymerization

The chemical structure of a polymer can be varied over a wide range in order to obtain a product with a particular combination of desirable properties. Copolymerization is used to change polymer properties and to allow the synthesis of an almost unlimited number of different products by variations in the nature and relative amounts of the two monomer units in the copolymer product [22–24].

Homopolymerization of the functional monomer produces a product with every segment carrying a functional group, while copolymerization is the reaction that joins two or more monomer units to give polymer that contains more than one type of structural unit in the chain with only a fraction of the segments carrying the required group depending on the molar ratio of comonomers employed. This potential for dilution of the functional groups along the macromolecular backbone allows sensitive control over the loading. In addition, the main advantages of copolymerization lie in the modified properties of the polymers obtained, e.g. a change in solubility or a crosslinked structure [25]. The required physical form of crosslinked polymers containing the functional groups can be achieved during the polymerization process itself by copolymerizing monomer with divinyl compounds.

1.2 CHEMICAL FUNCTIONALIZATION OF SYNTHETIC ORGANIC POLYMERS

The preparation of functional polymers by chemical modification is an important technique which has been used extensively both industrially to modify the properties of polymers for various technological applications and in the area of polymer-supported chemistry to prepare chemically reactive polymers [26–31]. The application of chemical modification processes to polymers makes it possible to create new classes of polymers which cannot be prepared by direct polymerization of the monomers owing to their instability or unreactivity and to modify the structure and physical properties of other commercial polymers to make them suitable for specific applications. For example, attempts to prepare linear poly(N-alkylethylenimines) directly by ring-opening polymerization of N-alkylethylenimines were unsuccessful but these products were recently prepared by the chemical modification of poly(N-acylethylenimines) [32]:

$$\overset{N}{\underset{R\diagup\diagdown O}{\big|\big|}} \longrightarrow \left(N-CH_2CH_2 \right)_n \underset{COR}{\big|} \longrightarrow \left(N-CH_2CH_2 \right)_n \underset{CH_2R}{\big|} \tag{1.1}$$

Generally, chemical reactions of polymers are of two types.

1. Reactions on the main chain include degradation reactions which are accompanied by a decrease in molecular weight or intermolecular reactions which result in three-dimensional structures (causing a molecular weight increase) and the synthesis of a graft or block copolymer.
2. Reactions on side chain units result in a change in chemical composition of the polymer without affecting the molecular weight. Starting with an already formed polymeric carrier with reactive groups and replacing them by the desirable functional groups through chemical reaction is the simplest and more frequently used technique for the preparation of high molecular weight polymers with functional groups. In this technique commercially available resins of high quality are normally employed and the desired functional groups are introduced by using standard organic synthesis procedures. The ease of chemical modification of a resin, and indeed the level of success in its subsequent application, can depend substantially on the physical properties of the resin itself.

While the chemical modification approach is attractive for its apparent simplicity and the fact that it ensures a product with a good physical form, it suffers one major drawback: the polymers cannot be purified after modification. Polymer reactions also have other disadvantages.

1. They must be carried out under mild conditions and the yield of all reactions must be quantitative because every undesirable group that is formed by a

side reaction will become a part of the polymer chain. Thus the derivatization reactions required must be as free of side reactions as possible.

2. It must be established that polymer degradation does not occur during the chemical modification, particularly for polymer chains which are sensitive for chemical reactions.

3. The polymers prepared in this way rarely have every repeat unit functionalized, and the distribution of the functional groups on the polymer matrix may not be uniform.

4. The functional groups attached to a polymer chain may have a quite different reactivity from the analogous small molecule because of the macromolecular environment. Thus more drastic reaction conditions may be required to reach a satisfactory conversion.

5. The density of reactive groups obtained is generally low, and thus these polymers are not satisfactory for the investigation of the reaction between the structure and the functionality.

6. The chemical and physical nature of the polymer often changes as a result of undesirable side reactions, such as dissolution, crosslinking, dehydrohalogenation etc.

7. The reactivity of a functional group may be low when it is directly attached to the backbone owing to steric hindrance by neighbouring side groups and depends mainly on the proper choice of a swelling or suspending solvent.

8. Rate constants of reactions often decrease as the degree of substitution increases which normally means that the overall substitution reaction cannot go to completion. This problem of decreasing reactivity as the substitution reaction proceeds on polymers has been overcome either by spacing the site group from the backbone via spacer groups or by the use of a copolymer composition.

9. The purification problem: frequently a large proportion of the functional groups have been modified in the final polymer product but it also contains some impurities in the form of unreacted groups or other functionalities resulting from side reactions.

10. Indeed, different methods of preparation may give rise to different functional group distributions.

11. A problem which is always encountered when working with crosslinked polymers is the difficult characterization of the products after reaction, since a number of analytical methods are not well suited for the study of insoluble materials.

1.2.1 Functionalization of polystyrene

Although many polymer types, including both aliphatic and aromatic organic as well as inorganic polymers, have been employed as a carrier for the functional group, the most widely used as support is the polystyrene matrix. Thus most recent work on the chemical modification of polymers has been centred on the introduction and modification of various functionalities on polystyrene. The uses of polymers other than polystyrene have met with limited success for reasons such as lack of reactivity, degradation of the polymer chain, or other unsuitable physical properties of the finished polymer. In principle, polystyrene fulfils the major requirements for a solid support because it has many advantages over other resins.

1. It undergoes facial functionalization through the aromatic ring by electrophilic substitution.
2. Compatibility: styrenic polymers are compatible with most organic solvents and so functional groups are easily accessible to the reagents and solvents.
3. Chemical stability: the aliphatic hydrocarbon backbone is resistant to attack by most reagents. Hence the polymer chains are not susceptible to degradative scission by most chemical reagents under ordinary conditions.
4. Mechanical stability: styrenic polymers are mechanically stable to the physical handling required in sequential synthesis.
5. Crosslinking: since the degree of crosslinking in the polymer influences its swelling nature and the pore dimension, the type and degree of crosslinking can easily be controlled during manufacture by regulating the concentration of divinylbenzene.
6. Polystyrene is readily available commercially.

Polystyrene, chloromethylated polystyrene and ring-lithiated polystyrene are used in the chemical modification of styrene resins for the preparation of new functional polymers because they provide a method of attaching a wide variety of both electrophilic and nucleophilic species. However, the use of commercial polystyrene beads in chemical modifications is often complicated by the presence of surface impurities such as suspending or stabilizing agents which remain from the polymerization process. Thus, cleaning of the beads to remove most of these impurities is necessary since surface contaminants can prevent the penetration of reagents into the swollen beads or lead to the need for more drastic reaction conditions.

1.2.2 Functionalization of condensation polymers

Although the mechanical properties of condensation polymers are often superior to those exhibited by polystyrene, little work has been done on the introduction of functional groups by chemical modification of condensation polymers. For example, chloromethylation of polymers containing oxyphenyl repeat units with chloromethylethylether at room temperature in the presence of $SnCl_4$ has been reported [33]:

$$\text{(1.2)}$$

The lithiation of condensation polymers with the aid of *n*-butyllithium has also been reported [34]. Poly(2,6-dimethyl-1,4-phenyl ether) was metallated to give both the ring (20%) and the alkyl group (80%) lithium product:

$$(1.3)$$

The mode of lithiation depends on the duration and temperature of the reaction.

1.2.3 Chemical modification under phase transfer catalysis

A large number of functional polymers have been prepared by chemical modification under classical conditions. However, many of these reactions carried out on crosslinked polymers proceed very slowly and produce a low degree of functionalization because of hindered diffusion of the reagents through the swollen gel and, in many cases, the heterogeneous nature of the reaction system. These difficulties may be alleviated by using specific solvents and/or catalysts.

Recently, phase transfer catalysis has been found to be a valuable tool in the preparation of crosslinked and linear polymers containing various functionalities [35–43]. The application of phase transfer catalysis to polymer functionalization involves the chemical modification of functional polymers in a two- or three-phase system. These reactions involve mainly nucleophilic displacements on polystyrene derivatives or reactions of polymers that have a reactive nucleophilic pendant group with various electrophiles. In addition to the ease of reaction and work up, it has generally been found that the phase-transfer-catalysed reactions afford better results than those carried out under classical conditions in terms of both polymer purity and functional yields. These simple, mild and economical methods have been used for the synthesis of functional polymers through, for example, chemical modifications of pendant chloromethyl groups in poly(chloromethylstyrene) by reaction with several inorganic salts [44] as well as the salts of organic compounds [45] in the presence of a typical phase transfer agent:

$$(1.4)$$

$$Y \equiv Br, SCN, SH, CH(CN)_2$$

1.2.4 Functionalization by grafting

A graft copolymer is a branched polymer in which the main backbone is chemically different from the side branches. The grafting technique has been successfully used for modifying the physical and chemical properties of various polymers by several methods.

(i) Radical chain transfer grafting. The radical chain transfer effect is more pronounced with polymer chains that contain labile atoms that are easily abstracted by the attacking free radical. Heating or irradiating a mixture of a linear polymer dissolved in an appropriate monomer and initiator results in transfer between the polymer chain and the radical formed from the initiator. A polymer radical can initiate polymerization of the monomer. The amount of grafting achieved by this effect is usually small and depends on the magnitude of the chain transfer constant of the polymer which is usually small. Thus, this method of grafting leads to a mixture of linear polymer and graft copolymer.

(ii) Polymeric initiator grafting. Polymeric initiator grafting involves the creation of initiator radicals, peroxide or azo groups in the polymer chain followed by polymerization of the monomer to be grafted to give the graft copolymer, e.g.

$$(1.5)$$

Anionic initiators have been used as sites for grafting in this technique:

$$(1.6)$$

(iii) Chemical reaction grafting. Chemical reaction grafting is formed by attaching polymers with functional groups at the chain ends onto other polymers with the aid of reactive functional groups present along the polymer chain.

(iv) Radiation grafting. Radiation grafting using a simultaneous method is a convenient one-step procedure for modifying polymers [46]; it is useful in particular for imparting wettability to hydrophobic polymer using hydrophilic monomers. For example, *p*-styryldiphenylphosphine has been grafted to poly-(vinyl chloride), polypropylene and crosslinked polystyrene beads at radiation dose levels that do not affect the properties of the resulting copolymer [46, 47]. This technique is valuable for monomers and polymers that are radiation sensitive to achieve the required grafting. The most commonly used energy sources are ionizing radiation, plasma gas discharge and ultraviolet light sources in the presence of photosensitizer [48, 49]. The technique involves irradiating a solution of polymer in monomer with radiation that results in radical formation on the primary polymer chain; the sites of radical formation become the points of initiation for the side chains. At the same time, the radiation initiates polymerization of the monomer and thus a mixture of graft copolymer and homopolymer is obtained. The predominant variables which influence the grafting yield include (a) the radiation dose and dose rate (time), (b) the concentration of monomer and sensitizer in the solvent, (c) the structure of both monomer and base polymer. However, for grafting the polymer surface a solvent is used. The requirements for an appropriate solvent are as follows.

1. It is a non-solvent of the base polymer.
2. Slight definitive interactions are necessary to provide reaction sites for grafting.
3. It must not swell the base polymer, i.e. a thin graft layer is obtained.
4. Good solvent–growing chain interactions assist the propagation of the graft chain outside the base polymer surface.
5. The solvent must be inert to the triplet excited state of the sensitizer.

1.2.5   Functionalization of membranes

Membranes containing functional groups, which dominate their choice and use as reactive materials, are made by (a) polymerizing styrene–divinylbenzene in sheet-shaped moulds followed by further chemical reactions for incorporation of the active species [50–55], (b) copolymerization of the functionalized monomer with divinylbenzene in thin film form and (c) mechanically incorporating powdered functionalized polymer into a sheet of some other extrudable or mouldable matrix.

1.3   FUNCTIONALIZATION OF BIOPOLYMERS

Polysaccharides-based supports prepared from cellulose, agarose, Sepharose and Sephadex are well known as gel-filtration media [56] in chromatographic procedures for the fractionation process. Some of these supports have been functionalized and used widely in applications such as affinity chromatography [57], enzyme immobilization [58] and ion exchangers [59, 60]. However, other supports have been employed in organic synthesis, e.g. in the binding of oxidizing and reducing anions as redox reagents [61], in the immobilization of boric acid for use in solid phase oligonucleotide synthesis [62, 63], in the support of homogeneous transition metal complexes for use as hydrogenation catalysts [64] and for the attachment of crown ethers as alkali metal complexing species [65].

A supported catalyst consisting of diethylamine groups chemically bonded to cellulose has been shown to split off carbon dioxide catalytically from α-bromocamphorcarboxylic acid at 40 °C and to catalyse the combination of benzaldehyde and HCN to form mandelonitrile [66]. Recently, a number of phosphonium and ammonium salts supported on cellulose have been synthesized and employed as phase transfer catalysts [67, 68].

In spite of the non-toxic and highly hydrophilic character of polysaccharides which is particularly effective for numerous hydrophilic situations, the main drawbacks to their wide application are that they are susceptible to microbial attack, show a high degree of adsorption on some substrates and have low capacities for functionalization. In addition, they show less mechanical and chemical stability than synthetic polymers. However, they have an advantage in applications in which the degradability of the main backbone is of importance in order to prevent long existence durations.

Cellulose primarily consists of a linear polymer of β-1,4-linked D-glucose units. Although cotton is the purest form of naturally occurring cellulose, it contains several impurities such as wax, pectins and colouring matter. These impurities can be removed under conditions that are not sufficient to bring about any fundamental change in the cellulose structure by extraction with organic solvents and hot alkali in the absence of oxygen and by leaching. Cellulose can be obtained commercially in various forms: microcrystalline, amorphous [69], beads [70], short and long fibred. Microcrystalline cellulose is generally crosslinked with a bifunctional reagent such as epichlorohydrin and is remarkably stable to chemical attack.

Sephadex is a three-dimensional network in which soluble dextran chains are crosslinked by glycerin ether bonds or reaction with epichlorohydrin in alkaline solution. Dextran is a branched chain polysaccharide composed of D-glucose units which are jointed mainly by means of α-1,6-glycosidic bonds and is branched by 1,2, 1,3 and 1,4 glycosidic linkages. Sephadex is stable to chemical attack by, for example, alkali and weak acids and can be heated without any change in properties to 110–120 °C.

Agarose is a linear polysaccharide consisting of alternating residues of D-galactose and 3,6-anhydro-L-galactose. It is freely soluble in boiling water and the polymer chains are held together by hydrogen bonds. It is less stable chemically than the covalently crosslinked dextran. The beaded derivatives of agarose have many of the properties of the ideal matrix and have been used successfully in numerous purification procedures. The uniform spherical shape of the gel particles is of particular significance. Further, they can readily undergo substitution reactions by an activation procedure and have a moderately high capacity for substitution.

The hydroxyl groups of polysaccharides have been utilized in chemical reactions to produce a variety of new functional groups which can in turn be utilized in further reactions. A broad variety of possible chemical structures are obtained by chemical modification of polysaccharide through esterification, etherification, oxidation or grafting. A number of chemical derivatives of polysaccharides have been prepared by the reaction of the activated form with the desired functional group. The chemical activation of the hydrophilic polysaccharide by cyanogen halides leads to the formation of cyclic imidocarbonates, which are responsible for most of the coupling capacity to primary amino groups, and carbamic acid esters, which are stable, inert and neutral.

$$
\underset{OH}{\overset{OH}{C}} \xrightarrow{\text{BrCN}} \underset{OH}{\overset{O-C\equiv N}{C}} \longrightarrow C\underset{O}{\overset{O}{\diamond}}C=NH
$$

$$\Big\downarrow RNH_2 \tag{1.7}$$

$$
\underset{OH}{\overset{OCONHR}{C}}
$$

Polysaccharides support matrices can also be activated by cyanuric chloride or dichloro-*syn*-triazine:

$$
C\text{—OH} + Cl\text{—triazine(R, Cl)} \longrightarrow C\text{—O—triazine(R, Cl)}
$$

$$\xrightarrow{R'NH_2} C\text{—O—triazine(R, NHR')} \tag{1.8}$$

$$R \equiv -OCH_2COOH, \quad -Cl, \quad -NHCH_2COOH$$

Furthermore, polysaccharides can be activated by oxidation with oxidizing agents such as sodium periodate [71] to produce reactive aldehyde or carboxylic acid groups for the attachment of other functional groups. Oxidation of polysaccharides to carboxylic acid groups can occur at two different locations: (a) by oxidation of the primary hydroxyl groups at the 6-position to produce monocarboxyl polysaccharide by NO_2; (b) by oxidation of two secondary hydroxyl groups at the 2- and 3-positions to produce dicarboxyl polysaccharide by a two-step conversion:

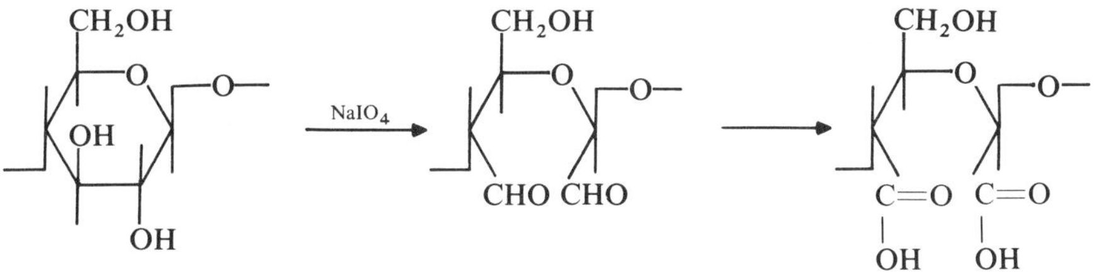

$$(1.9)$$

The first step involves the oxidation to dialdehyde groups by $NAIO_4$, HIO_4, $Pb(OAc)_4$ or $HCrO_3$, followed by their conversion to dicarboxylic groups by the action of chlorous acid or sodium chlorite. However, the main disadvantages of the oxidation method are (a) that the dialdehyde derivative is quite unstable to alkali which results in chain breakage and large losses of degree of polymerization and (b) it leads to a loss in strength of the polysaccharide, where the extent of strength loss is dependent on the amount of oxidizing agent and the conditions of oxidation used.

Acylation of polysaccharide hydroxyl groups with acryloyl chloride followed by subsequent derivatization techniques for the introduction of phase transfer catalysts on the cellulose backbone is outlined in Scheme 1.1 [67].

Grafting is another technique that has received considerable attention lately for modifying polysaccharides [72–77]. The main problem in this approach is the

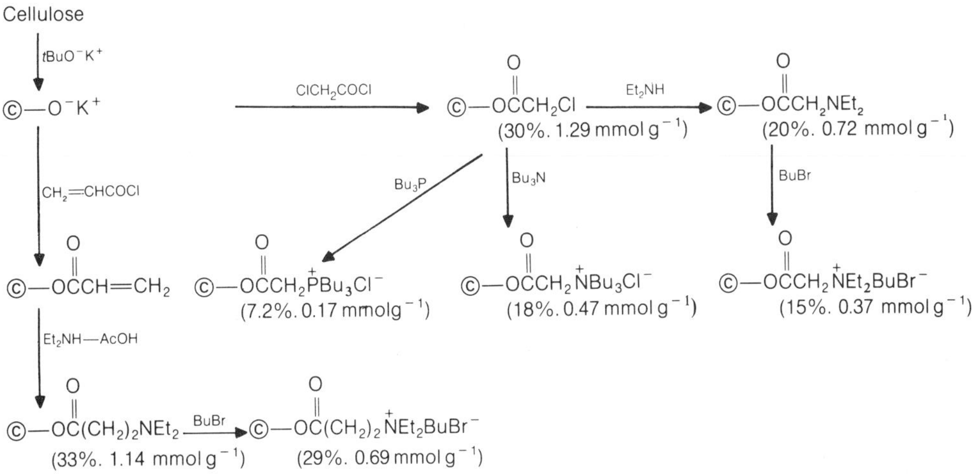

Scheme 1.1 Chemical functionalization of cellulose.

presence of homopolymers in the modified grafted polysaccharide which serve as an impurity and make characterization and evaluation of the grafted polysaccharides more difficult.

1.4 FUNCTIONALIZATION OF INORGANIC SUPPORTS

Modified silica, unlike synthetic polymers and polysaccharide polymers, is rigid and not subject to swelling. The choice of solvent is of little consequence to the physical and chemical behaviour of modified silicon as most of the functional groups are located on the surface. Physical adsorption of reagents and catalysts on an inorganic support, by hydrogen bonding between oxygen functions of the support surface and polar groups on the reagent or catalyst, have been used in many heterogeneous synthetic reactions [78]. A quite interesting approach to polymer supports has been introduced [79], which is based on chemical binding of reactive molecules to modified silica. Generally, modification of silica is achieved through the surface hydroxyl groups, owing to the small average pore diameter. Modification of the surface hydroxyl groups often leads to condensation of additional silica material in the pores. Several difficulties that arise in using organic polymers and that can be overcome by using modified inorganic support material include the following.

1. All reaction rates are unfavourably controlled by diffusion since functional groups are uniformly distributed throughout the organic polymeric resin. This distribution cannot be overcome by using resins with a lower concentration of functional groups.
2. The use of different solvents during the reaction and washing steps causes different swelling of resin particles, and thus affects reaction rates, yields and purity of the synthesized products [80, 81].
3. Variable swelling of the polymer particles by the different solvents is necessary in a solid phase synthesis but also causes some difficulties in the automation of the process; therefore batch procedures must be used instead of column procedures.

However, the properties of modified inorganic supports, such as the localization of reactive groups to the surfaces, the chemical and dimensional stabilities, the ease of filtration and the use in continuous-flow column operations, are used to overcome all these difficulties. Additionally, inorganic support materials offer several advantages over organic supports:

1. They prevent any ion exchange mechanism before or after the coupling step in multistep synthesis;
2. They have high mechanical strength;
3. They have high thermal stability;
4. High pressure operation can be used;

5. They are stable in solvents and acids;
6. They are resistant to microbial attack;
7. No special equipment is required for most procedures.

Despite these advantages, the main disadvantage of using modified silica as a support in organic synthesis is that the Si—Z—C bond ($Z \equiv O, N$) is highly polarized and thus very sensitive to attack by all reagents containing free hydroxyl groups, especially water, which results in removal of the synthesized molecule from the silicate polymers. This difficulty can be overcome by constructing a short aliphatic chain between the three-dimensional silicate network and the functional group by bonding organic molecules to the siliceous surface through Si—Z—Si—C bonds, which are more stable against an attack by electrophilic or nucleophilic agents than the Si—Z—C bonds are.

A significant difference between inorganic and organic matrices is the degree to which they can be functionalized. This represents one of the greatest drawbacks to the use of inorganic supports. Inorganic matrices have an upper limit of functional groups so that loadings of 1–2 meq/g are difficult to achieve, whereas organic matrices can carry up to 10 meq/g matrix. Thus, although the specific activity of inorganics may be lower, their potential for monocoordination and site isolation is greater than that of organic matrices. Nevertheless, limits to the range of applications of functionalized silica occur because of the chemical stability of the silica–oxygen bond in an alkaline medium.

The most useful technique for modifying silica involves the reaction of surface silanol groups with the organosilanes of formula X_3—Si—R—Y, which are able to improve the bonding between functional groups and silica particles markedly. Silane coupling agents either contain the desired functional group (reaction 1.10) or can be subject to later modification (reaction 1.11):

$$\text{Si}-\text{OH} + X_3 - \text{Si}-\text{R}-\text{Y} \longrightarrow \text{Si}-\text{O}-\text{Si}-\text{R}-\text{Y} \tag{1.10}$$

$$\text{Si}-(CH_2)_3-\text{Cl} \xrightarrow[\substack{KPPh_2 \\ LiCH_2-C_5H_4N \\ HNEt_2}]{} \begin{cases} \text{Si}-(CH_2)_3-PPh_2 \\ \text{Si}-(CH_2)_4-C_5H_4N \\ \text{Si}-(CH_2)_3-NEt_2 \end{cases} \tag{1.11}$$

Y is a reactive organic group, such as amino, mercapto, phosphino, vinyl, epoxy etc., which is bound via an alkyl or aryl chain to the silicon atom. X represents a hydrolysable group OR, Cl, NH or OCOR.

The activation of a silica surface and the successive covalent binding of active groups is normally realized by treatment of silica particles or gel with hydrochloric acid to afford a sufficient number of silanol groups to be reacted with different organosilane derivatives [82]. In general, functionalization of all surface hydroxyl groups is difficult to achieve and those remaining without modification can give rise to adsorption problems. Thus it is necessary to silylate the unmodified hydroxyl groups by reacting the modified support with excess hexamethylenedisilazane in order to minimize the adsorption of substrates or reagents. This type of modification has been employed successfully to anchor different functionalities. It has the advantage of being a one-step reaction, simple to perform under mild conditions, and a wide range of X_3—Si—R—Y compounds are available. These modification reactions have been used for the formation of chemically bonded layers of organic molecules on the surface of siliceous materials in the field of gas and liquid chromatography [83, 84].

The organosilanes X_3—Si—R—Y [85] can be prepared by hydrosilylation, i.e. by the addition of silane, $HSiX_3$, to an olefin derivative in the presence of a catalyst such as dipotassium hexachloroplatinate or palladium [86, 87]; the addition is anti-Markovnikov because of the unusual polarization in the silicon–hydrogen bond:

$$
\begin{array}{c}
\diagdown\!\!\!\!\diagdown\!\!\!\!-(CH_2)_{n-2}-\!\!\!/\!\!/ \quad + \quad X_3SiH \longrightarrow X_3Si-(CH_2)_n-\!\!\!/\!\!/ \\[1em]
\qquad\qquad\qquad\qquad\qquad\qquad\quad hv \downarrow Ph_2PH \\[0.5em]
\qquad\qquad\qquad\qquad X_3Si-(CH)_{n+2}-PPh_2
\end{array}
\qquad (1.12)
$$

(Ref. 88):

$$
\diagdown\!\!\!\!-(CH_2)_{n-2}-\!\!\!\bigcirc\!\!\!-CH_2Cl \xrightarrow{K_2PtCl_6} X_3Si-(CH_2)_n-\!\!\!\bigcirc\!\!\!-CH_2Cl
\qquad (1.13)
$$

(ref. 89). In addition, the organosilanes can be prepared by the reaction of Grignard reagent with $SiCl_4$ [86, 88, 89]:

$$
\begin{array}{c}
Me-\!\!\!\bigcirc\!\!\!-MgBr \quad + \quad SiCl_4 \longrightarrow Cl_3Si-\!\!\!\bigcirc\!\!\!-Me \\[1em]
\qquad\qquad\qquad\qquad\qquad\qquad\quad \downarrow NBS \\[0.5em]
\qquad\qquad\qquad Cl_3Si-\!\!\!\bigcirc\!\!\!-CH_2Br
\end{array}
\qquad (1.14)
$$

(ref. 90). The nature of X in X_3—Si—R—Y affects the extent of anchoring [90]. The concentration of functional groups anchored to silica decreases with increasing steric requirement of the hydrolysable group X. This is probably due to the reaction of a single hydrolysable group with a surface hydroxyl, leaving two groups free to block other sites from reacting with other X_3—Si—R—Y compounds.

Another type of silica-supported functional group has been prepared by coating silica gel with polystyrene which can in turn be further functionalized [91, 92]. For example, copolymers of styrene and divinylbenzenzene are prepared in the presence of azobisisobutyronitrile and silica gel. The product obtained is brominated and then reacted with $KPPh_2$ to give silica gel coated with phosphinated polystyrene [92]:

$$(1.15)$$

A totally different approach for functionalization of silica has been developed in which the functional group is built into trialkoxysilane and then polymerized to produce a non-linear polymer based on an Si—O—Si backbone [93]. For example, 2-(diphenylphosphine)ethyltriethoxysilane was refluxed in acetic acid with $Si(OEt)_4$ and a trace of concentrated HCl to give silica containing 11% phosphorus [94]:

$$(EtO)_3Si—CH{=}CH_2 + Ph_2PH \xrightarrow{\text{(t-BuO)}_2} (EtO)_3SiCH_2CH_2PPh_2$$

$$(1.16)$$

Furthermore, the inorganic polymer polyphenylsiloxane has been prepared by hydrolysis of $PhSiCl_3$ and subjected to further chemical functionalization [94–96]. After initial formation of a low molecular weight prepolymer, a ladder polymer is formed in the presence of KOH catalyst [94], the polymer is chloromethylated with an excess of chloromethyl methyl ether and $ZnCl_2$ catalyst, and phosphination occurs by reaction with $LiPR_2$ in tetrahydrofuran [95, 96]:

$$(1.17)$$

Other inorganic supports, such as *o*-alumina, molecular sieves (zeolites) and glass, which are essentially metal oxides, have hydroxyl groups on the surface that can be used as the point for attaching functional groups [97].

Although investigation of the employment of modified inorganic beads is limited, the fundamental simplicity of this technique seems attractive. The major applications of modified silica, as an example of inorganic polymers containing covalently attached functional groups, in solving some organic synthesis problems have been studied in several fields. In addition to the specific utilization of modified silica as a support material for liquid chromatography and for immobilization of biologically active materials [58, 98, 99], it has been used as a catalyst or reagent in the field of organic synthesis reactions [79]. It has also been successfully used as a support for peptide synthesis [87, 100], for oligonucleotide synthesis [101], for immobilizing transition metal catalysts [102] and other applications.

REFERENCES

1. Takemoto, K. and Mujata, M. (1980) *J. Macromol. Sci., Rev. Macromol. Chem.*, **C-18**, 83.
2. Moore, J.A. (ed.) (1977) *Macromolecular Synthesis*, coll. vol. 1, Wiley Interscience, New York.
3. Sandler, S.R. and Karo, W. (1974) *Polymer Synthesis*, vols 1 and 2, Academic Press, New York; (1980) *Polymer Synthesis*, vol. 3, Academic Press, New York.
4. Schildknecht, C.E. (1977) Cast polymerization, other bulk polymerizations. In *Polymerization Processes* (eds C.E. Schildknecht and I. Skeist), Wiley Interscience, New York, Chs 2, 4.
5. Munzer, M. and Trommersdorff, E. (1977) Polymerizations in suspension. In *Polymerization Processes* (eds C.E. Schildknecht and I. Skeist), Wiley Interscience, New York, Ch. 5.
6. Matsumoto, M., Takakura, K. and Okaya, T. (1977) Radical polymerization in solution. In *Polymerization Processes* (eds C.E. Schildknecht and I. Skeist), Wiley Interscience, New York, Ch. 7.
7. Barrett, K.E.J. (ed.) (1975) *Dispersion Polymerization in Organic Media*, Wiley, New York.
8. Hunt, H. (1949) US Patent 2 471 959; *Chem. Abstr.* **43**, 6002.
9. Tanimoto, S., Miyake, T. and Okano, M. (1974) *Synth. Commun.* **4**, 193.
10. Ogura, K., Kondo, S. and Tsuda, K. (1981) *J. Polym. Sci., Polym. Chem. Ed.*, **19**, 843.
11. Frechet, J.M.J., Farrall, M.J. and Nuyens, L.J. (1977) *J. Macromol. Sci. Chem.*, **A-11**, 507.
12. Hohenstein, X. (1950) US Patent 2 524 627; (1951) *Chem. Abstr.*, **45**, 903.
13. Screnson, W.R. (1965). *J. Chem. Educ.*, **42**, 8.
14. Sherrington, D.C., Graig, D.C., Dagleish, J., Domin, G., Taylor, J. and Meehan, G.V. (1977) *Eur. Polym. J.*, **13**, 73.
15. Harkins, W.D. (1947) *J. Am. Chem. Soc.*, **69**, 1428.
16. Smith, W.V. (1948) *J. Am. Chem. Soc.*, **70**, 3695.
17. Gardon, J.L. (1977) Emulsion polymerization. In *Polymerization Processes* (eds C.E. Schildknecht and I. Skeist), Wiley Interscience, New York, Ch. 6.
18. Gardon, J.L. (1977) Interfacial, colloidal and kinetic aspects of emulsion polymerization. In *Interfacial Synthesis*, vol. 1 (eds F. Millich and C.E. Carraher), Marcel Dekker, New York, Ch. 9.

19. Cooper, W. (1974) Emulsion polymerization. In *Reactivity, Mechanism and Structure in Polymer Chemistry* (eds A.D. Jenkins and A. Ledwith), Wiley Interscience, New York, Ch. 7.
20. Frechet, J.M.J., Darling, P. and Farrall, M.J. (1981) *J. Org. Chem.*, **46**, 1728.
21. Blackley, D.C. (1975) *Emulsion Polymerization*, Applied Science, London.
22. Ham, E. (ed.) (1964) *Copolymerization*, Wiley, New York.
23. Eastman, G.C. (1976) Copolymerization. In *Comprehensive Chemical Kinetics*, vol. 14A (eds C.H. Bamford and C.F.H. Tipper), Elsevier, Amsterdam, Ch. 4.
24. Cundall, R.B. (1963) Copolymerization. In *The Chemistry of Cationic Polymerization* (ed. P.G. Plesch), MacMillan, New York, Ch. 15.
25. Schultz, A.R. (1966) Crosslinking. In *Encyclopedia of Polymer Science and Technology*, vol. 4 (eds H.F. Mark, N.G. Gaylord and N.M. Bikales), Wiley Interscienc, New York, pp. 331–414.
26. Fettes, E.M. (ed.) (1964) *Chemical Reactions of Polymers*, Wiley Interscience, New York.
27. Frechet, J.M.J. and Farrall, M.J. (1977) In *Chemistry and Properties of Crosslinked Polymers* (ed. S.S. Labana), Academic Press, New York, p. 59.
28. Moore, J.A. (ed.) (1973) *Reactions on Polymers*, D. Reidel, Boston, MA.
29. Lenz, R.W. (1967) *Organic Chemistry of Synthetic High Polymers*, Wiley Interscience, New York, Ch. 17.
30. Loan, L.D. and Winslow, F.H. (1979) Reactions of macromolecules. In *Macromolecules: An Introduction to Polymer Science* (eds F.A. Bovey and F.H. Winslow), Academic Press, New York, Ch. 7.
31. Ceresa, R.J. (1978) The chemical modification of polymers. In *Science and Technology of Rubber* (ed. F.R. Eirich), Academic Press, New York, Ch. 11.
32. Kawakami, Y., Sugiura, T., Mizutani, Y. and Yamashita, Y. (1980) *J. Polym. Sci., Polym. Chem. Ed.*, **18**, 3009.
33. Daly, W.H., Chotiwana, S. and Nielsen, R. (1979) *Polym. Prepr., Am. Chem. Soc., Div. Polym. Chem.*, **20** (1), 835.
34. Chalk, A.J. and Hay, A.S. (1968) *J. Polym. Sci.*, **B-6**, 105.
35. Mathias, L.J. (1981) *J. Macromol. Sci. Chem.*, **A-15**, 853.
36. Frechet, J.M.J. (1981) *J. Macromol. Sci. Chem.*, **A-15**, 877.
37. Roeske, R.M. and Gesellchen, P.D. (1976) *Tetrahedron Lett.*, 3369.
38. Roovers, J.E.L. (1976) *Polymer*, **17**, 1107.
39. Farrall, M.J. and Frechet, J.M.J. (1978) *J. Am. Chem. Soc.*, **100**, 7998.
40. Frechet, J.M.J., deSmet, M. and Farrall, M.J. (1979) *Tetrahedron Lett.*, 137; (1979) *J. Org. Chem.*, **44**, 1774.
41. Nishikubo, T., Iizawa, T., Kobayashi, K. and Okawara, M. (1980) *Makromol. Chem. Rapid Commun.*, **1**, 765.
42. Gozdz, A.S. (1981) *Makromol. Chem. Rapid Commun.*, **2**, 443, 595.
43. Frechet, J.M.J. and Eichler, E. (1982) *Polym. Bull.*, **7**, 345.
44. Gozdz, A.S. and Rapak, A. (1981) *Makromol. Chem. Rapid Commun.*, **2**, 359.
45. Nishikubo, T., Iizawa, T. and Kobayashi, K. (1981) *Makromol. Chem. Rapid Commun.*, **2**, 387.
46. Garnett, J.L. (1979) *J. Rad. Phys. Chem.*, **14**, 79.
47. Garnett, J.L., Levot, R. and Long, M.A. (1981) *J. Polym. Sci., Polym. Lett. Ed.*, **19**, 23.
48. Hoffman, A.S. (1977) *Rad. Phys. Chem.*, **9**, 207; (1981) *Rad. Phys. Chem.*, **18** (1), 323; (1983) *Rad. Phys. Chem.*, **22**, 267.
49. Shen, M. and Bell, A.T. (eds) (1979) *Plasma Polymerization*, Am. Chem. Soc. Symp. Ser. 108, Washington, DC.
50. Juda, W. and McRae, W. (1960) US Patent 24 865.
51. Tsunoda, Y. and Seko, M. (1955) Jpn. Patent 5068–9.
52. Sprague, B.S. (1973) *J. Macromol. Sci. Phys.*, **8**, 157.

53. Miles, M.J. and Baer, E. (1970) *J. Mater. Sci.*, **14**, 1254.
54. Quynn, R.G. and Sprague, B.S. (1970) *J. Polym. Sci., A-2*, **8**, 1971.
55. Quynn, R.G. and Brody, H. (1971) *J. Macromol. Sci. Phys.*, **B-5** (4), 721.
56. Hjerten, S. (1964) *Biochem. Biophys. Acta*, **79**, 393.
57. Anonymous (1971) *Chem. Eng. News*, **49**, 86.
58. Suckling, C.J. (1977) *Chem. Soc. Rev.*, **6**, 215.
59. Peterson, E.A. (1970) *Cellulosic Ion Exchangers*, North-Holland, Amsterdam.
60. Lieser, K.H. (1979) *Pure Appl. Chem.*, **51**, 1503.
61. Perrier, D.M. and Benerito, R.R. (1976) *Appl. Polym. Symp.*, **29**, 213.
62. Weith, H.D., Wievers, J.L. and Gilham, P.T. (1970) *Biochemistry*, **9**, 4396.
63. Rosenberg, M., Wievers, J.L. and Gilham, P.T. (1972) *Biochemistry*, **11**, 3623.
64. Pracejus, H. and Bursian, M. (1972) DDR Patent 92031; (1973) *Chem. Abstr.*, **78**, 72591.
65. Djamali, M.G., Burba, P. and Lieser, K.H. (1980) *Angew. Makromol. Chem.*, **92**, 145.
66. Bredig, G. and Oerstuer, F. (1932) *Biochem. Z.*, **260**, 414.
67. Akelah, A. and Sherrington, D.C. (1981) *J. Appl. Polym. Sci.*, **26**, 3377; (1982) *Eur. Polym. J.*, **18**, 301.
68. Kise, H., Araki, K. and Seno, M. (1981) *Tetrahedron Lett.*, **22**, 1017.
69. Wadehra, I.L., Manley, R.S.J. and Goring, D.A.I. (1965) *J. Appl. Polym. Chem.*, **9**, 2634.
70. Peska, J., Stamberg, J. and Hradil, J. (1976) *Angew. Makromol. Chem.*, **53**, 73.
71. Jackson, E.L. (1944) In *Organic Reactions*, vol. 2, Wiley, New York, p. 341.
72. Rogovin, Z.A. (1971) *Polym. Sci. USSR (A)*, **13** (2), 497.
73. Faranone, G., Parasacco, G. and Cogrossi, C. (1961) *J. Appl. Polym. Sci.*, **5**, 16.
74. Arthur, J.C. (1970) *J. Macromol. Sci. Chem.*, **A-4** (5), 1057.
75. Stannett, V. (1970) *J. Macromol. Sci. Chem.*, **A-4**, 1177.
76. Krassing, H.A. and Stannett, V. (1970) *Adv. Polym. Sci.*, **4**, 111.
77. Immergut, E.H. (1965) Cellulose graft copolymers. In *Encyclopedia of Polymer Science and Technology*, vol. 3 (eds H.F. Mark, N.G. Gaylord and N.M. Bikales), Interscience, New York, p. 242.
78. McKillop, A. and Young, D.W. (1979) *Synthesis*, 401, 481.
79. Akelah, A. (1981) *Br. Polym. J.*, **13**, 107.
80. Bayer, E., Yung, G., Halasz, I. and Sebastian, I. (1970) *Tetrahedron Lett.*, 4503.
81. Bayer, E., Eckstein, H., Hagele, K., Konig, W., Bruning, W., Hagenmaier, H. and Parr, D.W. (1970) *J. Am. Chem. Soc.*, **92**, 1735.
82. Fritz, J.F. and King, J.N. (1976) *Anal. Chem.*, **48**, 570.
83. Kirkland, J.J. and Destefano, J.J. (1970) *J. Chromatogr. Sci.*, **8**, 309.
84. Abel, E.W., Pollard, Z.H., Uder, P.C. and Nickless, G. (1966) *J. Chromatogr.*, **22**, 23.
85. Noll, W. (1968) *Chemistry and Technology of Silicones*, Academic Press, New York and London, p. 582.
86. Oswald, A.A., Murrell, L.L. and Boucher, L.J. (1974) *Prepr., Am. Chem. Soc., Div. Petrol. Chem.*, **19**, 155.
87. Parr, W., Grohmann, K. and Hagele, K. (1974) *Liebigs Ann. Chem.*, 655.
88. Chvalovsky, V. and Bazant, V. (1951) *Coll. Czech. Chem. Commun.*, **16**, 580.
89. Parr, W. and Grohmann, K. (1971) *Tetrahedron Lett.*, 2633.
90. Boucher, L.J., Oswald, A.A. and Murrell, L.L. (1974) *Prepr., Am. Chem. Soc., Div. Petrol. Chem.*, **19**, 162.
91. Scott, R.P.W., Chan, K.K., Kucera, P. and Zolty, S. (1971) *J. Chromatogr. Sci.*, **9**, 577.
92. Arai, H. (1978) *J. Catal.*, **51**, 135.
93. Seidl, J., Malinsky, J., Dusek, K. and Heitz, W. (1967) *Adv. Polym. Sci.*, **5**, 114.
94. Conan, J., Bartholin, M. and Guyot, A. (1975–76) *J. Mol. Catal.*, **1**, 375.
95. Bartholin, M. Graillat, C., Guyot, A., Coudurier, G., Bandiera, J. and Naccache, C. (1977–78) *J. Mol. Catal.*, **3**, 17.

96.   Bartholin, M., Conan, J. and Guyot, A. (1977) *J. Mol. Catal.,* **2**, 307.
97.   Capka, M. and Hetflejs, J. (1974) *Coll. Czech. Chem. Commun.,* **39**, 154.
98.   Silman, I.H. and Katchalski, E. (1966) *Ann. Rev. Biochem.,* **35**, 873.
99.   Lindsey, A. (1970) *Rev. Macromol. Chem.,* **4**, 1.
100. Parr, W. and Grohmann, K. (1972) *Angew. Chem., Int. Ed. Engl.,* **11**, 314.
101. Smith, A.K., Basset, J.M. and Maitlis, P.M. (1977) *J. Mol. Catal.,* **2**, 223.
102. Bailey, D.C. and Langer, S.H. (1981) *Chem. Rev.,* **81**, 109.

2
Properties and characterization of functionalized polymers

There are a number of considerations in the choice of the functional polymers to be used in a specific application. Functionalized polymers must possess a structure which permits adequate diffusion of reagents into the reactive sites. This depends on the extent of swelling, compatibility, the effective pore size, the pore volume (porosity) and the chemical, thermal and mechanical stability of the resins under the conditions of a particular chemical reaction or reaction sequence. These in turn depend on the degree of crosslinking of the resin and the conditions employed during its preparation.

2.1 PHYSICAL FORMS

Macromolecules can be linear or crosslinked species. The latter are commonly referred to as resins. Each type possesses its own distinct advantages and disadvantages depending on the final utilization. Thus, in any application of functionalized polymer, the physical form of the polymer must be studied carefully in order to maximize the advantages of the polymer system while minimizing any potential problems.

2.1.1 Linear polymers

A linear polymer is a long-chain species in which the monomer molecules have been linked together in one continuous length. In the solid state the linear polymer molecules have a thread-like shape and occur in various conformations. They are usually present in crystalline or amorphous form. In the crystalline state the molecules are oriented in a regular manner with respect to each another. The degree of crystallinity depends on the structure of the polymer chains and the amount of chain flexibility, and can be increased by appropriate thermomechanical means. In the amorphous state a condition of maximal possible entropy determines the most probable shape. A linear polymer is capable of forming a molecular solution in a suitable solvent; an individual chain is not usually present as an extended chain but adopts a random coil conformation. The coil density is usually influenced by (a) the structure of the polymer chains, (b) the extent of solvation, (c) the molecular weight, (d) temperature and (e) ionic groups and their degrees of dissociation. Polymer coils readily expand in a good solvent and contract in a poor one.

The use of functionalized linear polymer is of growing interest especially when separation of the polymer after its application is not necessary or when the polymer must be soluble to permit working in a homogeneous phase to perform its function. Soluble polymer substrates are useful for chemical reactions both as a model system for solid phase reactions and as a soluble substrate for kinetic studies because they are not limited by diffusion control problems.

The advantages associated with the use of linear soluble polymers include the following.

1. Reactions can be carried out in homogeneous media, minimizing diffusion problems.
2. Functional groups are of equal accessibility.
3. Problems arising from the pore size distribution and reactions which involve substrates of large molecular size that are not able to penetrate all the pores of a crosslinked polymer can be overcome by the use of soluble polymers.
4. Reactions are not affected by the size of the polymer backbone and usually proceed to a high extent.
5. High conversions can be achieved which give yields comparable with those in the homogeneous phase.
6. Characterization at the various stages of the functionalization is easy.

In some applications the use of linear soluble polymers may give rise to some disadvantages, e.g. the separation of the polymer from low molecular weight contaminants may be difficult. Separation can be achieved by ultrafiltration, dialysis or precipitation. However, the recovery of the polymer by these methods may not be easy and is not quantitative. Moreover, low molecular weight species may be insoluble in the precipitating medium and thus complete removal of impurities from the precipitated polymer may not be achieved. Gel formation is yet another potential problem with the use of linear polymers.

2.1.2   Crosslinked polymers

If the polymers are formed in the presence of a crosslinking agent, or are subsequently crosslinked in a post-polymerization process, all the chains are effectively interconnected to form an infinite network. Such a system can no longer form a true molecular solution and may be regarded as insoluble in a strict thermodynamic sense. Crosslinked polymers exhibit considerable differences in properties depending on the degree of crosslinking and the method of preparation. They can be conveniently characterized in terms of their total surface area (internal and external), total pore volumes and average pore diameter. In general, the degree of crosslinking determines the solubility, extent of swelling, pore size, total surface area and mechanical stability of the polymer.

The use of crosslinked polymers in chemical applications is associated with some advantages, such as the following.

1. Since they are insoluble in all solvents, they offer the greatest ease of processing.
2. They can be prepared in the form of spherical beads which do not coalesce when placed in a suspending solvent and can be separated from low molecular weight contaminants by simple filtration and washing with various solvents.
3. Polymer beads with low degrees of crosslinking swell extensively, exposing their inner reactive groups to the soluble reagents.
4. More highly crosslinked resins may be prepared with very porous structures which allow solvents and reagents to penetrate the inside of the beads to contact reactive groups.

However, there are also a number of disadvantages arising from the use of crosslinked polymers.

1. From investigations of the reaction rates and kinetic course in solid phase synthesis it is found that the reaction sites within the polymeric matrix are not equivalent either chemically or kinetically, which makes quantitative conversions almost impossible.
2. The difficulty in accessibility of insoluble polymers appears to limit a more general application of these compounds in chemical reactions.
3. Factors such as rate of diffusion and pore size may restrict the reactions, especially in the case of larger substrates which may only be able to react at some of the more accessible sites located on the surface of the beads or within the larger pores.
4. It is often very difficult to control the loading of the resin accurately. However, it is possible to cause a reaction to occur at a fraction of the available sites by control of the swelling of the polymers. Such reactions on partially swollen resins give functional polymers in which the reactive sites are not distributed evenly throughout the bead but are concentrated in the more accessible sites only.
5. It is difficult to characterize adequately the structural changes which take place, since a number of analytical methods are not well suited for the study of insoluble materials.
6. Not all reagents can penetrate with ease into the crosslinked network.
7. The introduction of any functional group onto a resin may remove some of the original pore volume, whether this be in the form of permanent

macropores or in the form of gel porosity of solvent-swollen lightly crosslinked materials.

8. The generation of a polar environment in an originally non-polar support, and vice versa, by the introduction of appropriate functional groups can alter the solvent compatibility of the system significantly.
9. In some instances, ionic groups generated on a lightly crosslinked non-polar support can actually aggregate or cluster into charged nuclei, considerably increasing the rigidity of the resin matrix.
10. Crosslinked polymers exhibit considerable differences in properties depending on the degree of crosslinking and the method of preparation.

The following is a classification of the types of crosslinked polymers which are most frequently encountered with enhanced properties.

(a) Microporous or gel-type resins

Microporous or gel-type resins [1, 2] are generally prepared by suspension polymerization using a mixture of vinyl monomer and small amounts (less than 10%; in most cases less than 0.5%–2%) of a crosslinking agent containing no additional solvents. The growing polymer chains are solvated by non-incorporated monomer molecules, but as higher conversions are reached this solvation diminishes and finally disappears. The resulting nuclei tend to aggregate as the more extended portions of the polymer chains slowly collapse together, eventually forming a dense glass-like material in the form of spherical beads. The crosslinking sites are usually randomly distributed, producing a heterogeneous network structure. In the dry state the pores of a gel resin are small, and hence they are often referred to as microporous resins. However, on the addition of a good solvent extensive resolvation of the polymer chains causes considerable swelling with the formation of a soft gel and results in the reappearance of considerable porosity that depends on the degree of cross-linking.

Swellable polymers are found to offer advantages over non-swellable polymers. Of particular interest is their lower fragility, i.e. lower sensitivity to sudden shock, and their potential to achieve a higher loading capacity during functionalization. However, a decrease in crosslinking density will increase the swelling but will also result in soft gels which generally have low mechanical stability and readily fragment even under careful handling. Gels with lower density of crosslinking are difficult to filter and under severe reaction conditions can degrade to produce soluble linear fragments. In addition, gel-type resins that are lightly crosslinked may suffer considerable mechanical damage as a result of rapid and extreme changes in the nature of the solvating media and cannot be subjected to steady and high pressures.

Microporous resins with less than 1% crosslinking generally have low mechanical stability and readily fragment even under careful handling, while macroporous resins with more than 8% crosslinking are mechanically stable but unfortunately give rise to acute diffusional limitations that result in slow and incomplete reactions. In chemical applications, resins with a crosslink ratio of approximately 2% provide a satisfactory compromise, generally allowing adequate penetration by most reagents and yet retaining sufficient mechanical stability to provide ease of handling; they have found wide application as a result of these factors.

(b)   Macroporous and macroreticular resins

The mechanical requirements in industrial applications force the use of higher crosslinking densities for preparing resins with enhanced properties. Macroporous [3–11] and macroreticular [10, 12–16] resins are also prepared by suspension polymerization using higher amounts of the crosslinking agent but with the inclusion of an inert solvent as diluent for the monomer phase. The diluent can be either a good solvent for both the comonomers and the resulting polymer (non-crosslinked) or a good solvent only for the monomer and a precipitant for the polymer. As the polymer is solvated, a fully expanded network is formed with a considerable degree of small pore porosity. Poor solvents, however, lead to large pores. The proper pore size is achieved by a diluent composed of a mixture of two solvents, one of which is a good solvent for the polymer and the other is a precipitant. With macroporous resins the growing chains remain fully solvated in a good solvent during polymerization and do not collapse as the comonomer is consumed. Crosslink ratios of about 20% are most common, so that the matrix formed has sufficient mechanical stability in the solvent state and a large volume of solvent is retained. The resulting polymer, when the polymerization has been completed, contains cavities filled with the solvent; the pores may collapse partially when the solvent is removed because of the much larger extent of the solvated network during polymerization, but this collapse is reversible and if the polymer is placed again in a good solvent the initial macroporous structure is reformed. Macroporous resins will also absorb varying quantities of bad solvents and remain in a fully expanded form, i.e. removal of solvent yields a residual network with a permanent system of macropores.

When the solvent employed during polymerization is a good solvent for the monomer but a precipitant for the polymer (non-crosslinked), the term macroreticular is generally employed to describe the product. Macroreticular resin is non-swelling and macroporous, a rigid material with a high crosslinking ratio; it retains its overall shape and volume when the precipitant is removed. The method adopted for the synthesis of this type of resin consists essentially of the usual homogeneous solution phase process modified by inclusion of a non-solvent for the expected polymer. The ratio of non-solvent in the reaction mixture is critical and must be carefully adjusted to cause the crosslinked particles to precipitate at the desired stage of the polymerization. Control of particle size can be accomplished by adjusting the rate of stirring, but the nature of the solvent, non-solvent and crosslinker components mainly determines the physical characteristics of the final product.

The structure of these resins is quite different from that of the previous two. They have a large, definitive and permanent internal porous structure with an effective surface area larger than that of swollen beads. Macroreticular resins are generally much less sensitive to the choice of solvent and can absorb significant quantities of both solvents and non-solvents, which probably fill the available voids. In general, the whole structure is not susceptible to the dramatic changes observed with gel-type resins when the nature of the surrounding medium is changed. The dimensional stability of macroreticular resins makes them resistant to high pressure in column applications where better solvent flow rates can be achieved than would be the case with gel polymers. Macroreticular resins usually display negligible change in volume during their use. Moreover, they have further advantages in chemical applications. These include ease of filtration from the reaction medium and minimal effects of surface impurities. The main disadvantages of these resins include (a) a lower reactivity than their swellable counterparts, (b) a lower loading capacity, (c) their brittle nature (they may fracture under sudden stress during handling, with the formation of fine particles which are extremely difficult to work with) and (d) the difficulty in handling dry beads as a result of static electricity.

(c) Popcorn polymers

Popcorn polymers [17–22] are prepared by gently warming a mixture of vinyl monomer and a small amount of crosslinking agent, 0.1%–0.5%, in the absence of any initiators or solvents. Popcorn polymer is a white glassy opaque granular material, fully insoluble and porous, with a low density. It is not swellable in most solvents but is easily penetrated by small molecules. It has a reactivity comparable with that of solvent-swellable bead but is often more difficult to handle.

(d) Macronet polymers

Macronet polymers [23, 24] are three-dimensional crosslinked networks in which linear polymer chains are interconnected by a separate chemical reaction following polymerization. 'Macronets are usually produced in the presence of a reaction solvent such that the resulting material has a relatively floppy structure and is capable of reabsorbing large quantities of solvents.' As a result, it has the disadvantage of poor mechanical stability. Experience of the structure and loading of bound species accrues from a macronet species in which a linear polymer appropriately functionalized by copolymerization or chemical modification, is structurally analysed and characterized and is then crosslinked.

2.2 SOLVATION BEHAVIOUR

The solvent has a dominant influence on the physical nature and the chemical reactivity of immobilized molecules. An organic linear macromolecule can dissolve in an appropriate solvent to form a true molecular solution in which the concentration of polymer can be made to approach zero. Dissolving a polymer is a slow process that can take place if the polymer–polymer intermolecular forces can be overcome by strong polymer–solvent interactions in which the gel gradually disintegrates into solution.

In solution, the polymer chain generally exists as a random coil which may be highly expanded or tightly contracted depending on the thermodynamics of polymer–solvent interactions. Generally, a highly compatible or good solvent, where polymer–solvent contacts are highly favoured, will give rise to an expanded coil conformation, and as the solvating medium is made progressively poorer the coil contracts and eventually precipitation takes place. The conformations of randomly coiling mass occupy many times the volume of its segments alone. The random coil arises from the relative freedom of rotation associated with the chain bonds of most polymers and the large number of conformations accessible to the molecule. The ability of a given solvent to dissolve a linear polymer depends on (a) the chemical nature of the polymeric backbone, (b) the molecular weight, (c) the crystallinity, (d) the nature of the solvent, i.e. the polymer–solvent interaction forces, and (e) temperature.

However, the absence of solubility does not imply crosslinking. Other features, such as crystallinity, hydrogen bonding and a high molecular weight, give rise to sufficiently large intermolecular forces to hinder solubility. A crosslinked system can be solvated by a suitable solvent and remains macroscopically insoluble. In this case, swelling rather than solubility is the required property; the polymer can be solvated homogeneously only to a limited extent, beyond which addition of more solvent will not increase solvation. Swelling of resin beads is very important as it brings the polymer to a state of complete solvation and thus allows easy penetration of the network by molecules of the reagent. The crosslink ratio controls the behaviour of a resin in contact with a solvent and is inversely proportional to the degree of swelling. When a good solvent is added to a crosslinked polymeric network solvent molecules slowly diffuse into the polymer resulting in swelling and gelation and it becomes highly expanded and extremely porous. If the degree of crosslinking is low, then such gel networks can consist largely of solvent with only a small fraction of the total mass being polymer backbone. As the degree of crosslinking is increased, or if strong polymer–polymer intermolecular forces are present because of crystallinity or strong hydrogen bonding, then the ability of the network to expand in a good solvent is reduced and penetration of reagents to the interior may become impaired. With poor solvents, crosslinked matrices display little tendency to expand and movement of reagents within such an interior can become somewhat analogous to a diffusional process in the polymer solid. Solvent compatibility with the resin can be adjusted by mixing monomeric units in the polymer chain, i.e. by the use of copolymers.

Information on the degree of swellability of the polymers could be determined either from the measured density of the dry resin and the weight of imbided solvent using the centrifugation technique [25–29] or from the proportion of the specific gel bed volume to the bulk volume [30]. The volumetric swelling coefficient B can be calculated using Duesek's equation [31]:

$$B = \frac{\rho_{ap}}{\rho} + (w-1)\frac{\rho_{ap}}{\rho_{solv}} (mL\,mL^{-1}) \tag{2.1}$$

where ρ_{solv} is the solvent density $(g\,mL^{-1})$, w is the swollen polymer weight divided by the dried polymer weight and $w-1$ has the same meaning and value as the measured solvent uptake coefficient.

2.3 POROSITY

In the swollen state, a crosslinked polymer has a certain porosity in which the size and shape of the pores may continuously change owing to the solvating effect of a good solvent and hence the mobility of the polymer segments. Dry solid supports can be conveniently characterized in terms of their total surface area (internal and external), total pore volumes and average pore diameter. These physical parameters are not independent of each other but are generally interrelated by the simple geometrical equations

$$P = n\pi r^2 l \tag{2.2}$$

and

$$S = 2n\pi r l$$

where P is the pore volume, S is the surface area, r is the average pore radius, n is the number of pores, l is the average pore height and nl is the effective total pore length. If cylindrical pores with a large mean diameter are assumed, these equations illustrate that the number of pores is likely to be relatively small and that the total interior surface area is also restricted. Conversely, as the total interior surface area increases, the number of pores increases and the radius of each pore diminishes.

Gel-type supports usually have relatively small pore diameters and a large effective surface area which, in some circumstances, gives rise to high loading capabilities, up to approximately $10\ mmol\,g^{-1}$. Macroporous or macroreticular supports have large pore diameters but a relatively small surface area. Chemical modification of these species is thought to occur largely on the pore surfaces, and generally the highly crosslinked and entangled polymer chains are not readily available for functionalization. Loading capabilities are of the order of 3 mmol g^{-1}. A resin of high surface area (approximately $500\ m^2\,g^{-1}$) can be prepared by using a good solvent as a porogenic agent.

Surface areas are usually measured by N_2 adsorption–desorption isotherms. Total pore volumes can be obtained simply by measuring the volume uptake of an appropriate liquid. The total pore volume and pore size distribution depend upon the type and relative volume of the diluent, on the degree of crosslinking and on the reaction conditions. The pore volume PV of the polymers can be calculated using the equation

$$PV = \frac{1}{\rho_{ap}} - \frac{1}{\rho}(mL\,g^{-1})$$ (2.3)

where ρ_{ap} is the apparent density $(g\,mL^{-1})$ and ρ is the skeletal density $(g\,mL^{-1})$. These are measured by the picnometric technique. Electron microscopy and small angle X-ray scattering can be employed to measure pore diameters. The average pore diameter $\bar{D}$ can also be estimated according to the equation

$$\bar{D} = \frac{4PV}{S_{BET}} \times 10^4$$ (2.4)

The high porosity of the matrix has two desirable effects. It leads to good flow properties, and it does not hinder the penetration of molecules of high molecular weight. The polymer porosity $(\%P)$ can be calculated according to the equation [32]

$$\%P = 100\left(1 - \frac{\rho_{ap}}{\rho}\right)$$ (2.5)

It should be noted, however, that solvent–polymer interaction determines the porous structure of the networks decisively [33, 34].

2.4   PERMEABILITY

Permeability [35] is the measure of the ease with which an intact material can be penetrated by a given gas or liquid. This property is most important and most conveniently studied in the passage of matter through a membrane. Membranes are generally described as permeable, semipermeable or permselective depending upon the nature of the penetrants under consideration.

Membranes can be homogeneous or heterogeneous. A homogeneous membrane is defined as one which has uniform properties across all its dimensions, while a heterogeneous membrane has some anisotropy due to either molecular orientation during the manufacturing process or fillers, additives, voids or reinforcing materials. The permeability coefficient P is generally the proportionality constant between the flow of penetrant per unit area of membrane per unit time and the driving force per unit thickness of membrane. Permeation of relatively small molecules through a membrane may occur by one of the following processes.

(i) Flow through pores or capillaries in the non-homogeneous membrane

[36, 37]. The size of the permeant relative to pore size and the viscosity of the permeant are the controlling factors governing permeability. The simplest type of flow mechanism is viscous flow, in which the volume q of penetrant passing through a capillary of radius r and length, Δx in unit time is given by Poiseuille's equation:

$$q = \frac{\pi r^4 \Delta p}{8 \eta \Delta x} \qquad (2.6)$$

where η is the viscosity of the permeant and Δp is the pressure difference across the capillary. Accordingly, the permeability coefficient P corresponds to

$$P = \frac{\Phi \beta r^2}{8 \eta} \qquad (2.7)$$

where β is a tortuosity factor which increases the effective length from Δx to $\Delta x/\beta$ and Φ is the volume fraction of capillary in the membrane. For all penetrants that do not interact with the membrane, i.e. for which Φ and r are independent of the penetrant, the permeability coefficient is inversely proportional to the viscosity of the penetrant.

(ii) Diffuse flux of molecules dissolved in a membrane which has no pores or voids. In this process the penetrant dissolves and equilibrates in the membrane surface and then diffuses in the direction of lower chemical potential. If the boundary conditions on the two sides of the membrane are maintained constant, a steady state flux of the components will be established which can be described at every point within the membrane by Fick's first law of diffusion:

$$Q_i = -D_i \frac{dc_i}{dx} \qquad (2.8)$$

where Q_i is the mass flux (g cm^{-2} s^{-1}), D_i is the local diffusivity (cm^2 s^{-1}), c_i is the local concentration of component i (g cm^{-3}) and x is the distance through the membrane measured perpendicular to the surface. The measurement of permeability [38] is carried out by two basic methods: the transmission method and the sorption–desorption method.

In the transmission method a concentration gradient of the penetrant is applied across the membrane and the rate of transmission of penetrant passing through the membrane in unit time can then be determined by a number of techniques such as the refractive index method or interferometry, thermal conductivity, chemical analysis or colorimetry, gravimetric techniques, mass spectroscopy, gas chromatography and pressure–volume–temperature measurements of gases.

In the sorption–desorption method samples are rapidly brought into a liquid or vapour of known activity, and the diffusion coefficient and the solubility coefficient can be calculated from the rate of sorption and desorption and the equilibrium sorption value. This method is largely used for condensable vapour–polymer systems in which an appreciable amount of vapour is sorbed, and the solubility coefficient and diffusion coefficient are often dependent on concentration.

2.5 STABILITY

Chain degradation is generally possible both in the presence and in the absence of oxygen at higher temperatures. It may be caused by thermal, hydrolytic or mechanical effects. The mechanical stability of networks varies considerably from one material to another, and also depends on the nature of the mechanical stress and on the crosslink ratio. Lightly crosslinked materials are extremely fragile, particularly when in contact with a good solvent, and even conventional stirring techniques can cause considerable mechanical degradation of the support. Increased physical stability can be achieved with increased crosslinking but there always exists a balance between the required mechanical properties and the porosity of the network. Macroreticular resins can be employed in high pressure conditions and present some flexibility in the use of solvents. Gel-type resins are readily compressed and are not suitable for high pressure applications but can show marked mechanical resilience and ability to absorb shock because of their elastomeric properties in the swollen state. However, sudden dramatic shear will cause considerable damage. Similar effects can arise from osmotic shock if the nature of the solvent is changed dramatically, and rapid evaporation of solvents from the interior can also cause excessive rupture of the structure due to the sudden increase in volume.

2.6 REACTIVITY OF FUNCTIONALIZED POLYMERS

A functional group attached to a polymer chain may have a quite different reactivity from an analogous group on a small molecule because of its macromolecular environment. Thus, more drastic reaction conditions may be required to reach a satisfactory conversion. The design of a new reactive polymer must be planned by considering important factors which affect its activity. These include (a) the type of solvents and reagents to which the polymer must be subjected during the course of its functionalization or subsequent reactions and (b) the chemical behaviour of the support which depends on its physical form, crosslinking density, the flexibility of the chain segments and the degree of substitution.

Since the functional groups on the resin are not free to move, the surrounding low molecular weight substances must diffuse to the fixed reactive sites in the rigid-gel structure, essentially by using solvents with good swelling properties. The primary function of the solvent is to affect the degree of swelling of the polymer lattice, which is also an important factor in determining the chemical reactivity of immobilized molecules [39]. In fact, poorly swollen resin retards the rotation of unattached molecules imbibed in the matrix. The swollen polymer exhibits a high internal viscosity and the crosslinks restrict the long-range mobility of chain segments; thus the collision frequency of substituents attached to different chain segments is reduced substantially.

The role of a solvent in the application and reaction of a functionalized resin is complex. An ideal solvent should meet the following requirements.

1. It should interact with the polymer matrix to optimize the diffusional mobility of reagent molecules.
2. It should have the correct solvating characteristics to aid any chemical transformations being carried out.
3. It should not limit the reaction conditions which are to be applied.
4. It should enhance translucence rather than opacity.

Naturally, it is difficult to satisfy all these criteria simultaneously and the selection of a solvent often involves compromise.

Gel polymers are usually found to be slightly less reactive than linear polymers as reactions will be limited by diffusion of the reagent within the resin pores. The reaction yields can be affected by the degree of crosslinking. Highly crosslinked resins result in lower yields. Thus resins with very low degrees of crosslinking will be the most suitable as increased swelling will result in higher accessibility through enhanced diffusion properties. In addition, swellable polymers are found to offer the advantage of achieving higher loading capacity during functionalization. They are generally much less sensitive to the choice of solvents. Since the reactivity of these resins is not a function of swelling, swelling of the polymer matrix is not required prior to reactions. In this case, reaction can often be carried out in a variety of solvents without appreciable change in reaction rates. The rotational motion of the polymer chains and hence the mobility of the functional groups is a function of the polymer reactivity.

The reactivity of a functional group may be low when it is directly attached to the main chain; this may be a result of steric hindrance by the polymer backbone and neighbouring side groups. In addition to the microenvironment of the functional groups, surface impurities on the polymer beads have marked influence on the apparent lack of reactivity of a functionalized polymer. An additional cause for the apparent lack of reactivity may be that the structure of some of its functional groups is different from that which is assumed from the reaction sequence leading to it, i.e. the polymer may contain interfering functionalities introduced during its preparation or chemical modification.

The capacity of a polymer support is also important in terms of reactivity. A polymer with a very high capacity may only react partially due to a lack of accessibility of the functional sites. Since the size of the molecules which are attached to the polymer may increase during a synthesis and result in other changes such as variations in polarity of the medium, the accessibility of the polymer–substrate bond may become restricted and result in partial or difficult cleavage when the synthesis is complete. In contrast, a polymer with a very low capacity may not be useful for a synthesis on a practical scale. Furthermore, the reaction rate of the functional group depends on the nature of the functional group, the concentration of the low molecular weight species in solution in contact with the resin, the diffusion rate of the low molecular weight species, the diameter of the resin particles, the temperature of the reaction and the mixing rate.

REFERENCES

1. Millar, J.R., Smith, D.G., Marr, W.E. and Kressman, T.R.E. (1963) *J. Chem. Soc.*, 218.
2. Hoffmann, M. (1974) *Makromol. Chem.*, **175**, 613.
3. Tilak, M.A., Hollinden, C. and Stephen, C. (1971) *Org. Prep. Proc. Int.*, **3**, 183.
4. Grubbs, R.H., Gibbons, C., Kroll, L.C., Bonds, W.D. and Brubaker, C.H. (1973) *J. Am. Chem. Soc.*, **95**, 2373.
5. Seidl, J., Malinsky, J., Dusek, K. and Heitz, W. (1967) *Adv. Polym. Sci.*, **5**, 114.
6. Heitz, W. (1970) *J. Chromatogr.*, **53**, 37.
7. Corte, A. (1957) Ger. Patent 1 021 166.
8. Moore, J.C. (1964) *J. Polym. Sci.*, **A-2**, 835.
9. Lloyd, W.G. and Alfrey, T. (1962) *J. Polym. Sci.*, **62**, 301.
10. Kunin, R. Meitzner, E. and Bortnick, N. (1962) *J. Am. Chem. Soc.*, **84**, 305.
11. Barret, J.H. and Heights, C. (1974) US Patent 3 843 566.
12. Kunin, R., Meitzner, A.E. and Bortnick, N. (1962) *J. Am. Chem. Soc.*, **84**, 706.
13. Kunin, R., Meitzner, A.E., Oline, A.J., Fischer, A.S. and Frish, N. (1962) *Ind. Eng. Chem., Prod. Res. Dev.*, **1**, 140.
14. Beer, W., Kuhnle, D. and Funke, W. (1972) *Angew. Makromol. Chem.*, **23**, 205.
15. Obrecht, W., Seitz, U. and Funke, W. (1974) *Makromol. Chem.*, **175**, 3587; (1975) *Makromol. Chem.*, **176**, 2771.
16. Kuhnle, D. and Finke, W. (1970) *Makromol. Chem.*, **139**, 255; (1972) *Makromol. Chem.*, **158**, 135.
17. Letsinger, R.L., Kornet, M.J., Mahadevan, V. and Jerina, D.M. (1964) *J. Am. Chem. Soc.*, **86**, 5163.
18. Breitenbach, J.W. (1968) *Adv. Macromol. Chem.*, **1**, 139.
19. Immergut, E.H. (1953) *Makromol. Chem.*, **10**, 93.
20. Letsinger, R.L. and Hamitton, S.B. (1959) *J. Am. Chem. Soc.*, **81**, 3009.
21. Letsinger, R. L. and Kornet, M.J. (1963) *J. Am. Chem. Soc.*, **85**, 3045.
22. Shambhu, M., Theodorakis, M.C. and Digenis, G.A. (1977) *J. Polym. Sci., Polym. Chem. Ed.*, **15**, 525.
23. Davankov, V.A., Tsyurupta, M.P. and Rogozhin, S.V. (1976) *Angew. Makromol. Chem.*, **53**, 19.
24. Davankov, V.A., Rogozhin, S.V. and Tsyurupta, M.P. (1974) *J. Polym. Sci., Polym. Symp.*, **47**, 95.
25. Pepper, K.W., Reichenberg, D. and Hale, D.K. (1952) *J. Chem. Soc.*, 3129.
26. Gregor, H.P., Hoeschele, G.K., Potenza, J., Tsuk, A.G., Feinland, R., Shida, M. and Teyssie, P. (1965) *J. Am. Chem. Soc.*, **87**, 5525.
27. Greig, J.A. and Sherrington, D.C. (1978) *Polymer*, **19**, 163.
28. Stamberg, J. and Sevcik, S. (1966) *Coll. Czech. Chem. Commun.*, **31**, 1009.
29. Pepper, K. W. (1951) *J. Appl. Chem.*, **1**, 126.
30. Heitz, W. and Platt, K.L. (1969) *Makromol. Chem.*, **127**, 113.
31. Dusek, K. (1965) *Coll. Czech. Chem. Commun.*, **30**, 3804.

32. Bortel, F. (1965) *Przem. Chem.*, **44**, 255.
33. Alfrey, T. and Lloyd, W.G. (1962) *J. Polym. Sci.*, **62**, 159; (1967) US Patent 332 269.
34. Lloyd, W. G. and Alfrey, T. (1962) *J. Polym. Sci.*, **62**, 301.
35. Yasuda, H., Clark, H.G. and Stannett, V. (1968) *Encyclopedia Polym. Sci. Tech.*, **9**, 794.
36. Barrer, R.M. (1956) *Diffusion In and Through Solids*, Cambridge University Press, London.
37. Carmen, P.C. (1956) *Flow of Gases through Porous Medial*, Academic Press, London.
38. Stannett, V. and Yasuda, H. (1965) *Testing of Polymers*, Wiley, New York.
39. Regen, S.L. (1974) *J. Am. Chem. Soc.*, **96**, 5275; (1975) *J. Am. Chem. Soc.*, **97**, 3108.

Part Two   Chemical Applications

Since the mid-1960s synthetic macromolecules have been increasingly recognized as organic species capable of behaving as organic reactants and susceptible, under appropriate conditions, to all the chemical transformations of smaller organic species. The use of polymers as reactive molecules in organic synthesis was introduced by Merrifield in 1963, when he introduced his 'solid-phase technique' for the synthesis of peptides [1]. In this technique, an insoluble macromolecule was used as a protecting group, simultaneously providing an easy method for isolating and purifying the product of each condensation step. Since that announcement, functionalized polymers have found widespread application in organic synthesis and related chemical fields [2–36]. They have been employed as stoichiometric reagents, as catalysts and as substrate carriers in separations and in multiple synthesis. The use of functionalized polymers in organic synthesis and other chemical applications shows, as with any new technique, both advantages and disadvantages. The most important advantages in using a functionalized polymer over conventional low molecular weight species include the following.

1. The simplification of product work-up, i.e. separation and isolation from reaction side-products or byproducts after each reaction step, is the most practical advantage of the technique. In the case of crosslinked polymer resins simple filtration procedures can be used for isolation and washing, and the need for complex chromatographic techniques can be eliminated. With linear polymers, techniques such as precipitation, sedimentation and ultrafiltration can be employed.
2. Resins provide the possibility of automation in the case of repetitive stepwise synthesis, and the facility for carrying out reactions on a commercial scale in the same way as ion exchange resins. This property has industrial attractions.
3. Supported reagents may also be used more conveniently in large excess to increase reaction rates and hence to drive reactions to completion to increase

the reaction yields, without problems in the work-up procedures and separation.

4. Scarce and expensive materials can be efficiently retained when attached to a polymer and in principle, if appropriate chemistry is available, they can be recycled and reused many times. The regeneration of polymeric byproduct is very important economically.

5. The reactivity of an unstable reagent or catalyst may be decreased when supported on a resin and the corrosive action of, for example, protonic acids can also be minimized by this effective encapsulation. In addition, they make reactive explosive gas species suitably stable for safe storage over prolonged periods.

6. Since polymers are non-volatile, toxic and malodorous materials, they often lack many inconvenient properties of the corresponding low molecular weight reagents and can be rendered environmentally more acceptable.

7. In some cases, the chemical and steric structure of the polymer backbone provides a special microenvironment for reactions of the pendant reactive group. For example, the hydrophobic environment of the polymer protects water-sensitive catalysts from attack by atmospheric moisture. The nature of the backbone, e.g. the bulkiness of the substituents, the polarity, and the size and structure of the pores, imparts some specificity to the transition state of the reaction which may lead to a regioselective or stereospecific reaction.

8. A number of potentially important reactivity changes may be induced by the use of a functionalized polymer. The insolubility of crosslinked polymer restricts the interaction of functional groups and makes them inaccessible to each other, which leads to reactions not possible with the analogous soluble molecules. A high degree of crosslinking, a low level of functionalization, low reaction temperatures and the development of electronic charges near the polymer backbone tend to encourage this situation, which may be regarded as mimicking the solution condition of 'infinite dilution'. In these circumstances, intermolecular reaction of bound molecules is prevented and such attached residues can be made either to react intramolecularly or to react selectively with an added soluble reagent. Polymer-supported metal complexes with vacant coordination sites can be regarded as fulfilling this description, with the resin inhibiting the normal solution oligomerization processes of such species.

9. Under certain circumstances it is also possible to achieve the complementary state of 'high concentration' by heavily loading a flexible polymer matrix with one particular moiety in an attempt to force its reaction with a second polymer-bound species.

Balancing the above advantages, there are also a number of considerable disadvantages.

1. Probably the most important of these is the likely additional time and cost in synthesizing a supported reagent or catalyst. This may well be offset by the

potential advantages, and certainly in the case of regeneratable and recyclable species the objection essentially disappears.

2. The occurrence of predominantly slow reactions and poor yields, however, can seldom be accommodated and can be a problem. The lower yields of products than in homogeneous reactions is caused by the polymer steric hindrance on the attachment of reactants to the reactive site, by the incompatibility of the reactants and the polymer, and by absorption of the product to the resin. Appropriate choice of support and reaction conditions can overcome this.

3. With applications involving the assembly or modification of a polymer-bound substrate, the final cleavage step releasing the product into solution can be incomplete, or vigorous conditions employed in cleavage may result in degradation of the polymer.

4. The overall chemical and mechanical stability of the support can often be limiting, and in the case of strongly acidic and basic ion exchange resins limited temperature stability has hindered their widespread commercial application.

5. The ultimate capacity of a functionalized polymer and the reduction of the degree of functionalization during the regeneration processes are also restricted, and may be important in preparative organic chemistry involving stoichiometric quantities of supported reagents. One of the major differences between inorganic supports and synthetic polymers is the lower loading capabilities of the former which, while being suitable for the attachment of catalysts, is totally unsuitable for high capacity demands.

6. The product obtained at the end of the multistep synthesis may be contaminated with undesired materials, which may increase the difficulty of obtaining the product in a pure state.

7. Investigation of the reaction rates and kinetic course revealed that the reaction sites within the matrix are neither chemically and nor kinetically equivalent, making quantitative conversion almost impossible.

8. Difficulties in the characterization of reactions on polymers can also arise. These are maximized in the case of resins where the techniques relying on the formation of a true homogeneous solution can at best be rendered inadequate or insensitive and at worst be completely useless.

9. The polarity of the polymer may change after reactions.

10. Finally, with the use of functionalized polymers there always exists the additional chemical option of a side reaction with the polymer itself.

Despite these potential drawbacks, a marked increase in the preparation and investigation of chemically active species bound to polymer supports has evolved and considerable scope now exists for their exploitation in routine synthetic chemistry, biochemistry and other chemical applications. However, the polymeric support used in chemical reactions should have the following properties in order to be effective.

1. The support should be totally insoluble in common solvents to avoid losses of catalyst and to facilitate handling and purification. This is generally achieved by the use of crosslinked polymers, either rigid (non-swellable) or flexible (swellable).
2. It should be capable of functionalization to a high degree of substitution of reactive sites which can be used for specific purposes and the functional groups should be uniformly distributed in the polymer.
3. It should undergo straightforward reaction with the reagents and be free of any side reactions and of undesirable competing functionalities.
4. It should be compatible with the solvents and reagents used, which is sometimes achieved by grafting reactive functional groups to the polymer backbone by a spacer-arm. It can also be increased by the incorporation of a solvent of the same nature as the support.
5. It should have good mechanical stability during handling. Mechanical and thermal stabilities are generally governed by the intrinsic nature of the polymer and by the extent of crosslinking. For example, where the use of a polystyrene support is limited to 150 °C, a condensed polyaromatic polymer is thermally much more stable; also macroreticular polystyrenes possess a much more rigid network than conventional 2% divinylbenzene resins. The thermal stability concerns also the link between the support and the reactive part of the functionalized support. This property is needed to avoid leaching of the active group in a process or on recycling.
6. The polymeric byproduct should be capable of being regenerated after use via a simple procedure.
7. The backbone of the support must be inert towards the reagents and not impart steric restriction to the circulation of the reagent and of the product near the active sites.
8. With regard to physical form, small spherical crosslinked beads are often preferred.
9. Diffusion in a polymer depends on the nature of its porous structure, which in turn is usually determined by the extent of swelling of the polymer lattice. This is governed by the degree of crosslinking as well as the nature of the substrate and solvent used in the reaction.

REFERENCES

1.   Merrifield, R.B. (1963) *J. Am. Chem. Sci.*, **85**, 2149.
2.   Patterson, J.A. (1971) In *Biochemical Aspects of Reactions on Solid Supports*, (ed. G.R. Stark), Academic Press, New York, p. 189.
3.   Pittman, C.U. and Evans, C.O. (1973) *Chem. Tech.*, 650.
4.   Overberger, C.G. and Sannes, K.N. (1974) *Angew. Chem., Int. Ed. Engl.*, **13**, 99.
5.   Leznoff, C.C. (1974) *Chem. Soc. Rev.*, **3**, 65.
6.   Patchornik, A. and Kraus, M.A. (1975) *Pure Appl. Chem.*, **43**, 503.
7.   Crowley, J.I. and Rapoport, H. (1976) *Acc. Chem. Res.*, **9**, 135.
8.   Mathur, N.K. and Williams, R.E. (1976) *J. Macromol. Sci., Rev. Macromol. Chem.*, **C-15**, 117.

9. Patchornik, A. and Kraus, M.A. (1976) In *Encyclopedia of Polymer Science and Technology*, Suppl. vol. 1 (eds H.F. Mark, N.G. Gaylord and N.M. Bikales), Interscience, New York, p. 468.
10. Crosby, G.A. (1976) *Aldrichimica Acta*, **9**, 15.
11. Patchornik, A. and Kraus, M.A. (1976) *Pure Appl. Chem.*, **46**, 183.
12. Heitz, W. (1977) *Adv. Polym. Sci.*, **23**, 1.
13. Manecke, G. and Renter, P. (1978) *J. Polym. Sci., Polym. Symp.*, **62**, 227.
14. Leznoff, C.C. (1975) *Acc. Chem. Res.*, **11**, 327.
15. Manecke, G. and Storck, W. (1978) *Angew. Chem., Int. Ed. Engl.*, **17**, 657.
16. Hodge, P. (1978) *Chem. Br.*, **14**, 237.
17. Patchornik, A. and Kraus, M.A. (1979) *Chim. Ind. (Milan)*, **61**, 830.
18. Manecke, G. and Renter, P. (1979) *Pure Appl. Chem.*, **51**, 2313.
19. Vogl, O. (1979) *Pure Appl. Chem.*, **51**, 2409.
20. Daly, W.H. (1979) *Makromol. Chem., Suppl.*, **2**, 3.
21. Hodge, P. and Sherrington, D.C. (eds) (1980) *Polymer-supported Reactions in Organic Synthesis*, Wiley, London.
22. Sherrington, D.C. (1980) *Br. Polym. J.*, **12**, 70.
23. Mathur, N.K., Narang, C.K. and Williams, R.E. (1980) *Polymers as Aids in Organic Chemistry*, Academic Press, New York.
24. Kraus, M.A. and Patchornik, A. (1980) *J. Polym. Sci., Macromol. Rev.*, **15**, 55.
25. Akelah, A. (1981) *Synthesis*, 413.
26. Akelah, A. (1981) *Br. Polym. J.*, **13**, 107.
27. Akelah, A. and Sherrington, D.C. (1981) *Chem. Rev.*, **81**, 557.
28. Frechet, J.M.J. (1981) *Tetrahedron*, **37**, 663.
29. Geckeler, K., Pillai, V.N.R. and Mutter, M. (1981) *Adv. Polym. Sci.*, **39**, 65.
30. Cainelli, G., Manescalchi, F. and Contento, M. (1981) In *Organic Synthesis Today and Tomorrow* (eds B.M. Trost and C.R. Hutchinson), Pergamon, London, p. 19.
31. Sherrington, D.C. (1982) *Nouv. J. Chim.*, **6**, 661.
32. Guyot, A. and Bartholin, M. (1982) *Progr. Polym. Sci.*, **8**, 277.
33. Akelah, A. and Sherrington, D.C. (1983) *Polymer*, **24**, 1369.
34. Sherrington, D.C. (1984) *Br. Polym. J.*, **16**, 164.
35. Ford, W.T. (ed.) (1986) *Polymeric Reagents and Catalysts*, Am. Chem. Soc. Symp. Ser. 308.
36. Akelah, A. (1988) *React. Polym.*, **8**, 273.

3
Polymeric reagents

A polymeric reagent is a reactive organic group bound to a polymeric support, and chemical reactions can be carried out with these reactive groups. The active functional group of the support is consumed during the course of the reaction and used in stoichiometric quantities to achieve the required chemical modification of the added substrate:

$$\text{(P)}\!-\!X + \text{substrate} \longrightarrow \text{(P)}\!-\!Y + \text{product} \qquad (3.1)$$
polymeric reagent   polymeric byproduct

The desired reactive groups can be attached to polymeric carriers by physical adsorption or by chemical bonding. Physically adsorbed reagents are generally unsatisfactory since the components tend to dissociate in use, and they are therefore unsuitable for column or cyclical applications [1]. The chemical reagents that have been covalently attached to polymeric carriers and successfully used in organic synthesis are listed in Tables 3.1–3.9. In such reactions, a polymeric reagent is usually used in a large excess to give high yields of the desired product, which is usually obtained directly in the solution. A high degree of functionalization is desirable to reduce the amounts of polymer and solvent which are required in a given reaction. Thus, functionalized polymers with a high percentage of reactive groups and a low degree of crosslinking are the most suitable for use as polymeric reagents.

After reaction, the byproduct remains attached to the insoluble polymer and can be removed by simple filtration. Some of these polymeric byproducts can be regenerated, by a single-step synthesis, for repeated use either without or with some chemical change.

3.1 POLYMERIC PHOSPHINE REAGENTS

Phosphorus-containing polymers have a wide range of applications as flame retardants and as chelating agents, and they have recently received considerable

Table 3.1 Polymeric phosphine reagents

Functionalized polymer	Application	Reference
(PS)—PPh$_2$	Wittig reaction	2–12
	Synthesis of vinyl ethers and thioethers, —C=CH—X—R (Z ≡ O, S)	3
	Cleavage of aromatic disulphides	13
	Esterification	14
	Reduction of ozonides	15
(PS)—(CH$_2$)$_4$—P(n-Bu)$_2$	Reduction of disulphides	16
	Hydroformylation of alkenes	
	Michael addition	
(PS)—(CH$_2$)$_n$—PPh$_2$X$_2$		
(a) $n = 0$; X$_2 \equiv$ CCl$_4$	Conversion of alcohols to chlorides	17–23
	Conversion of acids to acid chlorides	18, 19, 22
	Conversion of secondary amides to imidoyl chlorides	24
	Conversion of acids and amines to amides	19, 22
	Conversion of primary amides or aldoximes to nitriles	24
(b) $n = 0$; X$_2 \equiv$ CBr$_4$	Conversion of alcohols to alkyl bromides	25
(c) $n = 0$; X$_2 \equiv$ Cl$_2$	Peptide synthesis	26–28
	Conversion of alcohols to chlorides, acids to acid chlorides and amides to imidoyl chlorides	18
(d) $n = 0$; X$_2 \equiv$ Br$_2$	Conversion of ureas to carbodiimides and amides to imidoyl bromides and ketenimines	29
	Conversion of epoxides to halohydrins	30
	Conversion of alcohols to alkyl bromides	25
	Conversion of ethers to alkyl bromides	31
(e) $n = 0$; X$_2 \equiv$ I$_2$, Br$_2$, Cl$_2$	Esterification	32
(f) $n = 0$; X$_2 \equiv \left(-S - \underset{N}{\bigcirc} \right)_2$ 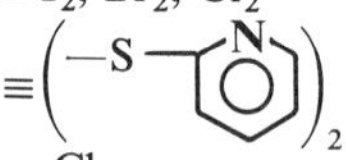	Peptide synthesis	33
(g) $n = 1$; X$_2 \equiv$ Cl$_2$	Conversion of acids to acid chlorides and amides to nitriles	18, 31
(PS)—(PPh$_2$)$_2$RhCl(CO)	Conversion of alkyl lithium and acid chlorides to ketones	34

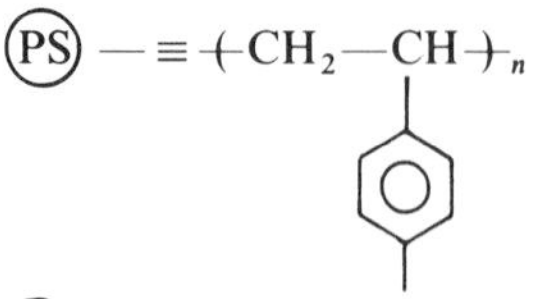

(PS) — ≡ +CH$_2$—CH+$_n$
       |
     [O]
       |

(PEG) — ≡ polyethylene glycol

(C)— ≡ cellulose

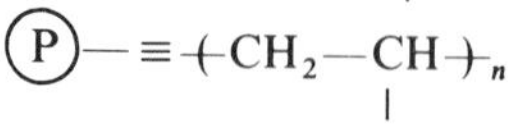

(P)— ≡ +CH$_2$—CH+$_n$
       |

(Si)— ≡ silica

interest for application in organic synthesis as reagents (Table 3.1) and as catalysts.

In spite of the widespread application of the Wittig reaction to olefin synthesis, a principal disadvantage of this reaction is the difficulty of separating the main product from the byproduct, triphenylphosphine oxide. In addition, the phosphine itself is a costly reagent. However, when the insoluble polymeric phosphine reagent 1 is used the byproduct remains attached to the polymer after the reaction and is readily separated from the desired product. In addition, the alkene products are frequently obtained in high purity and quantitative yields.

$$\textcircled{PS}-PPh_2 + R^1-CH-X \longrightarrow \textcircled{PS}-P^+-CH-R^1 \xrightarrow{:B^-} \textcircled{PS}-P=C-R^1$$

(3.2)

$$\downarrow R^3-\overset{\overset{R^4}{|}}{C}=O$$

$$\textcircled{PS}--PPh_2 \quad + \quad R^3-\overset{\overset{R^4}{|}}{C}=\overset{\overset{R^2}{|}}{C}-R^1$$

Moreover, the polymeric phosphine oxide byproduct 2 can readily be recycled and reused in further Wittig reactions [2, 10, 20].

$$\textcircled{PS}-\overset{\overset{O}{||}}{Ph_2} \xrightarrow{Cl_3SiH} \textcircled{PS}-PPh_2$$

(3.3)

The presence of lithium ions in conventional Wittig reactions, arising during generation of the ylide, leads to the formation of *trans*-olefins due to the preferred complexation of the *threo*-form of the betaine. Polymeric phosphonium reagents can give high yields of *cis*-olefins since inorganic lithium salts may be filtered off before the addition of the carbonyl compound [2].

In addition to its use in Wittig reactions, polymeric phosphine has also been used as a mild and efficient reagent for halogenation and dehydration or as condensing reagent. Polymeric phosphine reagents have been used as their dihalides (3, X ≡ Cl, Br) and with carbon tetrachloride (3, X ≡ CCl$_4$) for the cleavage of ethers [20] and for the conversion of carboxylic acids, alcohols and primary amides into the corresponding acid chlorides, alkyl halides and nitriles respectively. The recovered polymeric phosphine oxide is readily converted to the polymeric phosphine dihalide for reuse [11].

Recently, the polymeric phosphine dibromide 3, X ≡ Br, has been used for the preparation of some compounds which are sensitive to water, column chromatography and the elevated temperatures required in distillation. These include imidoyl bromide, carbodiimides and ketenimines [29]. These compounds were

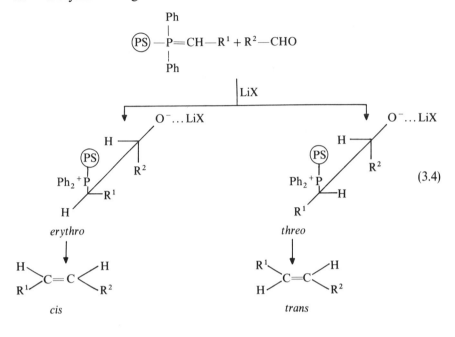

$$(3.4)$$

$$(3.5)$$

prepared in high yields from the corresponding urea (or thiourea) and N-substituted amides in the presence of polymeric reagent [15]. Polystyryldiphenyl-methoxymethyl-phosphonium chloride, **4**, has also been prepared and used for the synthesis of vinyl ethers and sulphide derivatives from carbonyl compounds [3]:

$$(3.6)$$

3.2   POLYMERIC SULPHONIUM SALTS

Thiols and thioethers are very useful synthetic reagents because of their low cost and high reactivity, which often allow the use of mild reaction conditions.

Table 3.2 Polymeric sulphonium salts

Functionalized polymer	Application	Reference
(PS)—(CH$_2$)$_n$—S$^+$MeR X$^-$		
(a) $n = 0$; R ≡ Me; X$^-$ ≡ MeSO$_4$$^-$	Epoxidation of aldehydes	36
(b) $n = 0$; R ≡ Me; X$^-$ ≡ BF$_4$$^-$		37
(c) $n = 0$; R ≡ Cl; X$^-$ ≡ Cl$^-$	Oxidation and selective mono-oxidation of alcohols	35, 38
(d) $n = 0$; R ≡ Me, Et; X$^-$ ≡ FSO$_3$$^-$	Epoxidation of aldehydes	39
(e) $n = 1$; R ≡ Me; X$^-$ ≡ I$^-$	Epoxidation of aldehydes	36, 40
(f) $n = 1$; R ≡ Me, Et; X$^-$ ≡ HCO$_3$$^-$	Peptide synthesis	41
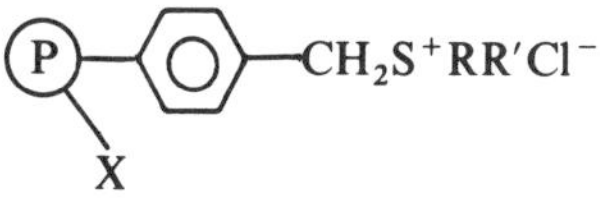 (P)—◯—CH$_2$S$^+$RR′Cl$^-$ with X below; X ≡ Ph, CONH$_2$		42–44
(PS)—S—CH$_2$$^-$ Li$^+$	Homologation of alkyl iodides and diiodides	38
(P)—COO—(CH$_2$)$_2$S$^+$MeR X$^-$		45
(a) R ≡ Me; X$^-$ ≡ MeSO$_4$$^-$		
(b) R ≡ CH$_2$COOH; X$^-$ ≡ Cl$^-$		
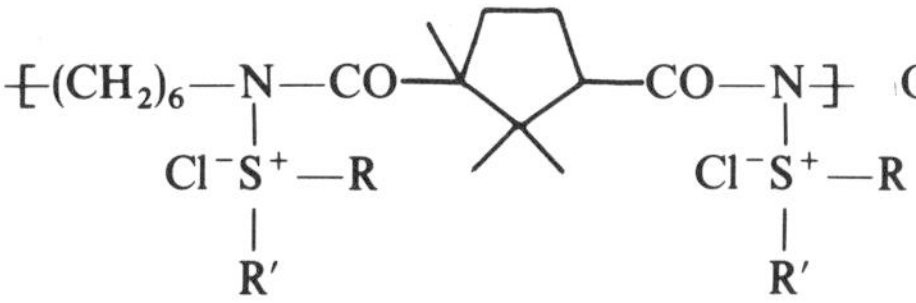 ⫲(CH$_2$)$_6$—N—CO-⟨⟩-CO—N⫲ with Cl$^-$S$^+$—R and R′ groups	Oxidation of alcohols	46
polymer with —O— linkage, Cl$^-$, CH$_2$S$^+$R$_2$		47

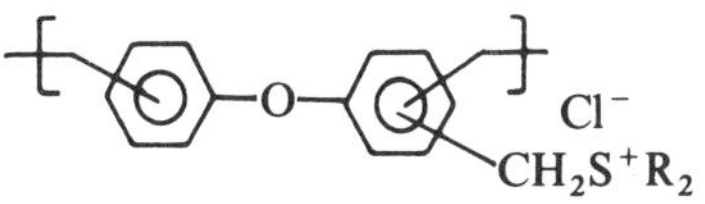

(PS) — ≡ +CH$_2$—CH+$_n$ (with phenyl substituent)

(P) — ≡ +CH$_2$—CH+$_n$

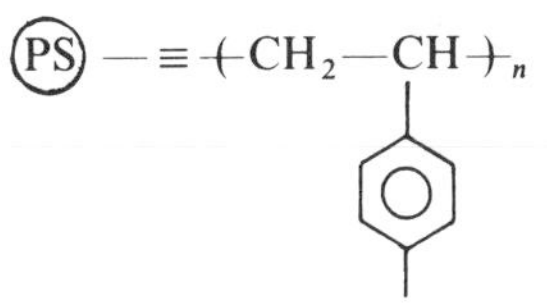

(PEG) — ≡ polyethylene glycol

(C) — ≡ cellulose

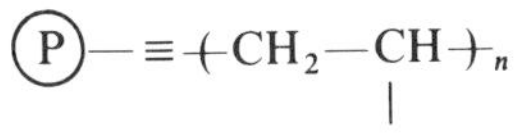

(Si) — ≡ silica

Unfortunately, the noxious odour of sulphide or thiol derivatives and the difficulty of removing sulphide byproducts from reaction mixtures detracts from their usefulness. A polymeric sulphide reagent may be used to overcome these disadvantages, and in addition in some applications it can be reused after washing since the original reagent is automatically reformed [35].

Even though polymeric sulphonium salts are onium salts and function in the same way as the polymeric ammonium and phosphonium salts, only few studies have been carried out on the application of these polymers (Table 3.2). Insoluble polymeric thiol disulphide systems have been successfully used in a number of oxidation–reduction processes [48, 49]:

$$R—S—S—R + 2H^+ + 2e \rightleftharpoons 2R—SH \tag{3.7}$$

By making use of the major advantages of polymeric sulphur reagents over monomeric reagents, benzaldehyde has been successfully converted to styrene oxide in 65% yield using a polymeric sulphonium methylide 5 [36]. The recovered polymeric methyl sulphide 6 was readily converted to the polymeric ylide 5 on treatment with methyl bromide.

$$\text{(PS)}—(CH_2)_nS^+—Me \ Br^- + Ph—CHO \longrightarrow Ph\overset{O}{\triangle} + \text{(PS)}—(CH_2)_nSMe_3 \tag{3.8}$$

Me

5a $n = 0$ **6a, 6b**
5b $n = 1$

Other reagents that have been used in the epoxidation of aldehydes are shown in Table 3.2. The Corey oxidation of primary and secondary alcohols and ketones can be achieved in high yields by using the polymeric thioanisole dichloride reagent **7** [35]:

$$\text{(PS)}—S^+—Cl \ Cl^- + R^1—CH—OH \xrightarrow{Et_3N} R^1—C=O + \text{(PS)}—S—Me \tag{3.9}$$

Me   R^2   R^2

7   **6a**

Depending on the fact that the reactive sites on the polymer maintain their separation during reaction, the selective mono-oxidation of diols to the monoaldehyde can also be achieved by using the polymeric thioanisole dichloride **7** [35]. The reagent used in this case contained a low concentration of methylthio groups (0.66 mmol g^{-1}), and one hydroxy group is able to react with the chlorosulphonium ion leaving the other hydroxy group unable to react with a second distant active site. Attempts to obtain a selective mono-oxidation of symmetrical diols, such as 1, 7-heptanediol, gave a mixture of the monoaldehyde and the dialdehyde.

$$HO—CH_2(CH_2)_5CH_2OH \xrightarrow{7} HO—CH_2(CH_2)_5CHO + OCH—(CH_2)_5CHO$$

$$\tag{3.10}$$

Moreover, the homologation of 1-iodooctane was achieved with the polymeric sulphonium ylide **8**, prepared by treatment of **6a** with n-butyllithium, and that of 1, 4-diiodobutane to its higher homologues has been achieved by using the same polymeric reagent under different conditions to give a mixture of products [38]:

$$\text{(PS)}—S—CH_2^- \ Li^+ + I—(CH_2)_4I \xrightarrow{NaI/MeI} I—(CH_2)_nI \tag{3.11}$$

8

Table 3.3 Polymeric halogenating reagents

Functionalized polymer	Application	Reference
P〈CH—CO / CH—CO〉N—X		
(a) X ≡ Br	Allylic and aromatic bromination	50–52
(b) X ≡ Cl	Aromatic chlorination	52, 53
(PS)—I·X$_2$		
(a) X ≡ F	Addition of F$_2$ to olefins to give —CF$_2$—C—	54, 55
(b) X ≡ Cl	Addition of Cl$_2$ to olefins	55–57
(P)—〈O〉N$^+$—R X$^-$		
(a) R ≡ H; X$^-$ ≡ Br$_3$$^-$	Bromination of olefins and ketones	57
(b) R ≡ H, Me; X$^-$ ≡ ICl$_4$$^-$	Halogenation of acetophenone	59
(P)—〈O〉N—X$_2$ R		
(a) R ≡ H; X$_2$ ≡ Br$_2$, Cl$_2$, I$_2$	Addition of halogen to olefins	57, 60, 61
	Bromination of alkyl benzenes and naphthalenes	62
(b) R ≡ menthyl; X$_2$ ≡ Br$_2$, I$_2$	Asymmetric halogenation of olefins and dienes	63
(PS)—CH$_2$N$^+$Me$_3$X$_3$$^-$		
(a) X$_3$$^-$ ≡ Br$_3$$^-$	α-bromination of carbonyl compounds; addition of Br$_2$ to alkenes and alkynes	64, 65
(b) X$_3$$^-$ ≡ BrCl$_2$$^-$	Chlorobromination of alkenes, alkynes and carbonyl compounds	65
(c) X$_3$$^-$ ≡ ICl$_2$$^-$	Chlorination of carbonyl and unsaturated compounds	65
(PS)—CH$_2$—P$^+$Ph$_3$ Br$_3$$^-$	Bromination of unsaturated and carbonyl compounds	66
(PS)—C=N—Br | Ph	Allylic bromination	67
(PS)—CH$_2$—NR$_2$·(PCl$_5$ or PBr$_3$)	Conversion of acids to acid chlorides and alcohols to alkyl chlorides or bromides	68
(PS)—CO—Cl	Conversion of acids to acid chlorides	69
(P)—〈O〉—O,O-PCl$_3$	Conversion of acids to acid chlorides and acetophenone to α-chlorostyrene	70

Table 3.3 (*Contd.*)

Functionalized polymer	Application	Reference

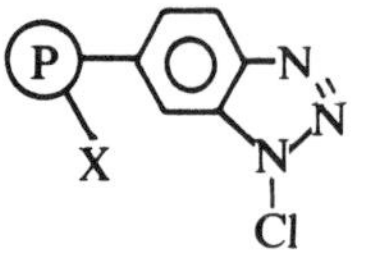

X ≡ COOH and —⟨O⟩—OH

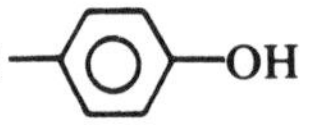

Chlorination of aromatic compounds — 71, 72

R ≡ CH$_2$Ph, CHMeCOOH,
   CMe$_2$CH$_2$SO$_3$H,
   (CH$_2$)$_m$—COOH
m = 1, 2, 3

— 73, 74

$+(CH_2)_6$—N—Z—R—Z—N$+$

Cl   Cl

Z ≡ CO, SO$_2$; R ≡ (CH$_2$)$_4$—, —⟨O⟩—

— 74

$+$N—CO—R$)_n$—
  |
  Cl

Chlorinating agent — 74, 75, 46

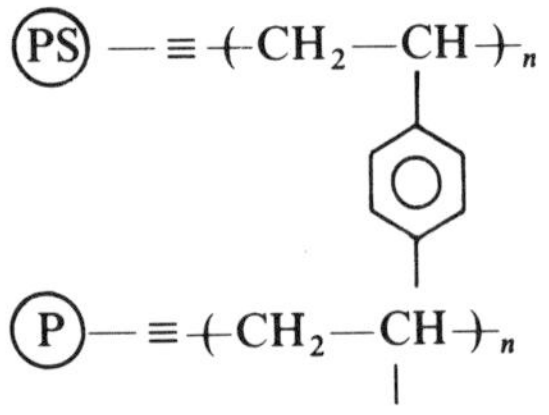

(PS) —≡ $+$CH$_2$—CH$+_n$
         |
       ⟨O⟩

(P) —≡ $+$CH$_2$—CH$+_n$
          |

(PEG) —≡ polyethylene glycol

(C)— ≡ cellulose

(Si)— ≡ silica

3.3 POLYMERIC HALOGENATING REAGENTS

Interesting but different polymeric halogenating reagents (Table 3.3) have been successfully used for the specific addition of halogen to olefins and for allylic halogenation. The success of these insoluble functionalized polymers is attributed to a combination of several advantages including changes in the specificity and reactivity of the functional group, the ease of removal of excess reagent and product separation, and the facility to regenerate the reagents. However, the reactivities of some of these halogenating polymers were found to be lower than those of the conventional reagents.

The reactions between alkylaromatic compounds and N-chlorosuccinimide carried out in the absence of solvent and free radical initiators lead to the formation of a mixture consisting of side chain and aryl chlorinated compounds. However, N-chloropolymaleimide, **9a**, was found to react specifically under the same conditions to yield only the corresponding aryl chloro-substituted products [53]. Dihalogen compounds were also successfully prepared in high yields by the addition of halogens to alkene using the polymer-supported aryl–iodine dihalogen reagents **10a**, **10b** [54, 56, 57]:

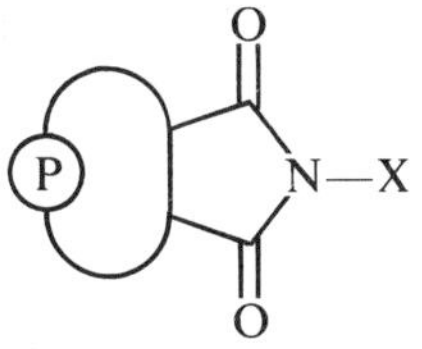

9a X = Cl
9b X = Br

$(PS)—IX_2$
10a X ≡ F
10b X ≡ Cl

Moreover, treatment of alkylaromatic and olefinic compounds with *N*-bromopolymaleimide (**9b**) led to the formation of several unexpected products [50, 51]. Treatment of **9b** with cumene in a 2.33:1 molar ratio in the presence of carbon tetrachloride as solvent and benzoyl peroxide as initiator gave α, β, β′-tribromocumene (48%), β-bromo-α-methylstyrene (15%) and α-bromomethylstyrene (13%). When the molar ratio of **9b** to cumene was 3.7:1, the yield of α, β, β′-tribromocumene increased to 85%. The formation of these unusual products shows that **9b** undergoes dehydrobromination after the allylic bromination.

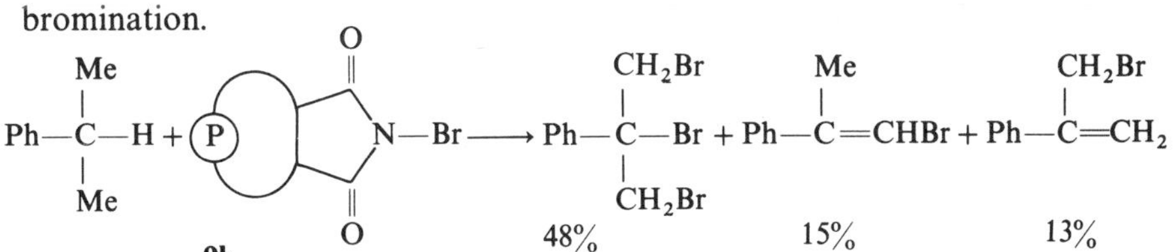

9b 48% 15% 13%

(13.12)

Recently, new polymeric halogenating reagents (**11**) have been prepared by modification of commercial anion exchange resins and have been used in the direct halogenation of various organic substrates such as alkenes, alkynes and α-carbonyl compounds [64, 65]:

$$(PS)—CH_2—N^+R_3X_3^- + R—CH_2C(=O)—R' \longrightarrow R—CH(X)—C(=O)—R' \quad (3.13)$$

11

3.4 POLYMERIC CONDENSING REAGENTS

Insoluble polymer-bound carbodiimide derivatives (**12**) have been prepared and used as condensing agents in the synthesis of peptides [76, 77]. Recently, peptide synthesis was carried out with various polymers containing the carbodiimide functional group, such as polymers containing the polyhexamethylene-carbodiimide [76] with **12a** [78] as condensing agent:

$(PS)—CH_2N{=}C{=}N—R$

12a R ≡ Et
12b R ≡ *i*-Pr

Other polymeric reagents (Table 3.4) were used in the Moffatt oxidation, in particular in the oxidation of highly sensitive alcohols and in the conversion of carboxylic acids to their anhydrides. Acid anhydrides were synthesized from the corresponding carboxylic acids by a simple procedure using the polymeric carbodiimide **12b** [82]. The products were obtained in high yield by simple filtration from the reaction mixture and evaporation of the solvent. The use of the polymeric carbodiimides has the advantage that the byproduct, urea, remains attached to the polymer **13** and is easily removed by filtration:

$$\text{(PS)}—CH_2N{=}C{=}N—C_3H_7\text{-}i + 2R—COOH$$

$$\mathbf{12b}$$

$$\longrightarrow \; \begin{array}{c} R—CO \\ R—CO \end{array}\!\!O \; + \text{(PS)}—CH_2NHCONH—Pr\text{-}i \quad (3.14)$$

$$\mathbf{13}$$

Polymer **13** can readily be converted back into the active polymeric reagent **12b** by treatment with tosyl chloride in the presence of triethylamine [82]. Although the Moffatt oxidation of alcohols is particularly useful with highly sensitive compounds, contamination of the products with urea constitutes the major problem of this method of synthesis. However, on using polymeric carbodiimides, good yields of the oxidation products are obtained with the elimination of the contamination problem [78]:

$$\begin{array}{c} R^1—CH—OH \longrightarrow R^1—C{=}O \\ \quad | \qquad\qquad\quad | \\ \quad R^2 \qquad\qquad\quad R^2 \end{array} \qquad\qquad (3.15)$$

Another polymeric condensing reagent incorporates sulphonyl chloride groups. For example, poly(3, 5-diethylstyrene) sulphonyl chloride has been prepared and used in oligonucleotide synthesis [88]. The use of this supported reagent has some advantages over the non-supported system, such as the elimination of emulsion problems and a reduction in the contamination of nucleotide product.

3.5 POLYMERIC REDOX REAGENTS

Although many useful procedures for oxidation processes have been reported, the main disadvantage of these is the relative difficulty in the preparation of the reagents and in the working-up of the reaction mixture. Polymeric redox systems were one of the earliest examples of polymeric reagents and alleviate these difficulties. The most important of the polymeric redox systems are the hydroquinone–quinone, thiol–disulphide, pyridine–dihydropyridine, polymeric dyes and polymeric metal complex systems [48, 49].

Aliphatic peroxy acids explode very readily on impact and analogous polymeric reagents based on polyacrylic acids behave similarly [95]. Aromatic

Table 3.4 Polymeric condensing reagents

Functionalized polymer	Application	Reference
$\text{(PS)}-(CH_2)_n-N=C=N-R$		
(a) $n = 0$; $R \equiv Et$		79–81
(b) $n = 1$; $R \equiv i\text{-Pr}$	Conversion of acids to anhydrides	82, 83
(c) $n = 1$; $R \equiv Et$	Peptide synthesis	78, 82
$\text{(PEG)}-N=C=N-R$	Dehydrating agent in peptide synthesis and oxidation of alcohols to carbonyl compounds	84
$+R-N=C=N)_n$	Peptide synthesis	76
(a) $R \equiv -(CH_2)_6-$		
(b) $R \equiv$		85
$\text{(PS)}-C\equiv C-NEt_2$	Conversion of acids to mixed anhydrides, esters and anhydrides	86
$\text{(PS)}-\overset{\overset{\displaystyle O}{\|}}{As}-Ph$	Synthesis of carbodiimides	87
$\overset{\|}{Ph}$		
	Peptide synthesis	88, 89
	Oligonucleotide synthesis	77
	Koenigs–Knorr glycoside synthesis	90
$\text{(PS)}-Z-NMe$	Esterification of acids, $R-COOH + R'-OH \rightarrow R-COOR'$	91
$Z-N-COMe$		
$\overset{\|}{Me}$		
$Z \equiv -CH_2-,$		
$-CH_2(NHCO(CH_2)_{10})_2-$		
$\text{(PS)}-(CH_2)_n-$	Dehydrobromination and esterification	92
$n = 1, 4, 7$		

Table 3.4 (*Contd.*)

Functionalized polymer	Application	Reference
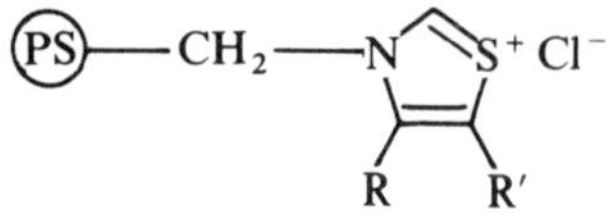 (PS)—CH$_2$—N⟨S⟩$^+$ Cl$^-$ (R, R')	Benzoin condensation	93
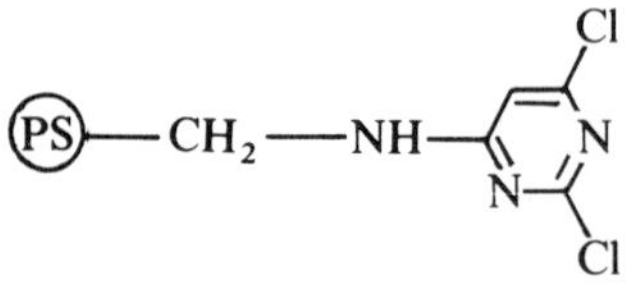 (PS)—CH$_2$—NH—⟨ring, Cl, N, N, Cl⟩	Dehydrating and desulphydrating agents	94
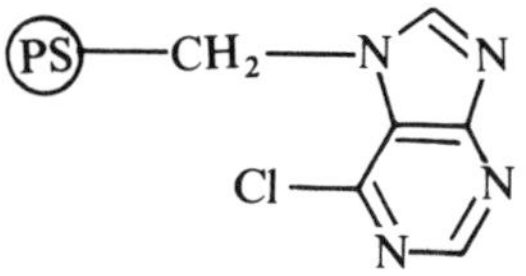 (PS)—CH$_2$—N⟨purine ring, Cl, N⟩	Synthesis of amides, carbamic acid esters and carbodiimides	94
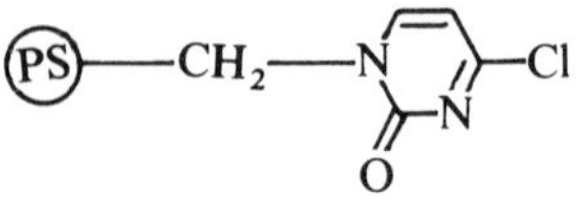 (PS)—CH$_2$—N⟨ring, Cl, N, O⟩		94

(PS) —≡ +(CH$_2$—CH)$_n$ with phenyl ring

(P) —≡ +(CH$_2$—CH)$_n$

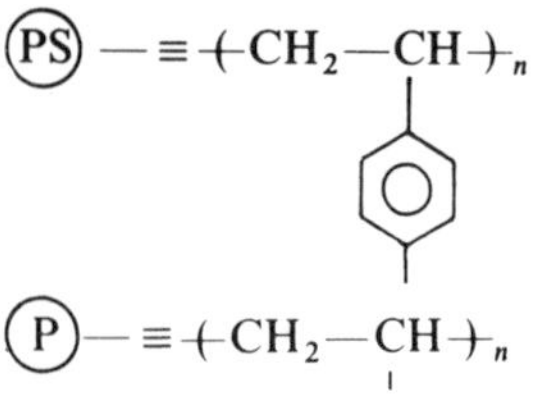

(PEG) —≡ polyethylene glycol

(C) —≡ cellulose

(Si) —≡ silica

peroxy acids in contrast are more stable and the polymer-supported analogues (**14**) of these have proved very useful indeed [96–98]:

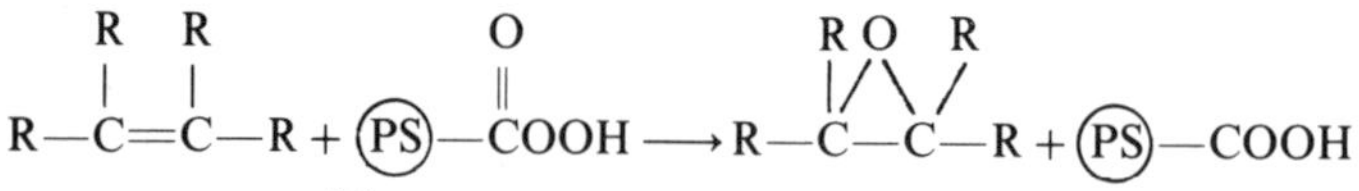

$$R-\underset{R}{\underset{|}{C}}=\underset{R}{\underset{|}{C}}-R + (PS)-\overset{O}{\overset{||}{C}OOH} \longrightarrow R-\underset{R}{\underset{|}{C}}\overset{O}{\overset{\diagup\diagdown}{-}}\underset{R}{\underset{|}{C}}-R + (PS)-COOH \quad (3.16)$$

$$\mathbf{14}$$

Chromic acid is a powerful oxidizing agent and a supported analogue **15** has been used for the oxidation of alcohols to carbonyl compounds in high yields [99, 100]. The reagent is also used for the conversion of alkyl halides to aldehydes and ketones:

$$(PS)-CH_2N^+R_3\,HCrO_4^- + R^2-\underset{R^3}{\underset{|}{CH}}-OH \longrightarrow R^2-\underset{R^3}{\underset{|}{C}}=O \quad (3.17)$$

$$\mathbf{15}$$

Organotin hydrides are well known as selective reducing agents for carbonyl compounds. The dihydrides are generally more reactive but less stable than the corresponding monohydrides. However, insoluble polymeric organotin dihydride **16** [101] has been prepared and used as a selective reducing agent for carbonyl compounds and alkyl halides:

$$\text{(PS)}-\overset{\displaystyle H}{\underset{\displaystyle H}{\text{Sn}}}-\text{Bu-}n$$

16

This polymeric reagent combines the advantages of the monomeric dihydride reagent, i.e. its high reactivity, and of the monohydride reagent, i.e. its selectivity, in addition to the advantages of a typical polymeric reagent: ease of operation and reaction work-up, avoidance of the odorous and toxic vapours characteristic of tin hydrides, and capability of regeneration.

The mechanism of the reduction with monomeric organotin hydrides is known to proceed by a two-step mechanism:

1. $$\ge\!\text{Sn}-\text{H}+-\overset{|}{\text{C}}\!=\!\text{O}\longrightarrow-\overset{|}{\text{CH}}-\text{O}-\text{Sn}\!\!\le$$

2. $$\ge\!\text{Sn}-\text{H}+-\overset{|}{\text{CH}}-\text{O}-\text{Sn}\!\!\le\longrightarrow-\overset{|}{\text{CH}}-\text{OH}+\ge\!\text{Sn}-\text{Sn}\!\!\le$$

However, reaction (2) cannot occur with the polymeric tin hydride reagent because of the restricted mobility of the polymer matrix. Thus the need to hydrolyse the intermediate tin alkoxide in order to isolate the reduction product can be used to advantage for the selective reduction of symmetrical dialdehydes. Terephthalaldehyde was reduced with polymeric tin reagent **16** to give 86% monoalcohol and 14% of dialcohol.

Recently, some polymeric redox resins with hydroquinone and catechol units as pendant groups were prepared and used for the oxidation of hydrazobenzene to azobenzene, as shown in Table 3.5.

Another interesting oxidizing reagent containing the amine oxide moiety has also been prepared by the chemical modification of polystyrene resin and has been used for the direct oxidation of alkyl halides to carbonyl compounds [141]. This polymeric *N*-oxide **17** eliminates the disadvantage associated with the use of soluble reagent and the polymeric byproduct can be recycled for reuse.

$$\text{R}-\overset{\displaystyle}{\underset{\displaystyle R}{\text{CH}}}-\text{X}+\text{(PS)}-\text{CH}_2-\overset{\displaystyle Me}{\underset{\displaystyle Me}{\text{N}^+}}-\text{O}^-\xrightarrow[-H_2O_2-]{}\text{(PS)}-\text{CH}_2-\text{NMe}_2+\text{R}-\overset{\displaystyle}{\underset{\displaystyle R}{\text{C}}}\!=\!\text{O}$$

$$\qquad\qquad\qquad\quad \textbf{17}\qquad\qquad\qquad\qquad\qquad\qquad\qquad\qquad\qquad (3.18)$$

In an attempt to eliminate the problem encountered by the regeneration of the polymeric byproduct, an interesting new method for oxidation based on a

Table 3.5 Polymeric redox reagents

Functionalized polymer	Application	Reference
(PS)—CH$_2$N$^+$Me$_3$ BH$_4^-$	Reduction of carbonyl compounds to alcohols	102, 103

(PS)—(CH$_2$)$_n$—S→BH$_3$
              |
              Me
 $n = 0, 1, 2$
	Reduction of ketones	104

(P)—⬡—O→N→BH$_3$	Reduction of carbonyl compounds	105–107
(P)—⬡—N→BH$_3$	Reduction of carbonyl compounds to alcohols	108, 109
(PS)—Sn(n-Bu)H$_2$	Reduction of carbonyl compounds Selective monoreduction of dials Reduction of alkyl halides to alkanes	101
(PS)—CH$_2$—N(pyrrolidine) BH$_3^-$ COONa Na$^+$	Reduction of carbonyl compounds	110
(C)—OBH$_3^-$ Na$^+$	Reduction of carbonyl compounds	110
(PS)—C(CH$_3$)$_2$OAlH$_4$	Reduction of ketones	111

Reagent	Application	Ref.
(PS)⎯⟨O–O⟩–OAlH$_4$, OAlH$_4$	Reduction of ketones	111
(PS)⎯CH$_2$⎯[anthracene] OH, OH	Reduction of quinones to quinols	112
(P)⎯[anthracene] OH ⎯⟨C$_6$H$_4$⟩SO$_3$H OH	Reduction of quinones to quinols	113
(P)⎯CO(CH$_2$)$_4$ SH SH	Reduction of disulphide bridges in peptides, proteins, oxidized form of glutathione and cystine	114
(PS)⎯CH$_2$⎯$^+$N⟨C$_6$H$_4$⟩CONH$_2$ Cl$^-$	Reduction of thionine, benzoquinone, methylene blue	61,115
(P)⎯⟨C$_6$H$_4$⟩X ⎯CH$_2$⎯N⟨ ⟩CONH$_2$ X = H, CONH$_2$; ⎯C$_6$H$_4$CH$_2$N$^+$Et$_3$ Cl$^-$, ⎯C$_6$H$_4$SO$_3$K, ⎯N=O	Reduction of various dyes, alloxan, ninhydrin	116–118
	Oxidation of alloxan, 2,6-dichloro phenol-indophenol, meacridinum iodide	
(P)⎯CO⎯⟨C$_6$H$_4$⟩N$^+$⎯Me X$^-$ X$^-$ ≡ I$^-$, SO$_4^{2-}$	Reducing agent	57

Table 3.5 (*Contd.*)

Functionalized polymer	Application	Reference
℗—N⁺(Me)⟨pyridine⟩ Br⁻	Redox reagent	57, 61, 119
℗—N(Br₂)⟨pyridine⟩	Conversion of thiols to disulphides	120
PS⟩—PPh₂, PPh₂ ⟩Cu⟨ H H ⟩B⟨ H H	Reduction of acid chlorides to aldehydes	121, 122
PS⟩—CH—NHCO—CH—CH₂, R, NH₂—M—S	Reduction of acetylene to ethylene	123
M ≡ Cu^{II}, Mo^{VI} —P—N⟨(CH₂)ₙ⟩N—P—N⟨(CH₂)ₙ⟩N— n = 2, 3	Desulphurization of polysulphide compounds	124
—(—CO—N—)ₙ— CH₂SH	Redox reagent	125

Reagent	Application	Ref.
Ⓟ—COOCH₂CH₂RuCl₂·NaBH₄	Reducing agent	126
ⓅS—PR₃·NiCl₂·NaBH₄	Reducing agent	127,128
Ⓟ—⬡N⁺—H ClCrO₃⁻	Oxidation of alcohols to carbonyl compounds	129
Ⓟ—⬡N⁺—H IO₄⁻ or IO₃⁻	Oxidation of alkyl halides to carbonyl compounds	99
Ⓟ—(⬡N⁺—H)₂ Cr₂O₇²⁻	Oxidation of quinols, catechols, glycols, phosphine	130
Ⓟ—⬡N·CrO₃ (Ph)	Oxidation of alcohols to carbonyl compounds	131–133
Ⓟ—⬡N·CrO₃	Oxidation of alcohols to carbonyl compounds	134
ⓅS—CH₂N⁺Me₃ HCrO₃⁻	Oxidation of alcohols and alkyl halides to carbonyl compounds	99,100
Ⓟ—⬡N⁺—H OBr⁻	Oxidation of alcohols to carbonyl compounds	135–137
Ⓟ—⬡N·(HBr or H₂SO₄)	Electrochemical oxidation of alkylbenzenes to carbonyl compounds	138
Ⓟ—[CH—CO / CH—CO]N—Br	Oxidation of alcohols	52
ⓅS—CH₂N⁺Me₃ IO₄⁻	Oxidation of phenols and sulphides	139

Table 3.5 (*Contd.*)

Functionalized polymer	Application	Reference
$\text{(PS)}-(CH_2)_n-\overset{\displaystyle O}{\underset{\displaystyle \,}{C}}\overset{\displaystyle Cl}{\underset{\displaystyle \,}{=}}N-R$ (a) $n=0$; $R \equiv$ Me, Ph (b) $n=1$; $R \equiv$ Ac	Oxidation of alcohols	139
$\text{(PS)}-CH_2-\overset{\displaystyle Cl}{\underset{\displaystyle \,}{N}}-\overset{\displaystyle O}{\underset{\displaystyle \,}{C}}-Me$	Oxidation of alcohols	139
$\text{(PS)}-COOOH$	Epoxidation of olefins Oxidation of thioethers	96, 140 97
$\text{(Si)}-(CH_2)_n-\!\!\bigcirc\!\!-COOOH$ $n = 0, 2$	Oxidation of tetrahydrothiophene	97
$\text{(P)}-COOOH$ $\quad\quad\; \searrow X$ $X \equiv H$ $X \equiv -C_6H_4-SO_3H$	Epoxidation of olefins Hydroxylation of olefins, conversion of acids to peroxy acids	95 98
$\text{(P)}-CH_2-\overset{\displaystyle Me}{\underset{\displaystyle Me}{N^+}}-O^-$	Oxidation of alkyl halides and tosylates to carbonyl compounds	141

Structure	Reaction	Ref.
$(PS)\!-\!Se\!-\!R$ $R \equiv Na, Cl, H, CN$	α,β-dehydrogenation of carbonyl compounds	142, 143
$(PS)\!-\!SeO_2H$	Oxidation of olefins, ketones, and alcohols	144
$(PS)\!-\!\overset{O}{\underset{\|}{Se}}\!-\!Ph$	Oxidation of β-Me-naphthalene to β-naphthaldehyde	142, 143
$(PS)\!-\!M\!=\!\!\langle\ \rangle\!-\!OMe$, $\quad M \equiv Se, Te$	Oxidation of thiols, phosphines, hydroquinones and thioketones	145
$(PS)\!-\!I(OAc)_2$	Oxidation of amines	56
$(P)\!-\!COO(CH_2)_n\!-\!CH_2$ ... flavin, $n = 1, 2$	Dehydrogenation of mandelate ester	146, 147
$\left((PS)\!-\!C\!\left(\!-\!\langle\ \rangle\!-\!NMe_2\right)_2\right)^{+} Cl^{-}$	Redox system	148
$(PS)\!-\!CH_2\!-\!$ quinone, $(PS)\!-\!CH_2\!-\!$ quinone	Oxidation of hydrazobenzene to azobenzene	149

Table 3.5 (*Contd.*)

Functionalized polymer	Application	Reference
	Oxidation of amines to ketones	150
	Redox reagent	151, 152
	Oxidation of alcohols to carbonyl compounds, hydrazo to azo, and sulphides to sulphoxides	71, 72
	Oxidation of alcohols to carbonyl compounds	153
	Redox system	154

Structure	Application	Reference
	Oxidation of thiols to sulphides	120
	Redox system	57
	Redox system	154
	Redox reagent	57
	Redox reagent	154
Z ≡ NH, S	Redox reagent	155, 156
	Redox reagent	157
	Dehydrogenation; oxidation of amines, cysteine, NADPH and vitamin C	158

Table 3.5 (*Contd.*)

Functionalized polymer	Application	Reference
$-(N=C)_n-$ with X side group; X ≡ Cl, OPh, NHPh	Peroxidation of alkylaromatic hydrocarbons	159, 160
$-(CO-R-CO-N-(CH_2)_6-N)_n-$ with Cl, Cl (a) R ≡ $-(CH_2)_4-$ Me (b) R ≡ (dimethyl cyclopentane structure, Me Me)	Oxidation of alcohols and thioethers Oxidation of tertiary amines Oxidation of thioethers	161–165 75 46
$-((CH_2)_m-CO-N)_n-$ with Cl, $m = 3, 6$	Oxidation of alcohols and diamines	46, 75 165–167
P$-(COO)_2$M·t-BuOOH ($O=M=O$)	Epoxidation of olefins	168
PS$-CH_2N[-(CH_2)_3-N=CH$-(phenol) $]_2$ Co^{2+}	Oxidation of 2,6-dimethyl phenol	169

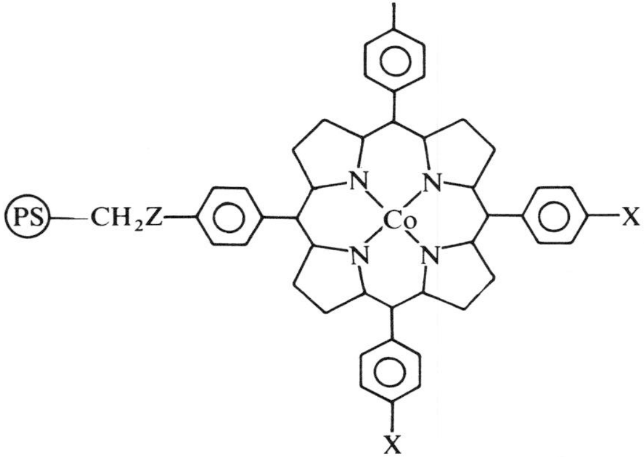

170   Oxidation of thiols to sulphides

171   Epoxidation of propene with
       t-BuOOH

172   Oxidizing agent

PS—CH₂Z

(a) X ≡ NH₂; Z ≡ NH
(b) X ≡ COOH; Z ≡ OCO

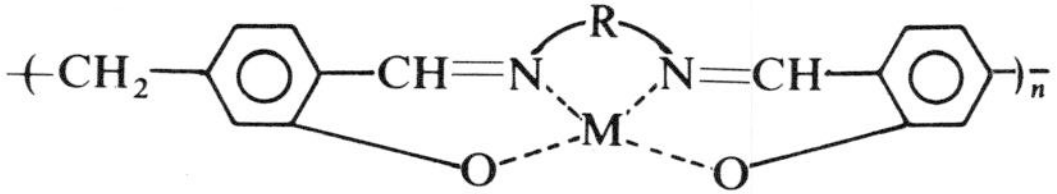

PS —NR₂·Mo(CO)₅

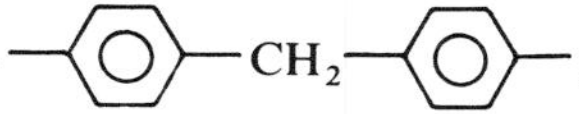

R ≡ —(CH₂)₄—,

M ≡ V, Mn

PS — ≡ (CH₂—CH)ₙ

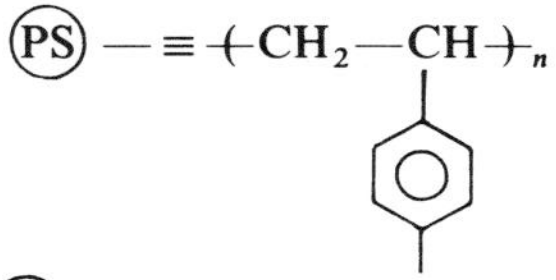

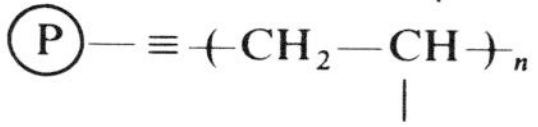

PEG — ≡ polyethylene glycol

C — ≡ cellulose

Si — ≡ silica

P — ≡ (CH₂—CH)ₙ

polymeric reagent electrochemically generated from poly(4-vinylpyridine) hydrobromide **18** was used for the oxidation of secondary alcohols to ketones under mild conditions and in high yields in the absence of electric current without any consumption of chemical oxidant and without producing any contaminating reduced product [135–137]. The polymeric byproduct **19** can be regenerated by electrochemical oxidation.

$$(3.19)$$

3.6   POLYMERIC PROTECTING GROUPS

There is no group capable of reacting selectively with only one functionality of a completely symmetrical bifunctional compound; however, functionalized insoluble polymers have been used recently with some success as selective monoblocking groups to facilitate the use of symmetrical bifunctional compounds in organic synthesis.

The utilization of functionalized insoluble polymers as protecting [173] groups for one of the functionalities of the unsymmetrical bifunctional compounds has been well known for many years in the synthesis of polypeptides, oligonucleotides and oligosaccharides. In these syntheses, the bifunctional substrate is coupled to the polymer through one of the functionalities and chemical modifications are carried out on the other free reactive functional group with appropriate reagents in successive steps; the completely modified product is cleaved from the polymeric protecting group at the end of the reaction sequence.

$$(3.20)$$

During the course of the synthesis the polymer-bound molecules are purified at each step by filtration and washing to remove byproducts and all soluble excess reagents used to increase the rate and the yield of the reaction. This technique has been used in polypeptide synthesis either for protecting an acid group of the N-protected amino acid or the amino group of the C-protected amino acid. The same concept has also been used for the protection of hydroxy groups in oligonucleotide and oligosaccharide syntheses.

The selectivity of functional polymer resins for mono-blocking depends on the 'infinite dilution' exerted by the polymer which can be achieved by using relatively highly crosslinked polymer with a low concentration of functional groups. In this type of reaction a large excess of a symmetrical bifunctional substrate is employed to ensure that only one of the functional groups reacts with the polymer-bound protecting group. This principle has been applied with success for selective mono-blocking of symmetrical diols [174–179], dialdehydes [180–182], diacids [183, 184] and diamines [185], allowing further chemical reactions at the free functional group end. For example a polymer containing a diol functional group, **20**, has been used as a mono-blocking agent for symmetrical dialdehydes [173, 180, 181] and some substituted benzaldehydes have been prepared by a variety of reactions on the free aldehyde group followed by acid cleavage from the polymer backbone.

(3.21)

Symmetrical diols have been shown to react with polymers containing acid chloride groups [179] or polymer-bound trityl chloride [174] to give mono-blocked diols capable of further reaction at the free alcohol end. This advantage of using insoluble functionalized polymers in the selective blocking of bifunctional compounds has been used for the preparation of insect sex attractants in high yields.

$$\text{(PS)}-COX + HO-A-OH \longrightarrow HO-A-OX \qquad (3.22)$$

Although cyclic boronate esters are formed readily and are stable to a number of reaction conditions, many are also quite sensitive to moisture and take up water with loss of phenylboronic acid. These properties of boronate esters coupled with the advantages of solid phase synthesis make a polystyrylboronic acid resin particularly attractive, since it provides a useful and easily removable diol-protecting group [186]. Thus glycoside derivatives have been synthesized in high yield using the polymeric boronic acid resin 21 as a diol-protecting group.

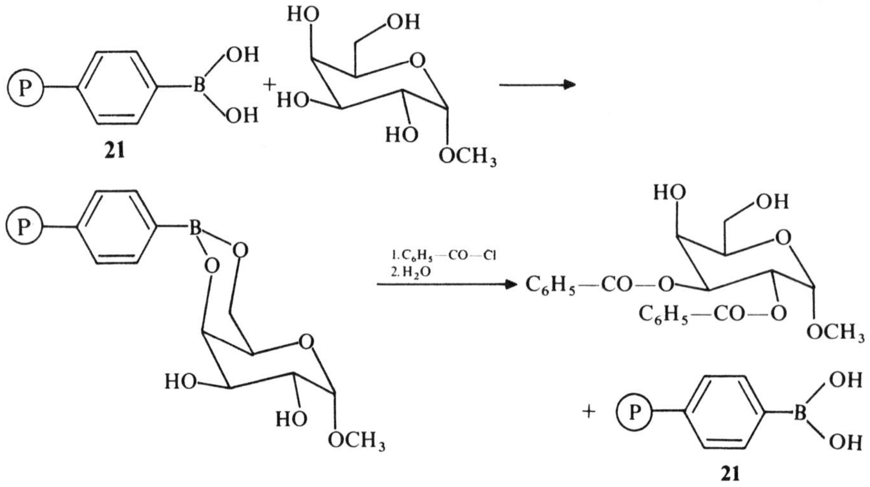

(3.23)

A number of functionalized polymers are also suitable for the mono-protection of symmetrical diacids and diamines. For example, a polymeric protecting group containing diazomethylene functionalities 22 has been prepared and used for the protection of carboxylic acids in the transformation of penicillin G1 (S7-oxide) into the corresponding cephalosporin [187]:

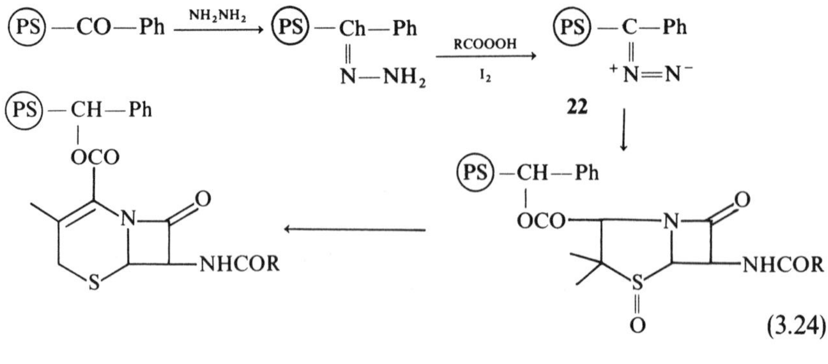

(3.24)

Table 3.6 Polymeric protecting groups

Functionalized polymer	Application	Reference
(PS)$-\underset{\underset{Ph}{\mid}}{\overset{\overset{Ph}{\mid}}{C}}-Cl$	Oligonucleotide synthesis	188
	Glycoside synthesis	189, 190
	Monoprotection of symmetrical diols	174–176, 190–193
	Monoprotection of triols and tetraols	174
(Si)$-\langle O \rangle-\underset{\underset{Ph}{\mid}}{\overset{\overset{Ph}{\mid}}{C}}-Cl$	Oligonucleotide synthesis	194
(PS)$-(CH_2)_n-COCl$	Monoprotection of diphenols and diols	140, 177, 191
(a) $n = 0$	Peptide and oligonucleotide synthesis	178
(b) $n = 1$	Monoprotection of symmetrical diols	179, 195
(PS)$-CHO$	Glycoside synthesis	196–199
(PS)$-B(OH)_2$	Glycoside synthesis	200
	Synthesis of carbohydrate derivatives	186, 201
	Partial acylation of acyclic polyols	202–204
(PS)$-CH_2Cl$	Monoprotection of dithiols	205
(PS)$-CH_2-Z-\underset{\underset{CH_2-OH}{\mid}}{CH_2CH}-OH$ $Z \equiv O, S$	Monoprotection of symmetrical aromatic diols	180, 181, 206–209
(PS)$-CH_2OCH_2-\underset{\underset{R}{\mid}}{C}\overset{-CH_2OH}{\underset{-CH_2OH}{}}$ $R \equiv H, Me$	Monoprotection of symmetrical aromatic diols	182
(PS)$-(CH_2)_n-OH$	Monoprotection of diacid chlorides	183, 184
$n = 1$	Alkylation of acids	5
$n = 2$	Synthesis of benzodiazopinone	210
(PS)$-\underset{\underset{O}{\parallel}}{\overset{\overset{O}{\parallel}}{S}}-CH_2CH_2OH$	Monoprotection of diacid chlorides	211
(P)$-OCO-CHR-NH_2$	Protection of acid during monoalkylation and dialkylation of NH_2 with RX	212

Table 3.6 (*Contd.*)

Functionalized polymer	Application	Reference

(PS)—CH$_2$OCH$_2$—CH—NH$_2$
           |
           R
 R ≡ Me, Et, PhCH$_2$
	Protection of carbonyl group in the synthesis of α-methylation of cyclohexanone	213

(PS)—C=N$^+$=N$^-$
     |
     Ph
	Transformation of penicillin to cephalosporin	187

(PS)—(CH$_2$)$_3$NHCO(CH$_2$)$_2$COOH	Oligonucleotide synthesis	214
(PS)—CH$_2$OCOCl	Peptide and oligonucleotide synthesis	215
(Si)—⟨O⟩—CH$_2$O(CH$_2$)$_n$CH =CH(CH$_2$)$_n$OH	Oligosaccharide synthesis	216
(PS)—CH$_2$OCOO—⟨O⟩—NO$_2$	Monoprotection of symmetrical diamines	185
NO$_2$ (PS)—CH$_2$O—⟨O⟩—CH$_2$OH MeO	Oligonucleotide synthesis	217
(Si)—R—NH$_2$ R ≡ (CH$_2$)$_3$,(CH$_2$)$_3$NH(CH$_2$)$_2$	Peptide synthesis	218, 219

(PS)—CH$_2$—N—CO—NH—NH$_2$
           |
           Me
	Oligonucleotide synthesis	220

NO$_2$ (P)—⟨O⟩—CH$_2$OR	Peptide synthesis	221, 222
NO$_2$ (PS)—CH$_2$NHCO—⟨O⟩—CH$_2$—Z—R Z ≡ O, NH	Peptide synthesis	221, 222
(PEG)—Z—CO—⟨O⟩—CH$_2$NHR NO$_2$	Peptide synthesis	223, 224

(PS)—COCH—OR
       |
       Me
	Peptide synthesis	223

(Si)—(CH$_2$)$_n$—⟨O⟩—CH$_2$OH	Peptide synthesis	97, 225–229

Another interesting functional polymer which can be used efficiently for the protection of symmetrical diamines contains benzyl-p-nitrobenzyl carbonate groups **23** [185]:

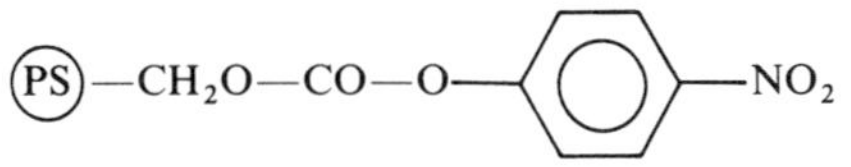

$$(PS)-CH_2O-CO-O-\!\!\!\bigcirc\!\!\!-NO_2$$

23

Other functionalized polymeric reagents used as selective mono-blocking groups are listed in Table 3.6.

3.7   POLYMERIC ACYLATION AND ALKYLATION REAGENTS

Several polymeric acylating reagents have been used in peptide syntheses, in which polymeric active esters of N-blocked amino acids were treated with C-blocked amino acids or oligopeptides. The product obtained in solution was separated and treated with a polymer containing N-blocked amino acid after removing the blocking group. The process was repeated to give the desired product:

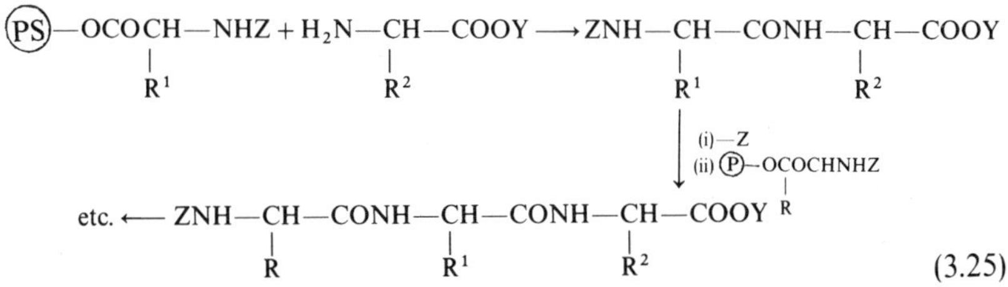

$$(PS)-OCOCH-NHZ + H_2N-CH-COOY \longrightarrow ZNH-CH-CONH-CH-COOY$$

with substituents R^1, R^2 (left side) and R^1, R^2 (right side)

$$\downarrow \begin{array}{l}(i)-Z \\ (ii)\,(P)-OCOCHNHZ\end{array}$$

$$\text{etc.} \longleftarrow ZNH-CH-CONH-CH-CONH-CH-COOY \overset{|}{R}$$

with substituents R, R^1, R^2

$$(3.25)$$

Other functional polymers have also been used in the acylation and alkylation of different substrates in general organic syntheses such as the acylation of amines, carboxylic acids and alcohols (Table 3.7). For example, the insoluble polymer containing the anhydride functional group **24** was used for the conversion of an amine or alcohol to amide or ester [237, 238].

$$(PS)-CO-O-CO-R \xrightarrow[\text{or R--OH}]{R-NH_2} \begin{array}{l}R-CONH-R' \\ \text{or } R-COOR'\end{array}$$

24

$$(3.26)$$

In conventional solution methods of C-alkylation and C-acylation of compounds

---

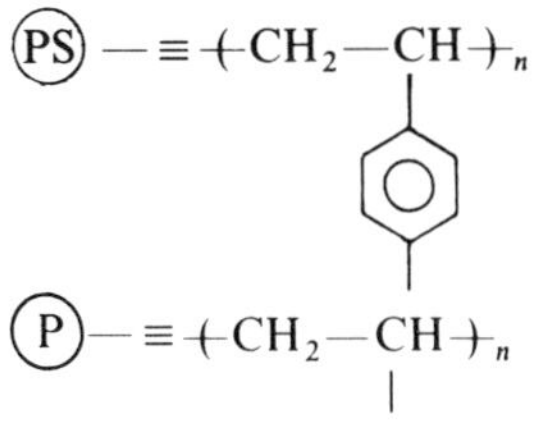

$$(PS)-\equiv \{CH_2-CH\}_n$$
(with pendant phenyl group)

$$(P)-\equiv \{CH_2-CH\}_n$$

$$(PEG)-\equiv \text{ polyethylene glycol}$$

$$(C)-\equiv \text{ cellulose}$$

$$(Si)-\equiv \text{ silica}$$

Table 3.7 Polymeric acylation and alkylation reagents

Functionalized polymer	Application	Reference
(PS) —$(CH_2)_n$OCOCHRR′		
$n = 1$	Conversion of acid chlorides and anhydrides to ketones	230–235
$n = 2$	Conversion of alkyl chloride to substituted acid derivatives	236
(PS) —CO—O—CO—R	Conversion of amines to amides and alcohols to esters	237, 238
(PS) —CH_2OCO—O—CO—R	Conversion of amines to amides and acids to anhydrides	239–241
(PS) —CO$(CH_2)_2$CO—O—CO—R	Conversion of amines to amides and alcohols to esters	237
(PS) —N=N—NHR	Conversion of acids to esters	242
(PS) —SO_2OCOCH$_3$	Acylation and alcohols and phenols	243
(PS) —CH_2NHCO$(CH_2)_4$— [cyclic SR SR]	Acylation of amines	244
$R \equiv COC_6H_4$—NO_2—p		
(P) —Z—CO$(CH_2)_4$— [cyclic AcS SAc]	Acylation of alcohols, phenols and thiols	245
$Z \equiv$ O, NH		
(PS) —CH_2—$\overset{O}{\underset{O}{\overset{\|}{\underset{\|}{S}}}}$—⟨O⟩—O—COR	Peptide synthesis	246–248
(PS) —$\underset{Me_3}{CHN^{+ -}}$$O_2$S—⟨O⟩—N=N—[ring OCOR]	Peptide synthesis	249
(P) —⟨O⟩—$(CH_2)_n$O—COR, $\underset{NO_2}{}$ $n = 0, 1$	Peptide synthesis. Acylation of 6-aminopenicillanic and 7-aminocephalosporanic acids	250–254
(PS) —Z—⟨O⟩—O—COR, $\underset{NO_2}{}$ (a) $Z \equiv CH_2, CH_2$OCO	Peptide synthesis	256, 257

Table 3.7 (*Contd.*)

Functionalized polymer	Application	Reference
(b) $Z \equiv CH_2$	Acylation of amines	258
(c) $Z \equiv CO$	Acylating reagent	255
$\text{(P)}-CONH(CH_2)_2-S-COR$	Acylation of amines	259
$\text{(P)}-COOCH_2CH_2O-\overset{\overset{O}{\parallel}}{\underset{\underset{O}{\parallel}}{S}}-\text{(O)}-R$ $R \equiv H, Me$	Alkylating agent	260
$\text{(PS)}-PR_2 \cdot Mo(CO)_6$	Acylation and alkylation of aromatic compounds	261
$\text{(PS)}-Mo(CO)_3$	Acylation and alkylation of aromatic compounds	261
$\text{(P)}\overset{CH-CO}{\underset{CH-CO}{\Big<}}N-O-COR$	Peptide synthesis	262–269
$\text{(PS)}-CH_2CH\overset{-CO}{\underset{CH_2-CO}{\diagdown}}N-O-COR$	Peptide synthesis	256
$\text{(P)}\overset{CH-CO}{\underset{CH-CO}{\Big<}}N-\text{(O)}-O-COR$ Ph $\quad$ NO_2	Peptide synthesis	270
$\text{(P)}-\text{(O)}-O-COR$ Me	Peptide synthesis	271
$\text{(PS)}-CH_2-\text{(O)}-OCO-R$ $R \equiv Me, Ph$	Alkylation of amines	272
$\text{(PS)}-CH_2-\text{(O)}-\begin{smallmatrix}N\\ \parallel N\\ N\end{smallmatrix}$ OCOR	Peptide synthesis	273, 274
$\text{(P)}-\begin{smallmatrix} \\ N=\end{smallmatrix}N-COCH_3$	Acylation of amines and alcohols	275

Table 3.7 (*Contd.*)

Functionalized polymer	Application	Reference
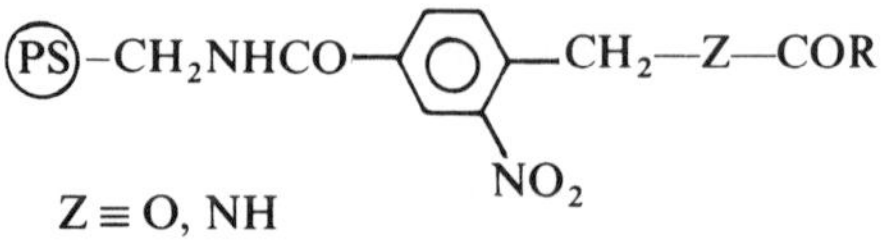 (PS)–CH$_2$NHCO–〈O〉–CH$_2$–Z–COR (NO$_2$) Z ≡ O, NH	Peptide synthesis	221, 222, 276
(P)–COOCH$_2$CH–CH$_2$O–Z–R (OH) Z ≡ CO, NH	Peptide synthesis	277
(PS)–CONH–O–COR	Peptide synthesis	266–268
(PS)–O–(–COCH$_2$N–)$_n$ (COR)		278
–(–CO(CH$_2$)$_4$CO–N–(CH$_2$)$_6$–N–)$_n$ (COR) (COR) R ≡ CF$_3$, CCl$_3$	R$_2$NH → R$_2$N–COCF$_3$ R–OH → R–OCOCF$_3$	279, 280
–(–R–CO–N–)$_n$ (COR′)		281
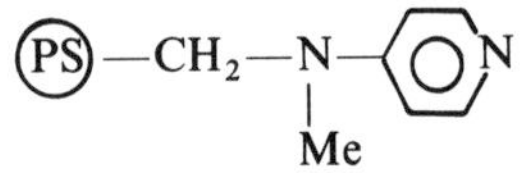 (PS)–CH$_2$–N–〈O N〉 (Me)	Acylation of alcohols	282

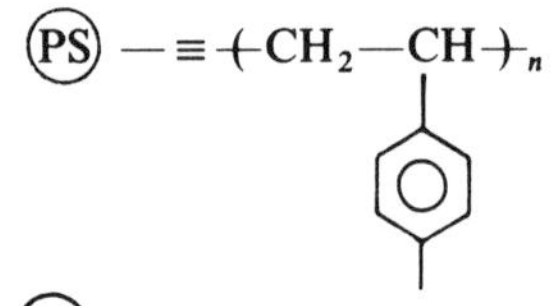

(PS) — ≡ –(CH$_2$–CH)$_n$– 〈O〉

(P) — ≡ –(CH$_2$–CH)$_n$–

(PEG) — ≡ polyethylene glycol

(C) — ≡ cellulose

(Si) — ≡ silica

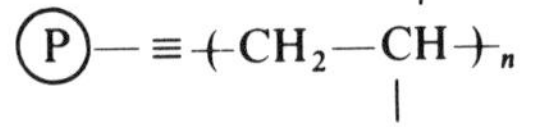

containing reactive methylene groups, there is always a possibility of formation of dialkylated or diacylated products, or of self-condensation as competing side reactions. Since the property of insolubility and hence the decreased chain mobilities of a resin-bound substrate in all common solvents can have the effect of isolating the reactive species formed on the polymer matrix from attaching each other and other components of the reaction mixture, the possibility of self-condensation and dialkylation or diacylation will be greatly reduced. Decreased chain mobility and degree of functionalization favour the formation of unsymmetrical or cross-condensation products with the elimination of the possibility of side-product formation. This approach has been used in the crossed alkylation and acylation of ester derivatives [230, 233, 236]. The reactive carbanion derivative of a bound ester is generated first and self-condensation with the carbonyl carbon atom of the unreacted ester is inhibited by the rigid matrix. The production of these stable mono-anions then allows reaction with acyl or alkyl halides to give selectively monoacylated or monoalkylated products, i.e.

$$\text{(PS)}-CH_2OCO-CH_2R \xrightarrow{:B} \text{(PS)}CH_2OCO-{}^-CH-R \xrightarrow{R'COCl} R'-COCH_2R$$

$$\downarrow R'X, H^+$$

$$\underset{\underset{R'}{|}}{R-CH-COOH}$$

(3.27)

3.8   POLYMER-BOUND NUCLEOPHILES

The use of anion exchange resins as basic catalysts in many synthetic organic reactions such as hydrolysis and condensation is one of the first examples of using functionalized polymers in organic chemistry. In addition, polymeric ammonium salts have been used as catalysts in phase transfer catalysed reactions. Anion exchange resins in which the bound negative ion is exchanged for a reactive nucleophile prior to use in a reaction have also been employed as polymer-supported anions in nucleophilic substitution reactions. There exists a close relationship between these systems and polymer-supported onium salts as phase

Table 3.8 Polymer-bound nucleophiles

Functionalized polymer	Application	Reference
(PS)—$CH_2N^+Me_3\ X^-$		
$X^- \equiv F^-$	Conversion of sulphonyl chloride tosylates, chlorides and bromides to fluorides	283, 284
	C—/O—alkylation, sulphenylation, Michael addition	285
$X^- \equiv Br^-, Cl^-, I^-$	Halogen exchange with alkyl halides	284
$X^- \equiv OCl^-$	Oxidation of diol	12
$X^- \equiv CN^-$	Conversion of alkyl halides to nitriles	12, 286–288
	Aldol condenstaion	289
	Cyanohydrin formation	290
	Benzoin condensation	291
$X^- \equiv SCN^-$	Conversion of alkyl halides to thiocyanates or isothiocyanates	287, 292
$X^- \equiv OCN^-$	Conversion of alkyl halides to ureas and urethanes	292
$X^- \equiv NO_2^-$	Conversion of alkyl halides to nitro-alkanes and nitrile esters	293–295
$X^- \equiv OCOONa^-$	Conversion of alkyl halides to alcohols	296
$X^- \equiv SCOMe$	Conversion of alkyl halides to thioacetates	297

Table 3.8 (*Contd.*)

Functionalized polymer	Application	Reference
$X^- \equiv SR^-$	Synthesis of sulphides, disulphides and thioesters	298, 299
$X^- \equiv OCOR^-$	Conversion of alkyl halides to esters	12, 300, 301
$X^- \equiv OH^-$	Condensation reactions	302
$X^- \equiv OAr^-$	Conversion of alkyl halides to ethers and synthesis of glycosides	295, 303–305
	Synthesis of aryloxyacetic acid	306
$X^- \equiv BH_4^-$	Reduction of aldehydes, hydroperoxides and solvent purification	307
$X^- \equiv SePh^-$	Synthesis of R—Se—Ph	308
$X^- \equiv O_2SPh^-$	Conversion of alkyl halides to alkyl sulphones	309
$X^- \equiv HFe(CO)_4^-$	Conversion of alkyl halides to aldehydes	310
$X^- \equiv H_3PO_2^{2-}, S_2O_3^{2-}, SO_3^{2-}, S_2O_4^{2-}$	Oxidation of alcohols to carbonyl compounds	311, 312
$X^- \equiv HCrO_4^-$	Oxidation of alcohols and alkyl halides to carbonyl compounds	99, 100
$X^- \equiv IO_4^-$	Oxidation of phenols and sulphides	12, 139
$X^- \equiv \underset{\underset{Y}{\mid}}{C}H—\underset{\underset{O}{\parallel}}{P}(OR)_2, Y \equiv CN, COOMe$	Conversion of carbonyl compounds and dioxolanes to olefins	235
$X^- \equiv BrO_2^-$	Oxidation of alcohols	313
$\text{(PS)}—\underset{\underset{Me}{\mid}}{S}^+—R\ FSO_3^-$ $R \equiv Me, Et$	Epoxidation of carbonyl compounds	39
$\text{(PS)}—CH_2—P^+(Ph)_3\ Br_3^-$	Bromination of carbonyl and unsaturated compounds	66
$\text{(PS)}—SO_3^-\ NH_4^+$	Conversion of carbonyl compounds to tetrahydropyrimidine derivatives 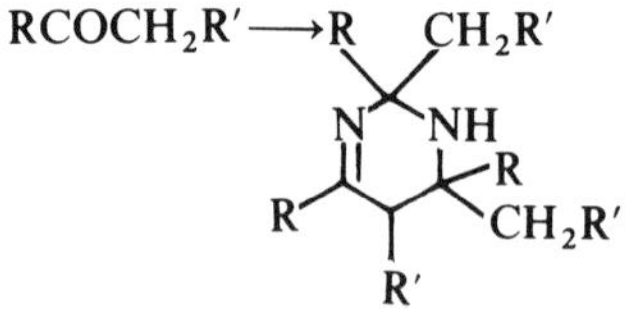	314

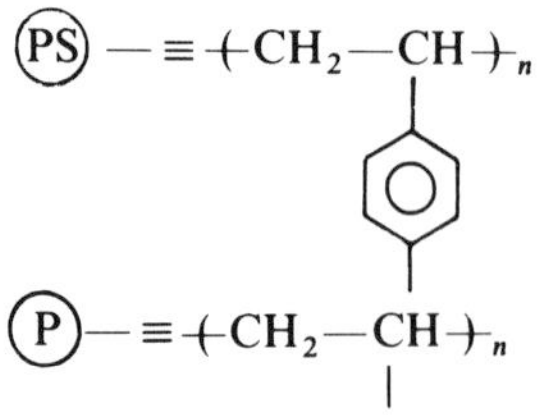

$$RCOCH_2R' \longrightarrow$$

R   CH₂R'

N   NH

     R

R'   CH₂R'

R'

$\text{(PS)}— \equiv +CH_2—CH+_n$

    [benzene ring with O]

$\text{(PEG)}— \equiv$ polyethylene glycol

$\text{(C)}— \equiv$ cellulose

$\text{(P)}— \equiv +CH_2—CH+_n$

$\text{(Si)}— \equiv$ silica

Table 3.9 Miscellaneous

Functionalized polymer	Application	Reference
(PS)—SO$_2$N$_3$	Diazo transfer to β-dicarbonyl compounds	315
	RRCH—CHO $\longrightarrow$ RRC=N$_2$	316
(PS)—CH$_2$NMe$_2$	Acid acceptor	317
(C)—O—C—NRR′ $\overset{\displaystyle \parallel}{\underset{\displaystyle \text{NH}}{}}$	Synthesis of N-alkylarginines, R^{2}R^{3}NH $\longrightarrow$ RR^1—N—C (=NH)NR^{2}R^3	318
(PS)—CH$_2$S—C=N—CN $\underset{\displaystyle \text{SR}}{\overset{\displaystyle \mid}{}}$	Conversion of amines to N-cyanoguanidines, (RNH)$_2$C= N—CN	319
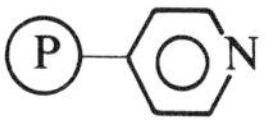 (P)—⟨◯⟩N	Acid acceptor	320, 321
(PS)—NH—C(=S)—NH—C$_6$H$_{13}$O$_5$ $\underset{\displaystyle \text{NCS}}{}$	Polymeric Girard reagent	322–325
(PS)—C=N$^+$=N$^-$ $\underset{\displaystyle \text{Ph}}{\overset{\displaystyle \mid}{}}$	Transformation of penicillin to cephalosporin	326
(PS)—(CH$_2$)$_n$—Li		
$n = 0$	Metallating agent for synthesis of organometallic polymers	327–328
$n = 1\text{–}4$	Metallating agent for synthesis of acids	329
	Metallating agent to halogen-containing compounds and CH-acidic substances	330
(PS)—HgClO$_4$	Conversion of silyl-protected piperazinediones into bicyclic derivatives	331
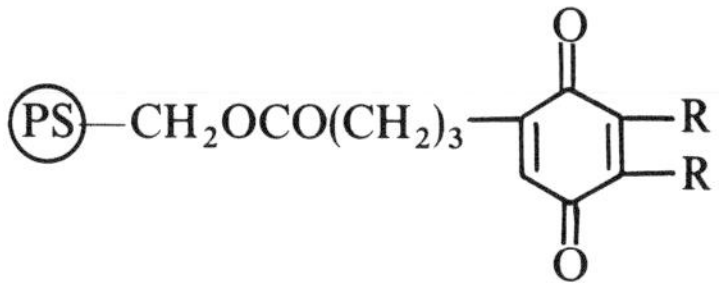 (PS)—CH$_2$OCO(CH$_2$)$_3$—[quinone]—R R R $\equiv$ H, CN	Synthesis of sensitive quinones (Thiele–Meisenheimer reaction)	332
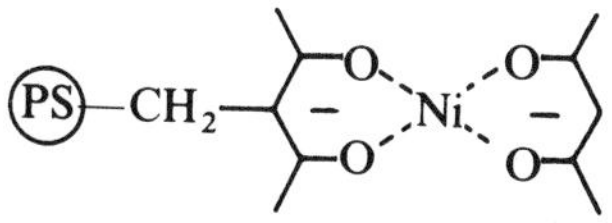 (PS)—CH$_2$—[Ni(O,O)$_2$ chelate]	Michael reaction	333

(PS) — $\equiv$ +(CH$_2$—CH)$_n$ ⟨◯⟩

(P) — $\equiv$ +(CH$_2$—CH)$_n$ |

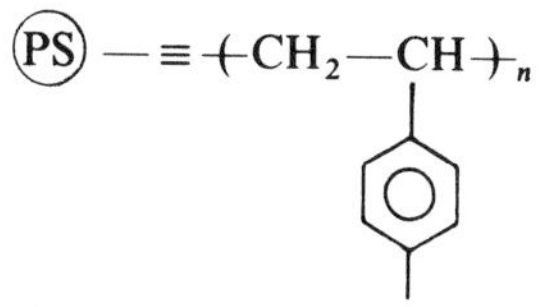

(PEG) — $\equiv$ polyethylene glycol

(C) — $\equiv$ cellulose

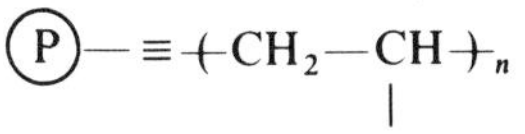

(Si) — $\equiv$ silica

transfer catalysts, where the anion exchange may be regarded as taking place *in situ* during reaction (section 4.5). Where the ion exchange is carried out as a separate process, the reagent can be isolated and dried before use under essentially anhydrous conditions. Most applications have employed halide ions as the nucleophile, but recently a bound benzenesulphinate, a thioacetate and a carbonate (25) and others have been used in the synthesis of sulphones, thioacetates and alcohols respectively (Table 3.8).

$$(PS)—CH_2N^+Me_3\,Z^- + R—X \longrightarrow R—Y + (PS)—CH_2N^+Me_3\,X^- \qquad (3.28)$$

25 $Z \equiv {}^-O_2SPh$ $Y \equiv SO_2Ph$
       ${}^-O—CO—ONa$ OH
       ${}^-SCOMe$ SCOMe

Aldehydes were obtained in high yield from alkyl halides in the presence of tetracarbonylhydridoferrate anions. However, drying of the reagent and separation of iron-containing byproducts from the organic compounds are difficult problems which limit the usefulness of the usual procedure in solution. A polymer containing tetracarbonylhydridoferrate anions (26) was used to eliminate the problems of the solution procedures [310]:

$$(PS)—CH_2N^+Me_3\,HFe(CO)_4{}^- + R—X \longrightarrow R—CHO \qquad (3.29)$$
               26

Other polymer-supported anions used in nucleophilic substitution reactions are listed in Table 3.8.

3.9 MISCELLANEOUS

Several reagents do not fall conveniently into any of the previous categories and these are listed in Table 3.9. Polymer-bound tosylazide 27 [316] is an important example. This has a useful activity but has greater thermal stability than its monomeric analogue. It can be handled with safety and provides improved yields in diazo transfer reactions to several β-dicarbonyl compounds:

$$(PS)—SO_2N_3 + R^1—CO—CH_2—CO—R^2 \longrightarrow R^1—CO—\underset{\underset{N_2}{\|}}{C}—CO—R^2$$
    27

or

$$R^3—CO—\underset{\underset{R^4}{|}}{CH}—CHO \longrightarrow R^3—CO—\underset{\underset{R^4}{|}}{C}=N_2 \qquad (3.30)$$

REFERENCES

1. McKillop, A. and Young, D.W. (1979) *Synthesis*, 402, 461.
2. Heitz, W. and Michels, R. (1973) *Liebigs Ann. Chem.*, 227.
3. Akelah, A. (1982) *Eur. Polym. J.*, **18**, 559.
4. Clarke, S.D., Harrison, C.R. and Hodge, P. (1980) *Tetrahedron Lett.*, 1375.
5. Camps, F., Castells, J., Font, J. and Vela F. (1971) *Tetrahedron Lett.*, 1715.

6. Heitz, W. and Michels, R. (1972) *Angew. Chem., Int. Ed. Engl.*, **11**, 298.
7. McKinley, S.V. and Rakshys, J.W. (1972) *J. Chem. Soc. Chem. Commun.*, 134.
8. Camps, F., Castells, J. and Vela, F. (1974) *An. Quim.*, **70**, 374.
9. Castells, J., Font, J. and Virgili, A. (1979) *J. Chem. Soc., Perkin Trans.*, *1*, 1.
10. Bernard, M. and Ford, W.T. (1983) *J. Org. Chem.*, **48**, 326.
11. Bernard, M., Ford, W.T. and Nelson, E.C. (1983) *J. Org. Chem.*, 3164.
12. Hodge, P., Khoshdel, E. and Waterhouse, J. (1984) *Markromol. Chem.*, **185**, 489.
13. Amos, R.A. and Fawcett, S.N. (1984) *J. Org. Chem.*, **49**, 2637.
14. Amos, R.A., Emblidge, R.W. and Havens, N. (1983) *J. Org. Chem.*, **48**, 3598.
15. Ferraboschi, P., Gambera, C., Azadani, M.N. and Santaniello, E. (1986) *Synth. Commun.*, **16**, 667.
16. Kim, B., Kodomari, M. and Regen, S.L. (1984) *J. Org. Chem.*, **49**, 3233.
17. Regen, S.L. and Lee, D.P. (1975) *J. Org. Chem.*, **40**, 1669.
18. Relles, H.M. and Schluenz, R.W. (1974) *J. Am. Chem. Soc.*, **96**, 6469.
19. Hodge, P. and Richardson, C. (1975) *J. Chem. Soc. Chem. Commun.*, 622.
20. Sherrington, D.C., Graig, D.C., Dagleish, J., Domin, G., Taylor, J. and Meehan, G.V. (1977) *Eur. Polym. J.*, **13**, 73.
21. Harrison, C.R. and Hodge, P. (1978) *J. Chem. Soc. Chem. Commun.*, 813.
22. Harrison, C.R., Hodge, P., Hunt, B.J., Khoshdel, E. and Richardson, G. (1983) *J. Org. Chem.*, **48**, 3721.
23. McKenzie, W.M. and Sherrington, D.C. (1982) *J. Polym. Sci., Polym. Chem. Ed.*, **20**, 431.
24. Harrison, C.R., Hodge, P. and Rogers, W.J. (1977) *Synthesis*, 41.
25. Hodge, P. and Khoshdel, E. (1984) *J. Chem. Soc.*, **2**, 194.
26. Appel, R., Struver, W. and Willms, L. (1976) *Tetrahedron Lett.*, 905.
27. Appel, R. and Willms, L. (1977) *J. Chem. Res. (S)*, 84; *(M)*, 0901.
28. Appel, R. and Willms, L. (1981) *Chem. Ber.*, **114**, 858.
29. Akelah, A. and El-Borai, M. (1980) *Polymer*, **21**, 255.
30. Caputo, R., Ferreri, C., Noviello, S. and Palumbo, G. (1986) *Synthesis*, 499.
31. Michels, R. and Heitz, W. (1975) *Makromol. Chem.*, **176**, 245.
32. Caputo, R., Corrada, E., Ferreri, C. and Palumbo, G. (1986) *Synth. Commun.*, **16**, 1081.
33. Horiki, K. (1976) *Tetrahedron Lett.*, 4103.
34. Pittman, C.U. and Hanes, R.M. (1977) *J. Org. Chem.*, **42**, 1194.
35. Crosby, G.A., Weinshenker, N.M. and Uk, H.S. (1981) *J. Am. Chem. Soc.*, **19**, 843.
36. Tonimoto, S., Horikawa, J. and Oda, R. (1967) *Kogyo Kagaku Zasshi*, **70**, 1269; (1968) *Chem. Abstr.*, 69406-h; (1969) *Yuki Gosei Kagaku Shi*, **27**, 989; (1970) *Chem. Abstr.*, 32306-g.
37. Ogura, K., Kondo, S. and Tsuda, K. (1981) *J. Polym. Sci., Polym. Chem. Ed.*, **19**, 843.
38. Crosby, G.A. and Kato, M. (1977) *J. Am. Chem. Soc.*, **99**, 278.
39. Farrall, M.J., Durst, T. and Frechet, J.M.J. (1979) *Tetrahedron Lett.*, 203.
40. Tanimoto, S., Miyake, T. and Okano, M. (1974) *Synth. Commun.*, **4**, 193.
41. Dorman, L.C. and Love, J. (1969) *J. Org. Chem.*, **34**, 158.
42. Sexsmith, D.R. and Frazza, E.J. (1965) US Patent 3 216 979.
43. Rassweiler, J.H. and Sexsmith, D.R. (1962) US Patent 3 060 156.
44. Hatch, M.J., Mayer, F.J. and Lloyd, W.D. (1969) *J. Appl. Polym. Sci.*, **13**, 721.
45. Bailey, F.E. and Combe, E.M. (1970) *J. Macromol. Sci. Chem.*, **A-4**, 1293.
46. Yamaguchi, H. and Schulz, R.C. (1976) *Makromol. Chem.*, **177**, 3441.
47. Geyer, G.R. (1966) US Patent 3 248 279.
48. Cassidy, H.G. and Kuhn, K.A. (1965) *Oxidation Reduction Polymers*, Wiley Interscience, New York.
49. Lindsey, A.S. (1970) *Rev. Macromol. Chem.*, **4**, 1.
50. Yaroslavsky, C., Patchornik, A. and Katchalski, E. (1970) *Tetrahedron Lett.*, 3629.

51. Yaroslavsky, C., Patchornik, A. and Katchalski, E. (1970) *Isr. J. Chem.*, **8**, 37.
52. Yanagisawa, Y., Akiyama, M. and Okawara, M. (1969) *Kogyo Kagaku Zasshi*, **72**, 1399; *Chem. Abstr.*, **71**, 113410-t.
53. Yaroslavsky, C. and Katchalski, E. (1972) *Tetrahedron Lett.*, 5173.
54. Zupana, M. and Pollak, A. (1975) *J. Chem. Soc. Chem. Commun.*, 715.
55. Sket, B., Zupan, M. and Zupet, P. (1984) *Tetrahedron*, **40**, 1603.
56. Hallensleben, M.L. (1972) *Angew. Makromol. Chem.*, **27**, 223.
57. Okawara, M., Oiji, Y. and Imoto, E. (1962) *Kogyo Kagaku Zasshi*, **65**, 1652 and 1658; (1963) *Chem. Abstr.*, **58**, 8051-e.
58. Frechet, J.M.J., Farrall, J.M. and Nuyens, L.J. (1977) *J. Macromol. Sci. Chem.*, **A-11**, 507.
59. Sket, B. and Zupan, M. (1984) *Tetrahedron*, **40**, 2865.
60. Zabicky-Zissman, J.Z., Oren, I. and Katchalski, E. (1972) US Patent 3 700 610.
61. Lloyd, W.G. and Durocher, T.E. (1963) *J. Appl. Polym. Sci.*, **7**, 2025.
62. Sket, B. and Zupan, M. (1986) *J. Org. Chem.*, **51**, 929.
63. Chiellini, E., Callaioli, A. and Solaro, R. (1985) *React. Polym.*, **3**, 357.
64. Cacchi, S. and Caglioti, L. (1979) *Synthesis*, 64.
65. Bongini, A., Cainelli, G., Contento, M. and Manescalchi, F. (1980) *Synthesis*, 143; (1980) *J. Chem. Soc. Chem. Commun.*, 1278.
66. Akelah, A., Hassanien, M. and Abdel-Galil, F. (1983) *Polym. Prepr., Am. Chem. Soc., Div. Polym. Chem.*, **24**(2), 467; (1984) *Eur. Polym. J.*, **20**, 221.
67. Manecke, G. and Stark, M. (1975) *Makromol. Chem.*, **176**, 285.
68. Cainelli, G., Contento, M., Manescalchi, F. and Piess, L. (1983) *Synthesis*, 306.
69. Hallensleben, M.L. (1973) *Angew. Makromol. Chem.*, **31**, 143.
70. Manecke, G. and Wallis, J. (1983) *Markromol. Chem. Rapid Commun.*, **4**, 119.
71. Manecke, G. Ehrenthal, E., Finck, W. and Wunsch, F. (1978) *Isr. J. Chem.*, **17**, 257.
72. Manecke, G. and Reuter, P. (1978) *J. Polym. Sci., Polym. Symp.*, **62**, 227.
73. Kaczmar, B.U. and Traser, S. (1976) *Makromol. Chem.*, **177**, 1991.
74. Schuttenberg, H. and Schulz, R.C. (1971) *Makromol. Chem.*, **143**, 153.
75. Sato, T. and Schulz, R.C. (1970) *Makromol. Chem.*, **180**, 299.
76. Wolman, Y., Kivity, S. and Frankel, M. (1967) *J. Chem. Soc. Chem. Commun.*, 629.
77. Rubinstein, M. and Patchornik, A. (1972) *Tetrahedron Lett.*, 2881; (1975) *Tetrahedron*, **31**, 1517.
78. Ito, H., Takamatsu, N. and Ichikizaki, I. (1975) *Chem. Lett.*, 577.
79. Kamogawa, H., Nanosawa, M., Uehara, S. and Osawa, K. (1979) *Bull. Chem. Soc. Jpn*, **52**, 533.
80. Nutt, R.F. (1979) US Patent 4 102 877; (1979) *Chem. Abstr.*, **90**, 39283.
81. Kondo, K., Ohbe, Y., Inaki, Y., and Takemoto, K. (1975) *Technol.. Rep. Osaka Univ.*, **25**, 487; (1976) *Chem. Abstr.*, **84**, 106103.
82. Weinshenker, N.M. and Shen, C.M. (1972) *Tetrahedron Lett.*, 3281, 3285.
83. Adam, W. and Yany, F. (1977) *Anal. Chem.*, **49**, 676.
84. Mutter, M. (1978) *Tetrahedron Lett.*, 2843.
85. Thomas, N.W., Beradinelli, F.M. and Edelman, R. (1979) US Patent 4 128 599; *Chem. Abstr.*, **90**, 88361.
86. Moore, J.A. and Kennedy, J.J. (1978) *J. Chem. Soc. Chem. Commun.*, 1079.
87. Smith, C.P. and Temme, G.H.J. (1983) *J. Org. Chem.*, **48**, 4681.
88. Brown, J. and Williams, R.E. (1971) *Can. J. Chem.*, **49**, 3765.
89. Williams, R.E., Brown, J. and Lauren, D.R. (1972) *Polym. Prepr., Am. Chem. Soc., Div. Polym. Chem.*, **13** (2), 823.
90. Nishino, J., Kondo, S., Yura, T., Tamaki, K. and Sakaguchi, Y. (1974) *Kobunshi Ronbunshu*, **31**, 244; (1974) *Chem. Abstr.*, **81**, 120123-m.
91. Shinkai, S., Tsuji, H., Hara, Y. and Manabe, O. (1981) *Bull. Chem. Soc. Jpn*, **54**, 631.
92. Tomoi, M., Kato, Y. and Kakiuchi, H. (1984) *Makromol. Chem.*, **185**, 2117.

93. Castells, J., Dunach, E., Geijo, F., Pujol, F. and Segura, P.M. (1978) *Isr. J. Chem.*, **17**, 278.
94. Kondo, K., Murakami, M. and Takemoto, K. (1983) *Makromol. Chem.*, **184**, 497.
95. Tagaki, T. (1967) *J. Polym. Sci., Polym. Lett. Ed.*, **B-5**, 1031; (1974) *J. Polym. Sci., Polym. Lett. Ed.*, **B-12**, 681; (1975) *J. Appl. Polym. Sci.*, **19**, 1649.
96. Harrison, C.R. and Hodge, P. (1974) *J. Chem. Soc. Chem. Commun.*, 1009; (1976) *J. Chem. Soc., Perkin Trans. 1*, 605, 2252.
97. Greig, J.A., Hancock, R.D. and Sherrington, D.C. (1980) *Eur. Polym. J.*, **16**, 293.
98. Helfferich, F. and Luten, D.B. (1964) *J. Appl. Polym. Sci.*, **8**, 2899.
99. Cardillo, G., Orena, M. and Sandri, S. (1976) *Tetrahedron Lett.*, 3985.
100. Cainelli, G., Cardillo, G., Orena, M. and Sandri, S. (1976) *J. Am. Chem. Soc.*, **98**, 6737.
101. Weinshenker, N.M., Crosby, G.M. and Wong, J.Y. (1975) *J. Org. Chem.*, **40**, 1966.
102. Gibson, H.W. and Bailey, F.C. (1977) *J. Chem. Soc. Chem. Commun.*, 815.
103. Sansoni, B. and Sigmund, O. (1961) *Naturwissenschaften*, **48**, 598.
104. Crosby, G.A. (1975) US Patent 3 928 293; (1976) *Chem. Abstr.*, **84**, 106499-u.
105. Hallensleben, M.L. (1973) *Z. Naturforsch.*, **28-b**, 540.
106. Hallensleben, M.L. (1974) *J. Polym. Sci., Polym. Symp.*, **47**, 1.
107. Domb, A. and Avny, Y. (1985) *J. Macromol. Sci. Chem.*, **A-22**, 167.
108. Menger, F.M., Shinozaki, H. and Lee, L.C. (1980) *J. Org. Chem.*, **45**, 2724.
109. Cernia, E. and Gasparini, F. (1975) *J. Appl. Polym. Sci.*, **19**, 917.
110. Akelah, A. and Sherrington, D.C. (1983) *Polymer*, **24**, 147.
111. Hodge, P. (1980) In *Polymer Supported Reactions in Organic Synthesis*, (eds P. Hodge and D.C. Sherrington), Wiley, London, p. 121.
112. Kemp, D.S. and Reczek, J. (1977) *Tetrahedron Lett.*, 1031.
113. Manecke, G., Kossmehl, G., Hartwich, G. and Gawlik, R. (1968) *Angew. Makromol. Chem.*, **2**, 86.
114. Gorecki, M. and Patchornik, A. (1973) *Biochim. Biophys. Acta*, **303**, 36.
115. Lindsey, A.S., Hunt, S.E. and Savill, N.G. (1966) *Polymer*, **7**, 479.
116. Endo, T., Kawasaki, H. and Okawara, M. (1979) *Polym. Prepr., Am. Chem. Soc., Div. Polym. Chem.*, **20**(1), 617.
117. Tsuchida, E., Hasegawa, E. and Ohno, H. (1977) *J. Polym. Sci., Polym. Chem. Ed.*, **15**, 417.
118. Endo, T. and Okawara, M. (1979) *J. Polym. Sci., Polym. Chem. Ed.*, **17**, 3667.
119. Adams, C.E. and Kimberlin, C.N. (1959) US Patent 2 899 396; (1960) *Chem. Abstr.*, **54**, 1770.
120. Christensen, L.W. and Heacock, D.J. (1978) *Synthesis*, 50.
121. Amaratunga, W. and Frechet, J.M.J. (1981) *Polym. Prepr., Am. Chem. Soc., Div. Polym. Chem.*, **22**(1), 151.
122. Frechet, J.M.J. and Amaratunga, W. (1982) *Polym. Bull.*, **7**, 361.
123. Huche, M., Savignac, M. and Pourcelot, G. (1981) *Nouv. J. Chim.*, **5**, 109.
124. Harpp, D.J., Adams, J., Gleason, J.G., Mullins, D. and Steliou, K. (1978) *Tetrahedron Lett.*, 3989.
125. Cairns, T.L., Gray, H.W., Scheider, A.K. and Schreiber, R.S. (1949) *J. Am. Chem. Soc.*, **71**, 655.
126. Hines, L., O'Driscoll, K.F. and Rempel, G.L. (1975) *J. Catal.*, **38**, 435.
127. Lecolier, S. (1974) Fr. Patent 2 270 238.
128. Howell, I.V. and Hancock, R.D. (1975) Br. Patent 1 408 013.
129. Frechet, J.M.J., Warock, J. and Farrall, M.J. (1978) *J. Org. Chem.*, **43**, 2618.
130. Harrison, C.R. and Hodge, P. (1982) *J. Chem. Soc., Perkin Trans. 1*, 509.
131. Frechet, J.M.J., Darling, G. and Farrall, M.J. (1980) *Polym. Prepr., Am. Chem. Soc., Div. Polym. Chem.*, **21**(2), 272.
132. Frechet, J.M.J., Darling, G. and Farrall, M.J. (1981) *J. Org. Chem.*, **46**, 1728.

133. Brunelet, T., Gelbard, G. and Guyot, A. (1981) *Polym. Bull.*, **5**, 145.
134. Brunelet, T. and Gelbard, G. (1983) *Nouv. J. Chim.*, **7**, 483.
135. Yoshida, J., Nakai, R. and Kawabata, N. (1980) *J. Org. Chem.*, **45**, 5269.
136. Yoshida, J., Hashimoto, J. and Kawabata, N. (1982) *J. Org. Chem.*, **47**, 3575.
137. Yoshida, J., Sofuku, H. and Kawabata, N. (1983) *Bull. Chem. Soc. Jpn*, **56**, 1243.
138. Yoshida, J., Ogura, K. and Kawabata, N. (1984) *J. Org. Chem.*, **49**, 3419.
139. Hodge, P. (1980) In *Polymer Supported Reactions in Organic Synthesis*, (eds P. Hodge and D.C. Sherrington), Wiley, London, p. 88.
140. Frechet, J.M.J. and Haque, K.E. (1975) *Macromolecules*, **8**, 130.
141. Frechet, J.M.J., Darling, G. and Farrall, M.J. (1980) *Polym. Prepr., Am. Chem. Soc., Div. Polym. Chem.*, **21**(2), 270.
142. Michels, R., Kato, M. and Heitz, W. (1976) *Makromol. Chem.*, **177**, 2311.
143. Kato, M., Michels, R. and Heitz, W. (1976) *J. Polym. Sci., Polym. Lett. Ed.*, **14**, 413.
144. Taylor, R.T. and Flood L.A. (1983) *J. Org. Chem.*, **48**, 5160.
145. Hu, N.X., Aso, Y., Otsubo, T. and Ogura, F. (1986) *Bull. Chem. Soc. Jpn*, **59**, 879.
146. Kondo, K., Fujita, T. and Takemoto, K. (1973) *Makromol. Chem.*, **174**, 7.
147. Kondo, K., Takahashi, T. and Takemoto, K. (1977) *J. Polym. Sci., Polym. Lett. Ed.*, **15**, 77.
148. Braun, D. (1959) *Makromol. Chem.*, **33**, 181.
149. Iwabuchi, S., Nakahira, T., Fukushima, Y., Saito, O. and Kojima, K. (1981) *J. Polym. Sci., Polym. Chem. Ed.*, **19**, 785.
150. Daly, W.H. and Kaufman, D.C. (1973) *Polym. Prepr., Am. Chem. Soc., Div. Polym. Chem.*, **14**, 1181.
151. Kaneko, M., Sugai, M. and Yamada, A. (1978) *Makromol. Chem.*, **179**, 2431.
152. Kamogawa, H., Kato, M. and Sugiyama, H. (1968) *J. Polym. Sci., A-1*, **6**, 2967.
153. Yoneda, F., Sakuma, Y. and Matsushita, U. (1978) *Heterocycles*, **9**, 1763.
154. Manecke, G. and Kossmehl, G. (1964) *Makromol. Chem.*, **80**, 22; **70**, 112.
155. Kamogawa, H. (1964) *Seni Kogyo Shikensho Iho*, **68**, 61.
156. Kamogawa, H., Larkin, J.M., Toei, K. and Cassidy, H.G. (1964) *J. Polym. Sci.*, **A-2**, 3603.
157. Ushakov, S.N., Klimova, O.M., Karchmarchik, O.S. and Smul'skaya, E.M. (1962) *Dokl. Akad. Nauk SSSR*, **143**, 231.
158. Manecke, G., Bahr, C. and Reich, C. (1959) *Angew Chem.*, **71**, 646.
159. Lunin, A.F., Sakaharov, M.M., Shostakova, L.M. and Geebov, L.S. (1974) *Chem. Abstr.*, **81**, 169295-h.
160. Sakharov, M.M., Shestakova, L.M. and Lunin, A.F. (1973) *Kinet. Katal.*, **14**, 808.
161. Schuttenberg, H., Klump, G., Kaczmar, U., Turner, S.R. and Schulz, R.C. (1973) *J. Macromol. Sci. Chem.*, **A-7**, 1085.
162. Kaczmar, R.C. (1973) *Angew. Chem., Int. Ed. Engl.*, **12**, 430.
163. Sato, Y., Kunieda, N. and Kinoshita, M. (1972) *Chem. Lett.*, 1023; (1977) *Makromol. Chem.*, **178**, 683.
164. Schuttenberg, H. and Schulz, R.C. (1971) *Angew. Makromol. Chem.*, **18**, 175.
165. Schuttenberg, H., Klump, G., Kaczmar, U., Turner, S.R. and Schulz, R.C. (1972) *Polym. Prepr., Am. Chem. Soc., Div. Polym. Chem.*, **13**, 866.
166. Schuttenberg, H. and Schulz, R.C. (1971) *Angew. Chem., Int. Ed. Engl.*, **10**, 856.
167. Hahn, K. and Schulz, R.C. (1976) *Angew. Makromol. Chem.*, **50**, 53.
168. Ivanov, S., Boeva, R. and Tanielyan, S. (1979) *J. Catal.*, **56**, 150.
169. Drago, R.S., Gaul, J., Zombeck, A. and Straub, D.K. (1980) *J. Am. Chem. Soc.*, **102**, 1033.
170. Rollamann, L.D. (1975) *J. Am. Chem. Soc.*, **97**, 2132.
171. Ivanov, S.D., Boeva, R. and Tanielyan, S. (1976) *React. Kinet. Catal.*, **5**, 297.
172. Sawodny, W., Grunes, R. and Reitzl, H. (1982) *Angew. Chem., Int. Ed. Engl.*, **21**, 775.
173. Leznoff, C.C. (1978) *Acc. Chem. Res.*, **11**, 327.

174. Frechet, J.M.J. and Nuyens, L.J. (1976) *Can. J. Chem.*, **54**, 926.
175. Fyles, T.M. and Leznoff, C.C. (1976) *Can. J. Chem.*, **54**, 935.
176. Leznoff, C.C., Fyles, T.M. and Weatherston, J. (1977) *Can. J. Chem.*, **55**, 1143.
177. Leznoff, C.C. and Dixit, D.U. (1977) *Can. J. Chem.*, **55**, 3351.
178. Letsinger, R.L. and Mahadevan, V. (1965) *J. Am. Chem. Soc.*, **87**, 3526; (1966) *J. Am. Chem. Soc.*, **88**, 5319.
179. Leznoff, C.C. and Wong, J.Y. (1972) *Can. J. Chem.*, **50**, 2892.
180. Leznoff, C.C. and Wong, J.Y. (1973) *Can. J. Chem.*, **51**, 3756.
181. Wong, J.Y., Manning, C. and Leznoff, C.C.(1973) *Angew. Chem., Int. Ed. Engl.*, **13**, 666.
182. Leznoff, C.C. and Greenberg, S. (1976) *Can. J. Chem.*, **54**, 3824.
183. Leznoff, C.C. and Goldwasser, J.M. (1977) *Tetrahedron Lett.*, 1875.
184. Goldwasser, J.M. and Leznoff, C.C. (1978) *Can. J. Chem.*, **56**, 1562.
185. Dixit, D.M. and Leznoff, C.C. (1977) *J. Chem. Soc. Chem. Commun.*, 798; (1978) *Isr. J. Chem.*, **17**, 248.
186. Seymour, E. and Frechet, J.M.J. (1976) *Tetrahedron Lett.*, 1149; Frechet, J.M.J., Nuyens, L.J. and Seymour, E. (1979) *J. Am. Chem. Soc.*, **101**, 432.
187. Chapman, P.H. and Walker, D. (1975) *J. Chem. Soc. Chem. Commun.*, 690.
188. Melby, L.P. and Strobach, D.R. (1967) *J. Am. Chem. Soc.*, **89**, 450.
189. Frechet, J.M.J. and Haque, K.E. (1975) *Tetrahedron Lett.*, 3055.
190. Frechet, J.M.J. (1975) *Polym. Prepr., Am. Chem. Soc., Div. Polym. Chem.*, **16**, 255.
191. Fyles, T.M., Leznoff, C.C. and Weatherston, J. (1977) *Can. J. Chem.*, **55**, 4135; (1978) *Can. J. Chem.*, **56**, 1031.
192. Leznoff, C.C. and Fyles, T.M. (1976) *J. Chem. Soc. Chem. Commun.*, 251.
193. Svirskaya, P. and Leznoff, C.C. (1984) *J. Chem. Ecology*, **10**, 321.
194. Koster, H. (1972) *Tetrahedron Lett.*, 1527.
195. Wong, J.Y. and Leznoff, C.C. (1973) *Can. J. Chem.*, **51**, 2452.
196. Ayres, J.T. and Mann, C.K. (1965) *Polym. Lett.*, **3**, 505.
197. Frechet, J.M.J. and Schuerch, C. (1971) *J. Am. Chem. Soc.*, **93**, 492.
198. Hanessian, S., Ogawa, T., Guindon, Y., Kamennof, J.L. and Roy, R. (1975) *Carbohydr. Res.*, **38**, C-15.
199. Frechet, J.M.J. and Pelle, G. (1975) *J. Chem. Soc. Chem. Commun.*, 225.
200. Farrall, M.J. and Frechet, J.M.J. (1976) *J. Org. Chem.*, **41**, 3877.
201. Frechet, J.M.J., Nuyens, L.J. and Seymour, E. (1979) *J. Am. Chem. Soc.*, **101**, 432.
202. Frechet, J.M.J. (1976) *Polym. Prepr., Am. Chem. Soc., Div. Polym. Chem.*, **17**, 515.
203. Frechet, J.M.J. and Seymour, E. (1978) *Isr. J. Chem.*, **17**, 253.
204. Bulten, N.P., Hodge, P. and Thrope, F.G. (1981) *J. Chem. Soc., Perkin Trans. 1*, 1963.
205. Farrall, M.J. and Frechet, J.M.J. (1978) *J. Am. Chem. Soc.*, **100**, 7998.
206. Leznoff, C.C. and Sywanyk, W. (1977) *J. Org. Chem.*, **42**, 3203.
207. Xu, Z.H., McArthur, C.R. and Leznoff, C.C. (1983) *Can. J. Chem.*, **61**, 1405.
208. Chou, T.S. and Shih, Y.E. (1984) *J. Chin. Chem. Soc.*, **31**, 87.
209. Hodge, P. and Waterhouse, J. (1983) *J. Chem. Soc., Perkin Trans. 1*, 2319.
210. Camps, F. Castells, J. and Pi, J. (1974) *An. Quim.*, **70**, 848.
211. Frechet, J.M.J. (1980) In *Polymer Supported Reactions in Organic Synthesis*, (eds P. Hodge and D.C. Sherrington), Wiley, London, p. 312.
212. Szabo, L., Coppens, E., Clauder, O. and Pecher, J. (1977) *Bull. Soc. Chim. Belg.*, **86**, 35.
213. McArthur, C.R., Worster, P.M., Jiang, J.L. and Leznoff, C.C. (1982) *Can. J. Chem.*, **60**, 1936.
214. Ogilvie, K.K. and Nemer, M.J. (1980) *Tetrahedron Lett.*, 4159.
215. Letsinger, R.L. and Kornet, M.J. (1963) *J. Am. Chem. Soc.*, **85**, 3045.
216. Eby, R. and Schuerch, C. (1975) *Carbohydr. Res.*, **39**, 151.

217. Zehavi, U. and Patchornik, A. (1973) *J. Am. Chem. Soc.*, **95**, 5673.
218. Watanabe, N. (1981) *Chem. Lett.*, (10), 1331.
219. Wachter, E., Hofner, H. and Machleidt, W. (1975) *In Solid Phase Methods in Protein Sequence Analysis*, (ed. R. Laursen), Pierce Chem. Co., Reckford, IL, p. 31.
220. Rubinstein, M., Amit, B. and Patchornik, A. (1975) *Tetrahedron Lett.*, 1445.
221. Rich, D.H. and Gurawara, S.K. (1973) *J. Chem. Soc. Chem. Commun.*, 610; (1975) *Tetrahedron Lett.*, 301.
222. Rich, D.H. and Gurawara, S.K. (1975) *J. Am. Chem. Soc.*, **97**, 1575.
223. Tjoeng, F.S., Tong, E.K. and Hodges, R.S. (1978) *J. Org. Chem.*, **43**, 4190.
224. Tjoeng, F.S., Staines, W., Pierre, S.S. and Hodges, R.S. (1977) *Biochim. Biophys. Acta*, **490**, 489.
225. Parr, W. and Grohmann, K. (1972) *Angew. Chem., Int. Ed. Engl.*, **11**, 314.
226. Waddell, T.G. and Leyden, D.E. (1981) *J. Org. Chem.*, **46**, 2406.
227. Bayer, E., Yung, G., Halasz, I. and Sebastian, I. (1978) *Tetrahedron Lett.*, 4503.
228. Parr, W. and Grohmann, K. (1971) *Tetrahedron Lett.*, 2633.
229. Parr, W., Grohmann, K. and Hagele, K. (1974) *Liebigs Ann. Chem.*, 655.
230. Kraus, M.A. and Patchornik, A. (1971) *Isr. J. Chem.*, **9**, 269; (1971) *J. Am. Chem. Soc.*, **93**, 7325.
231. Chang, Y.H. and Ford, W.T. (1981) *J. Org. Chem.*, **46**, 3756.
232. Chang, Y.H. and Ford, W.T. (1981) *Polym. Prepr., Am. Chem. Soc., Div. Polym. Chem.*, **22**(1), 157.
233. Kraus, M.A. and Patchornik, A. (1974) *J. Polym. Sci., Polym. Symp.*, **47**, 11.
234. Patchornik, A. and Kraus, M.A. (1970) *J. Am. Chem. Soc.*, **92**, 7587.
235. Cainelli, G., Contento, M., Manescalchi, F. and Regnoli, R. (1980) *J. Chem. Soc., Perkin Trans. 1*, 2516.
236. Camps, F., Castells, J., Ferrando, M. and Font, J. (1971) *Tetrahedron Lett.*, 1713.
237. Shambhu, M.B. and Digenis, G.A. (1973) *Tetrahedron Lett.*, 1627.
238. Letsinger, R.L., Kornet, M.J., Mahadevana, V. and Jerina, D.M. (1964) *J. Am. Chem. Soc.*, **86**, 5163.
239. Martin, G.E., Shambhu, M.B., Shakhshir, S.R. and Digenis, G.A. (1978) *J. Org. Chem.*, **43**, 4571.
240. Shambhu, M.B. and Digenis, G.A. (1974) *J. Chem. Soc. Chem. Commun.*, 619.
241. Martin, G.E., Shambhu, M.B. and Digenis, G.A. (1987) *J. Pharm. Sci.*, **67**, 110.
242. Adaway, T.J. and Harwood, H.J. (1977) *Polym. Prepr., Am. Chem. Soc., Div. Polym. Chem.*, **18**(1), 661.
243. Aug, T.L. and Harwood, J. (1973) *J. Macromol. Sci. Chem.*, **A-7**, 1079.
244. Nambu, Y., Endo, T. and Okawara, M. (1980) *J. Polym. Sci., Polym. Chem. Ed.*, **18**, 2793.
245. Okawara, M., Fujimoto, A. and Endo, T. (1978) *Isr. J. Chem.*, **17**, 264.
246. Marshall, G.R. and Liener, I.E. (1970) *J. Org. Chem.*, **35**, 867.
247. Wieland, T. and Birr, C. (1966) *Angew. Chem., Int. Ed. Engl.*, **5**, 310.
248. Flanigan, E. and Marshall, G.R. (1970) *Tetrahedron Lett.*, 2403.
249. Wieland, T. and Birr, C. (1967) *Chimia*, **21**, 581.
250. Fridkin, M., Patchornik, A. and Katchalski, E. (1965) *Isr. J. Chem.*, **3**, 69; (1965) *J. Am. Chem. Soc.*, **87**, 4646; (1966) *J. Am. Chem. Soc.*, **88**, 3164; (1968) *J. Am. Chem. Soc.*, **90**.
251. Patchornik, A., Fridkin, M. and Katchalski, E. (1970) Ger. Patent 1 913 486; *Chem. Abstr.*, **72**, 66932-y.
252. Bodenszky, M. and Funk, K.W. (1973) *J. Org. Chem.*, **38**, 1296.
253. Barton, M.A., Limieux, R.U. and Savoie, J.Y. (1973) *J. Am. Chem. Soc.*, **95**, 4501.
254. Skylarov, L.Y. and Shashkova, I.V. (1969) *Zh. Obsch. Khim.*, **39**, 2778.
255. Cohen, B.J., Hafeli, H.K. and Patchornik, A. (1984) *J. Org. Chem.*, **49**, 922.

256. Kalir, R., Fridkin, M. and Patchornik, A. (1974) *Eur. J. Biochem.*, **42**, 151.
257. Pause, G.T. and Laufer, D.A. (1970) *Tetrahedron Lett.*, 4181.
258. Rebek, J. and Trend, J.E. (1979) *J. Am. Chem. Soc.*, **101**, 737.
259. Gosselet, M., Sebille, B. and Buvet, R. (1979) *Eur. Polym. J.*, **15**, 1079.
260. Peska, J., Stamberg, J., Schmidt, P. and Benes, M.J. (1974) *J. Polym. Sci., Polym. Symp.*, **47**, 19.
261. Tsonis, C.P. and Farona, M.F. (1976) *J. Organomet. Chem.*, **114**, 293.
262. Rogozhin, S.V., Davidovich, Y.A., Andreav, S.M. and Yurtanow, A.A. (1973) *Dokl. Akad. Nauk SSSR, Ser. Khim.*, **212**, 108.
263. Fridkin, M., Patchornik, A. and Katchalski, E. (1972) *Biochem.*, **11**, 466.
264. Andreav, S.M., Tsiryapkin, V.A., Samoilova, N.A., Mironova, N.V., Davidovich, Y.A. and Rogozhin, S.V. (1977) *Synthesis*, 303.
265. Laufer, D.A., Chapman, T.M., Marlborough, D.F., Vaidya, V.M. and Blout, E.R. (1968) *J. Am. Chem. Soc.*, **90**, 2696.
266. Okawara, M., Akiyama, M. and Narita, M. (1969) *J. Polym. Sci., A-1*, **7**, 1299.
267. Narita, M., Teramoto, T. and Okawara, M. (1972) *Bull. Chem. Soc. Jpn*, **45**, 3149.
268. Akiyama, M., Yanagisawa, Y. and Okawara, M. (1969) *J. Polym. Sci., A-1*, **7**, 1905.
269. Akiyama, M., Shimizu, K. and Narita, M. (1976) *Tetrahedron. Lett.*, 1015.
270. Teramoto, T., Narita, M. and Okawara, M. (1977) *J. Polym. Sci., Polym. Chem. Ed.*, **15**, 1369.
271. Manecke, G. and Haake, E. (1968) *Naturwissenschaften*, **55**, 343.
272. Huang, X., Chan, C.C. and Zhou, Q.S. (1982) *Synth. Commun.*, **12**, 709.
273. Kalir, R., Warshawsky, A., Fridkin, M. and Patchornik, A. (1975) *Eur. J. Biochem.*, **59**, 55.
274. Heusel, G., Bovermann, G., Gohring, W. and Jung, G. (1977) *Angew. Chem., Int. Ed. Engl.*, **16**, 642.
275. Manecke, G. and Stark, M. (1984) *Makromol. Chem.*, **185**, 847.
276. Schally, A.V., Mittler, J.G. and White, W.F. (1971) *Biochem. Biophys. Res. Commun.*, **43**, 393.
277. Svec, F., Hradil, J., Coupek, J. and Kalal, J. (1975) *Angew. Makromol. Chem.*, **48**, 135.
278. Beyermann, H.C., DeLeer, E.W.B. and Van Vossen, W. (1972) *J. Chem. Soc. Chem. Commun.*, 929.
279. Schulz, R.C. and Schuttenberg, H. (1976) *Angew. Chem.*, **88**, 848.
280. Gunster, E.J. and Schulz, R.C. (1978) *Makromol. Chem.*, **179**, 2583, 2587; (1979) *Makromol. Chem.*, **180**, 1891.
281. Atherton, E., Clive, D.L.J. and Sheppard, R.C. (1975) *J. Am. Chem. Soc.*, **97**, 6584.
282. Tomoi, M., Akada, Y. and Kakiuchi, H. (1982) *Makromol. Chem. Rapid Commun.*, **3**, 537.
283. Borders, C.L., MacDonell, D.L. and Chambers, J.L. (1972) *J. Org. Chem.*, **37**, 3549.
284. Cainelli, G., Manescalchi, F. and Panunzio, M. (1976) *Synthesis*, 472.
285. Miller, J.M., So, K.H. and Clark, H. (1978) *J. Chem. Soc. Chem. Commun.*, 466.
286. Gordon, M., DePamphilis, M.L. and Griffin, C.E. (1963) *J. Org. Chem.*, **28**, 698.
287. Hodge, P. (1980) In *Polymer Supported Reactions in Organic Synthesis*, (eds P. Hodge and D.C. Sherrington), Wiley, London, p. 92.
288. Gordon, M. and Griffin, C.E. (1962) *Chem. Ind.*, 1079.
289. Durr, G. (1956) *Compt. Rend., Ser. C.*, **242**, 1630.
290. Schmidle, C.J. and Mansfield, R.C. (1952) *Ind. Eng. Chem.*, **44**, 1388.
291. Castells, J. and Dunach, E. (1985) *Chem. Lett.*, 1859.
292. Cainelli, G. and Manescalchi, F. (1979) *Synthesis*, 141.
293. Yamada, R., Noguchi, T., Urata, Y. and Okabe, K. (1969) *Mem. Def. Acad. Math. Phys. Chem. Eng. Yokosuka*, 667; (1970) *Chem. Abstr.*, **73**, 13813-z; (1973) *Mem. Def. Acad. Math. Phys. Chem. Eng. Yokosuka*, 13; (1974) *Chem. Abstr.*, **81**, 3281-a.

294. Hodge, P. (1980) In *Polymer Supported Reactions in Organic Synthesis*, (eds P. Hodge and D.C. Sherrington), Wiley, London, p. 95.
295. Gelbard, G. and Colona, S. (1977) *Synthesis*, 113.
296. Cardillo, G., Orena, M., Porzi, G. and Sandri, S. (1981) *Synthesis*, 793.
297. Cardillo, G., Contento, M., Manescalchi, F. and Musatto, M.C. (1981) *Synthesis*, 302.
298. Cainelli, G., Contento, M., Manescalchi, F. and Plessi, L. (1982) *Gazz. Chim. Ital.*, **112**, 461.
299. Cainelli, G., Contento, M., Manescalchi, F., Plessi, L. and Panunzio, M. (1983) *Gazz. Chim. Ital.*, **113**, 523.
300. Cardillo, G. and Manescalchi, F. (1975) *Synthesis*, 723.
301. Astle, M.J. and Zaslowsky, J.A. (1952) *Ind. Eng. Chem.*, **44**, 2867.
302. Shimo, K. and Wakamatsu, S. (1963) *J. Org. Chem.*, **28**, 504.
303. Rowe, R.J., Kaufmann, K.L. and Piantadosi, C. (1958) *J. Org. Chem.*, **23**, 1622.
304. Collins, M. and Laws, R.J. (1973) *J. Chem. Soc., Perkin Trans. 1*, 2013.
305. Iversen, T. and Johansson, R. (1979) *Synthesis*, 823.
306. Deshmukh, J.G., Jagdale, M.H., Mane, R.B. and Salunkhe, M.M. (1986) *Synth. Commun.*, **16**, 479.
307. Cook, M.M., Wagner, S.E., Mikulski, R.A., Demko, P.R. and Clements, J.G. (1978) *Polym. Prepr., Am. Chem. Soc., Div. Polym. Chem.*, **19**(2), 415.
308. Weber, J.W., Faller, P., Kirsch, G. and Schneider, M. (1984) *Synthesis*, 1044.
309. Manescalchi, F., Orena, M. and Savoia, D. (1979) *Synthesis*, 445.
310. Cainelli, G., Manescalchi, F. and Ronchi, A.U. (1979) *J. Org. Chem.*, **43**, 1598.
311. Regen, S.L. (1977) *J. Org. Chem.*, **42**, 875.
312. Manecke, G., Bahr, C. and Reich, C. (1959) *Angew. Chem.*, **71**, 160.
313. Kageyama, T. and Yamamoto, T. (1981) *Makromol. Chem.*, **182**, 705.
314. Bernardus, A.O.A. (1975) *US Patent 3 904 625*.
315. Roush, W.R., Feitler, D. and Rebek, J. (1974) *Tetrahedron Lett.*, 1391.
316. Durr, H., Hauck, G., Bruck, W. and Kober, H. (1981) *Z. Naturforsch.*, **36-b**, 1149.
317. Carpino, L., Williams, J.R. and Lopusinski, A. (1978) *J. Chem. Soc. Chem. Commun.*, 450.
318. Pundak, S. and Wilchek, M. (1981) *J. Org. Chem.*, **46**, 808.
319. Zupan, M., Sket, B., Vodopivec, J., Zupet, P., Molan, S. and Japelj, M. (1981) *Synth. Commun.*, **11**, 147.
320. Hallensleben, M.L. and Wurm, H. (1976) *Angew. Chem., Int. Ed. Engl.*, **15**, 163.
321. Greber, G. and Merchant, S. (1969) *Int. Symp. on Macromolecular Chemistry, Budapest, Kinetics and Mechanism of Polyreactions*, vol. 5, p. 131.
322. Dowling, L.M. and Stark, G.R. (1969) *Biochemistry*, **8**, 4728.
323. Frankhauser, P., Fries, P., Stahala, P. and Brenner, M. (1974) *Helv. Chim. Acta*, 271.
324. Laursen, R.A. (1966) *J. Am. Chem. Soc.*, **88**, 5344.
325. Schellenberger, A., Jeschkeit, H., Henkel, R. and Lehmann, H. (1967) *Z. Chem.*, **7**, 191.
326. Chapman, P.H. and Walker, D. (1975) *J. Chem. Soc. Chem. Commun.*, 690.
327. Braun, D., Daimon, H. and Becker, G. (1963) *Makromol. Chem.*, **62**, 183.
328. Braun, D. (1967) *Angew. Chem.*, **73**, 197.
329. Braun, D. and Seeling, E. (1964) *Chem. Ber.*, **97**, 3098.
330. Hallensleben, M.L. (1973) *Angew. Makromol. Chem.*, **31**, 147.
331. Dung, J.C., Armstrong, R.W. and Williams, R.M. (1984) *J. Org. Chem.*, **49**, 3416.
332. Perry, G.J. and Sutherland, M.D. (1982) *Tetrahedron*, **38**, 1471.
333. Fei, C.P. and Chan, T.H. (1982) *Synthesis*, 467.

4
Polymeric catalysts

A polymeric catalyst is a conventional catalytic species supported on a macromolecular backbone which is used in catalytic quantities relative to reaction substrates and can often be reused many times without loss of activity because the polymeric byproduct has the same structure as the polymeric catalyst. The performance of a polymeric catalyst is influenced by the physical and chemical properties of the support. In addition, its efficiency is related to other factors as (a) the properties of both catalyst and reactants since these determine the activation energy of the reaction, (b) the distribution coefficient of the reactants between the solution and catalyst, and (c) the nature of the solvent. The rate of the catalysed reaction depends on the rates of diffusion of substrates and products within the polymer matrix which, in turn, will depend on the ratio of the molecular size of the reactants to the micropore size of the polymer and consequently will be related to the degree of crosslinking and degree of swelling of the polymer. In other cases, the chemical reaction at the surface or within the polymer granules determines the rate, which will depend on the concentration of active sites available at the surface layer of the granule.

Heterogenation of homogeneous catalysts by attaching them to polymers leads to improved stability and selectivity since the catalysts produced combine the advantages of the polymer and the homogeneous catalyst. However, the increased experimental convenience of a polymeric catalyst is offset by a significant reduction in reactivity associated with, for example, the diffusional limitations imposed by resin supports. The selectivity of the polymer-attached catalysts towards substrates of different sizes can be controlled by the loading of the catalyst on the polymer support and also by the solvent used to swell the polymer. In general, the selectivity increases with decreasing swelling ratio of the polymer. In a polar solvent, it depends on the separation of the functionality. In cases where diffusion controls the reaction rate, the polymers can exhibit specificity or selectivity depending on the molecular size of the substrate and micropores of the polymer.

A wide range of catalysts has been supported in this way, ranging from ion exchange resins to the highly specifc enzymic catalysts.

4.1   ION EXCHANGE RESIN CATALYSTS

A wide range of functionalized polymers has been developed primarily for use in ion exchange processes and particularly for the removal of contaminant ions from aqueous solutions. The acid and base forms of ion exchange resins probably represent the earliest examples of synthetic polymeric catalysts [1, 2] and the use of these materials in this way has been reviewed previously [3–7].

The use of ion exchange resins as acidic and basic catalysts incorporates several advantages such as (a) elimination of the huge amount of waste material that is linked with large size production in industrial processes, (b) simplification of the work-up process because of the lack of extensive separation steps, (c) elimination of corrosion problems and (d) reduction in the amount of process chemicals. However, the high cost and lower thermal stability limit their commercial applications on an industrial scale. In general, the maximum working temperature for acid resins is approximately $125\,°C$ whilst basic resins are less stable and can be employed without loss of activity only up to $60\,°C$. Acidic resins can be stored in the proton or alkali metal ion forms, but strongly basic resins are most conveniently stored as the chloride ion form since their hydroxide forms tend to absorb CO_2 from the atmosphere and lose their activity.

The commercial availability of a series of resins ranging from strongly acid systems based on sulphonated polystyrene (1) to strongly basic systems involving the hydroxyl salts of bound ammonium ions (2) has facilitated the study of a great variety of reactions. Virtually all organic syntheses involving catalysis by homogeneous acids or bases have also been carried out using appropriate polymeric catalysts. These species are currently being exploited commercially.

Table 4.1   Ion-exchange resin catalysts

Functional polymer	Application	Reference
(P)—(O)—SO₃H	Hydrolysis of	
	Esters	8–10
	Enamines	11
	Amides, peptides and proteins	12–22
	Glycosylamines	23, 24
	Carbohydrates	25–27
	Esterification of	
	α–amino acids, fatty acids and glucose	28–32
	Acetal and ketal formation	33–36
	Cyclization	37–40
	Condensation	40–42
	Dehydration	43–48
	Alkylation	49–51, 52
	Rearrangement	53–55

Table 4.1 (*Contd.*)

Functionalized polymer	Application	Reference
	Decarboxylation	56
	Amide formation	57
	Hydration	25, 53, 58, 59
	Prins reaction	60
	Etherifications	61
(PS)—SO_3H		
X		
$X \equiv SO_3Ag$	Hydrolysis of esters	62
$X \equiv Me(CH_2)_{15}\overset{+}{N}Me_3$	Hydrolysis of esters	63–65
$X \equiv (PhCH_2)_3—N^+Me$	Hydrolysis of esters	63–65
$X \equiv OH$	Hydrolysis of dextrin	66
$X \equiv COOH$	Hydrolysis of amylose and sucrose	67, 68
(P)—COOH	Hydrolysis	69
	Esterification	70
(PS)—$CH_2NHCH_2PO_3H_2$	Acylation of furan	71
(PS)—$CH_2—\overset{+}{N}R_3\,OH^-$	Hydrolysis	72–81
	Dehydrohalogenation	75, 82
	Condensation	41, 83–89, 90–96
	Hydration	77, 78, 97, 98
	Cyclization	98
	Esterification	95, 99
	Alkylation	100
	Elimination	98, 101, 102
(P)—⟨○⟩N	Acylation	103
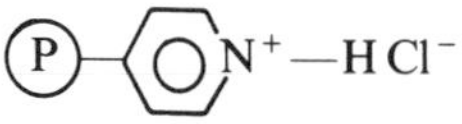 (P)—⟨○⟩N$^+$—$H\,Cl^-$	Acetalization of carbonyl compounds and esterification of acids	104
(PS)—$CH_2(NHCO(CH_2)_{10})_2$—N—⟨○⟩N		
Me		
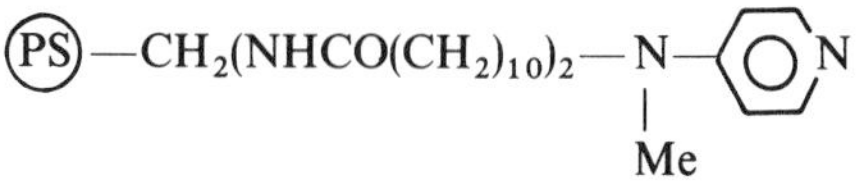	Esterification of acids	105

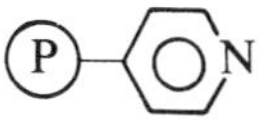

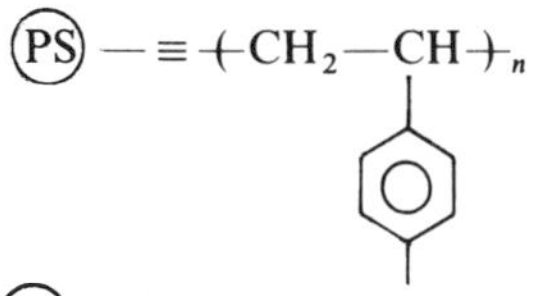

(PS) — $\equiv$ $(CH_2—CH)_n$ ⟨○⟩   (C)— $\equiv$ cellulose

(P)— $\equiv$ $(CH_2—CH)_n$ (Si)— $\equiv$ silica

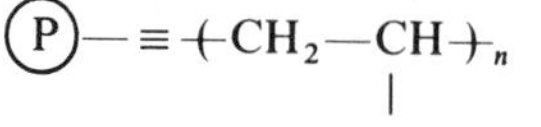

Nafion—H $\equiv$ $(CF_2—CF_2)_m(OCF_2—\overset{\overset{\displaystyle CF_3}{|}}{C})_n$

$O(CF_2)_2—SO_3H$

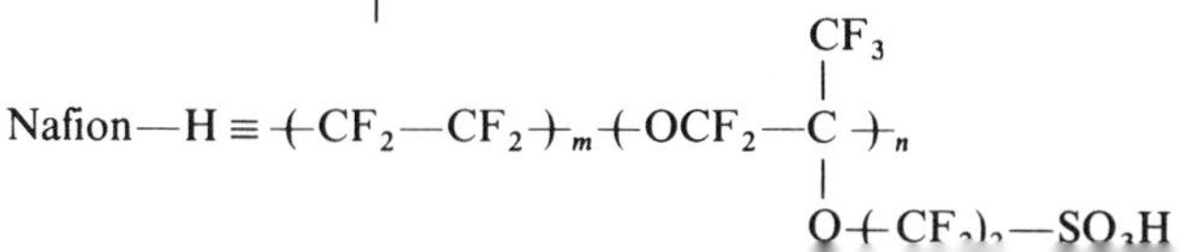

The major groups of organic reactions which are catalysed by ion exchange resins include acetal and ketal formation, alkylation, condensation, cyanoethylation, dehydration, epoxidation, esterification, etherification, hydration, hydrolysis, cyclization, isomerization and polymerization, as listed in Table 4.1. More recently, conventional anion exchange resins have been employed as supported phase transfer catalysts (section 4.5).

$$ (P) \!-\! \langle\bigcirc\rangle \!-\! SO_3H \qquad\qquad (P) \!-\! \langle\bigcirc\rangle \!-\! CH_2N^+R_3 \ ^-OH $$

1   2

4.2 POLYMERIC LEWIS ACID AND SUPERACID CATALYSTS

Polymeric Lewis acids have been prepared either by impregnation in polystyrene resin or by interaction with the functional group of the resin. The most frequently used resin is polystyrene; its hydrophobic nature protects the water-sensitive Lewis acid from hydrolysis by atmospheric moisture until it is placed in an appropriate solvent where it can be used in a chemical reaction. In addition, the presence of the polymer mediates the effect of the strong Lewis acid catalyst, producing higher yields of the desired product and lower yields of the competing higher molecular weight side-products. However, such catalysts are somewhat thermally unstable and degrade during their use.

The impregnation of a polystyrene resin with aluminium chloride by use of a suitable solvent produces a tightly bound complex of insoluble polymer and anhydrous Lewis acid on removal of the solvent [106]. Furthermore, this can be used very successfully as a mild moisture-insensitive catalyst for reactions requiring an acid catalyst in compounds with a sensitive secondary functional group. The impregnated Lewis acid is readily made available by swelling the resin with an appropriate solvent, while in the dry state the acid is protected from hydrolysis. The polymer appears to attenuate the activity of the catalyst to some extent and in some reactions undesirable complex condensation side-products are virtually eliminated. Polymeric aluminium chloride has been used as a mild effective catalyst for the formation of ethers, esters and acetals as shown in Table 4.2.

When a Lewis acid is introduced into a resin which is already functionalized with strong protonic acid groups, new acidic functions are generated [112, 113] with extremely high proton donor strengths and with an acidity approaching that of superacids. For example, polymeric superacids have been prepared by reacting $AlCl_3$ with sulphonic acid ion exchange resins. Such resins are capable of protonating paraffins, but where polystyrene is the base polymer the catalysts are somewhat unstable and are degraded during use. In an attempt to produce a more stable solid superacid, a commercially available perfluorinated copolymer containing perfluoroalkane–sulphonic acid groups 3 in combination with $AlCl_3$

Table 4.2 Polymeric Lewis acid and superacid catalysts

Functional polymer	Application	Reference
(P)—⟨O⟩·(BF$_3$ or AlCl$_3$)	Cyclization	107
	Acetal formation	108, 109
	Esterification	106
(PS)—SO$_3$H·MX$_3$	Alkylation of olefins	110, 111
MX$_3 \equiv$ BF$_3$, AlCl$_3$	Isomerization	112, 113, 114
MX$_3 \equiv$ FeCl$_3$, SnCl, TiCl, SbF$_5$	Esterification	115
(P)—⟨O⟩—O···BF$_3$ with Me below	Isomerization, etherification and rearrangement	116
Nafion—H	Transalkylation	117
	Acetal and thioacetal formation	118, 119
	Dehydration	120, 121
	Rearrangement	122
	Acylation	123, 124
	Alkylation	125, 126
	Hydration	127
	Etherification	128
	Esterification	129
	Nitration	130
	Epoxide ring opening	131
	O-silylation	132
	Protection and deprotection	132
	Cleavage of acetals and ketals	133, 134
	Fries rearrangement	135
	Rupe rearrangement	136
	Chlorination of alkenes	137
Nafion—Hg^{2+}	Nitration of aromatics	138
Cr^{3+} or Ce^{4+}	Oxidation of alcohols	139
Pd^{2+}	Carbomethylation of alkenes	140
Rh	Alkylation of benzene	140
	Carbonylation of aromatics	140
(P)—⟨O⟩—CH$_2$NR$_2$·(Pd$_5$ or PBr$_3$)	Conversion of acids to acid halides and alcohols to alkyl halides	141
(P)—⟨O⟩N·BF$_3$	Acetal formation	108

(PS)— ≡ ⟨CH$_2$—CH⟩$_n$ with ⟨O⟩   (C)— ≡ cellulose

(P)— ≡ ⟨CH$_2$—CH⟩$_n$ (Si)— ≡ silica

Nafion—H ≡ ⟨CF$_2$—CF$_2$⟩$_m$⟨OCF$_2$—C⟩$_n$ with CF$_3$ above C and O⟨CF$_2$⟩$_2$—SO$_3$H below have been prepared and used as a strong and efficient acid catalyst [142–144].

$$+(CF_2CF_2)_m—CF—CF_2+_n$$
$$|$$
$$(O—CF_2—CF+_zO—CF_2—CF_2—SO_3H$$
$$|$$
$$CF_3$$

$$\mathbf{3}\, m = 5\text{–}13.5$$
$$n = 1000$$
$$z = 1,2,3...$$

Where a perfluorinated copolymer is used much higher stability results from the inertness of its backbone to most reagents [145]. The enhanced acidity of the sulphonic group of Nafion-H is due to the electron-attracting fluorine atoms. However, problems arising from the morphology of the polymer and the relatively high cost seem to inhibit their industrial application as catalysts. Various systems have been developed recently and have been used as highly active catalysts in many organic syntheses such as the cracking and isomerization of hydrocarbons, nitration, rearrangement, Friedel–Craft reactions, esterification and other reactions as shown in Table 4.2.

4.3 POLYMERIC PHOTOSENSITIZER CATALYSTS

Several workers [146] have examined the possibility of attaching dyes to polymeric supports and using these as indicators in redox reactions, since they are accompanied by a distinct colour change, or as donor–acceptor substrates in biochemical reactions. Because of the variety of properties required in various biological and technological applications [147] polymeric dyes have also been prepared. For example, a water-insoluble chromophore was attached to an amine homopolymer via a portion of the amine groups. The remainder were then converted into solubilizing groups such as sulphonates [148], making the whole polymeric dye water soluble. The use of such polymeric dyes as non-toxic food dyes that do not absorb from the gastrointestinal tract has been described recently [149–153] (Chapter 8).

The applications of polymeric photosensitizers in photochemistry for the excitation of molecules into the excited state have received considerable interest owing to their potential use in solar energy storage [154] (Chapter 9). In these processes the polymeric sensitizer is excited by high energy into the triplet state and can then activate the substrate molecule while the sensitizer itself returns to the ground state.

Polymer-bound rose bengal [155] has been synthesized and used to sensitize the generation of singlet molecular oxygen for photo-oxidations of olefins. The singlet oxygen exhibits three modes of reactions with olefins; 1,4-cycloaddition with conjugated dienes to yield cyclic peroxides, an ene-type reaction to form allylic hydroperoxides, and 1,2-cycloaddition to give 1,2-dioxetanes which cleave thermally to carbonyl-containing products. Some examples of all three reaction types carried out with polymeric rose bengal as a sensitizer are shown in Table 4.3.

Table 4.3 Polymeric dyes and photosensitizer catalysts

Functional polymer	Application	Reference
(PS)—CH₂OCO—	Photosensitized oxidation	155
(PS)—CH₂	Photosensitized oxidation	155, 156
(PS)—CH₂OCO—⟨O⟩—COPh	Cycloaddition and dimerization of olefins	157, 158
(P)—COOR R ≡ Me and—	Photosensitized oxidation and reduction	159
(P)—⟨O⟩—CO—R R ≡ Me, Ph, (PS)—	Photosensitizer	160–165
(P)—CONH—R COO(CH₂)₂—OCOCH = CH·Ph R ≡	Photosensitizer and photosensitive	166
(P)—⟨O⟩—CO—⟨O⟩—OR R ≡ H, Me, Et, n-Bu, n-Oct, n-C₁₂H₂₃	Ultraviolet absorbers	167, 168
(P)—	Ultraviolet stabilizer	169

Table 4.3 (*Contd.*)

Functional polymer	Application	Reference
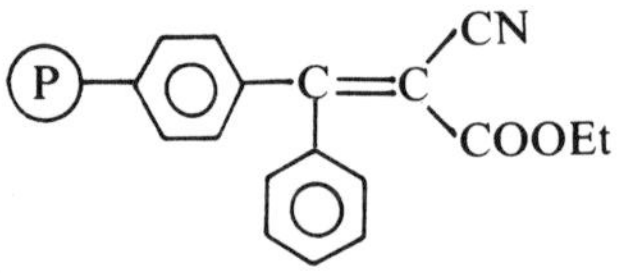	Ultraviolet stabilizer	170
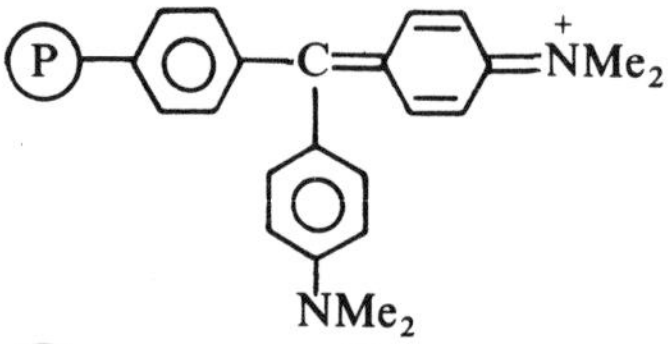	Photoresponse	171
	Soluble dye	148
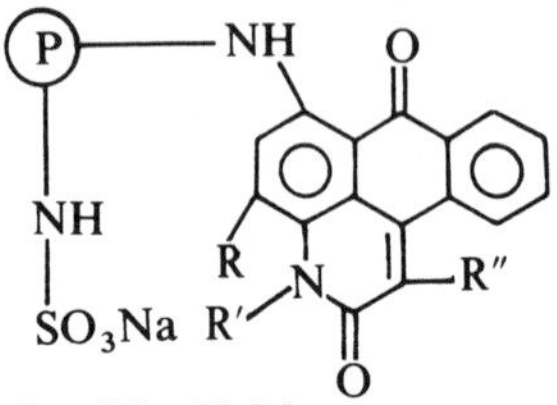 $R \equiv R' \equiv H, Me;$ $R'' \equiv H, COOMe, COOEt$	Soluble dye	148
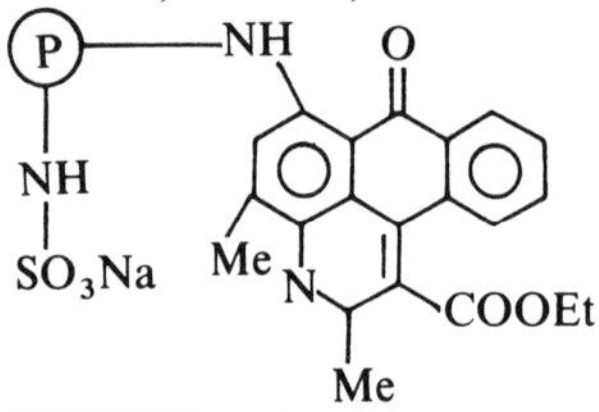	Soluble dye	148

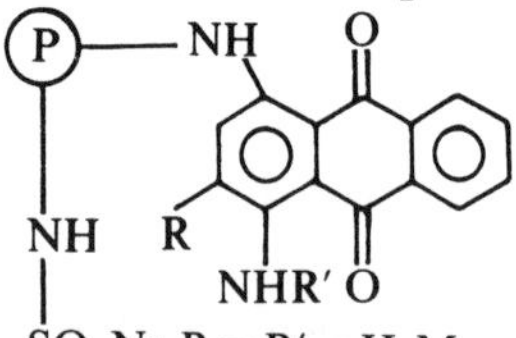

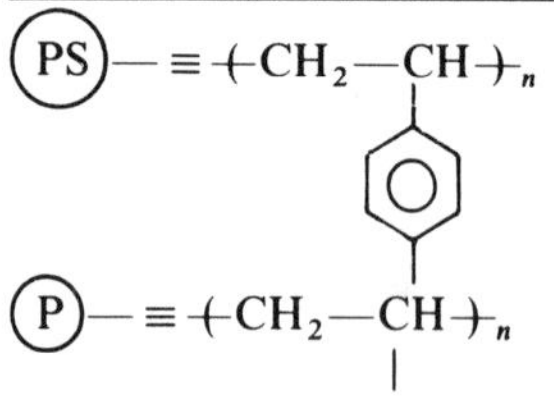

$\text{PS} - \equiv +CH_2-CH+_n$ (phenyl substituent)

$\text{C} - \equiv$ cellulose

$\text{P} - \equiv +CH_2-CH+_n$

$\text{Si} - \equiv$ silica

$$\text{Nafion}-H \equiv +CF_2-CF_2+_m+OCF_2-\underset{\underset{O+(CF_2)_2-SO_3H}{|}}{\overset{\overset{CF_3}{|}}{C}}+_n$$

The advantages of using an insoluble photosensitizer lie in the fact that the dye can be readily separated from the product and reused. Supported systems such as these must be translucent. The main problem in using polymers as supports for the sensitizers is the tendency of the polymer to crosslink or degrade through hydrogen abstraction from the backbone. Recently this problem has been overcome by using a soluble benzoylated phenyl Teflon, which has a good stability to ultraviolet sensitizer. Some examples of polymeric photosensitizers applied as heterogeneous catalysts are shown in Table 4.3.

4.4 POLYMERIC HYDROLYSING AND DECARBOXYLATING SYSTEMS

Several synthetic polymers with an essential functional structure of enzymes have been prepared and used to promote enzyme-type reactions [172–175]. Because the imidazole moiety of histidine is present in the active sites of several hydrolysing enzymes, many synthetic polymers containing imidazole and other pendant groups have been prepared and used to catalyse the hydrolysis of esters and amides. The polymeric catalysis is attributed to specific adsorption of the substrate and cooperative interaction of imidazole with other groups of the active centre on the substrate. After hydrolysis the product is eliminated from the polymer active centre which is then available for further catalysis. The esterolytic activity of various synthetic polymers containing imidazole and other active pendant groups are listed in Table 4.4.

The decarboxylation of carboxylic acids is a unimolecular reaction and is usually free from acid and base catalysis; the rate constant can be extremely solvent dependent. In particular, decarboxylation of 6-nitrobenzisoxazole-3-carboxylate has been widely investigated; it is slowest in aqueous solution and enhanced in aprotic or dipolar media.

Polysoaps with the characteristics of conventional micelles and polyelectrolytes are reactive polymers with a polar or charged moiety. They are excellent at activating many anionic reagents such as hydroxamate, thiolate and azide and can effectively catalyse reactions which are sensitive to the polarity of their environment. Polysoaps prepared by quaternization of polymeric amines with long-chain alkyl halides have been reported to be effective in the decarboxylation reaction and to enhance the rate of the reaction. For example, the reaction rate in the presence of fully quaternized polyethylenimines containing dodecyl

$$HOCO-CO-CH_2COO^- \longrightarrow HOCO-COCH_3$$

Table 4.4 Polymeric hydrolysing and decarboxylating systems

Functionalized polymer	Application	Reference
(a) X ≡ H (homopolymer)	Hydrolysis of 3-nitro-4-dodecanoyloxybenzoate, p-nitrophenylacetate and other ionic esters	176–186
(b) X ≡ COOH	Hydrolysis of m-MeOOC—C$_6$H$_4$—N$^+$Me$_3$ I$^-$	187–190
(c) X ≡ SO$_3^-$		188
(d) X ≡ SH	Hydrolysis of p-nitrophenylacetate	191
(e) X ≡ OH		187,188,192
(f) X ≡ p—HO—C$_6$H$_4^-$	Hydrolysis of phenyl ester derivatives	192
(g) X ≡ CON(Me)OH	Hydrolysis of phenyl ester derivatives	193
(h) X ≡ —CONH$_2$ and —CONPhOH		194,195
(k) X ≡	Hydrolysis of phenyl esters	196
(l) X ≡		197
	Hydrolysis of phenyl esters	198

Structure	Reaction	Refs.
P—CO—CH—CH$_2$—(imidazole), $X = NH_2$, COOH		199, 200
P—CONH—CH$_2$—(phenyl)—(imidazole), $X \equiv H$, $X \equiv COOH$	Hydrolysis of p-acetoxybenzoate / Hydrolysis of m-Me(CH$_2$)$_n$COO—C$_6$H$_4$—N$^+$Me$_3$ I$^-$	201 / 202
P—CONH—CH—CH$_2$—(imidazole), COOH, pyrrolidinone	Hydrolysis of phenyl esters	203
P—(CH$_2$)$_2$NHCH$_2$—(imidazole), (CH$_2$)$_2$NH—C$_{12}$H$_{25}$	Hydrolysis of p-nitrophenylacetate / Hydrolysis of 4-nitro-catechol sulphate	204, 205 / 206–209 / 210–213
P—(phenyl)—NH—(imidazole), $X = H$, COOH	Hydrolysis of 3-nitro-4-dodecanoyloxybenzoate	214–217

Table 4.4 (*Contd.*)

Functionalized polymer	Application	Reference
ⓟ–imidazole with OH, CON–CH₃	Saponification of active esters $MeCOO-\!\!\bigcirc\!\!-NO_2 \longrightarrow HO-\!\!\bigcirc\!\!-NO_2$ Solvolysis of *p*-acetoxybenzoate	201
ⓟ–CONH– benzimidazole (NH, N)	Hydrolysis of phenyl esters	185
ⓟ–imidazole (Me), X (a) X ≡ H (b) X ≡ –N pyrrolidinone (=O) (c) X ≡ –N⁺(Me)–N–Me I⁻ (d) X ≡ CONH₂ + CONPhOH	Hydrolysis of 3-nitro-4-acetoxybenzoate Hydrolysis of 3-nitro-4-acetoxybenzoate Hydrolysis of *p*-nitrophenylacetate	218 219 194, 195

(e) X ≡ — pyrazole —N—CH₂CO—N—CH₂Ph (with O⁻ and Me)   220

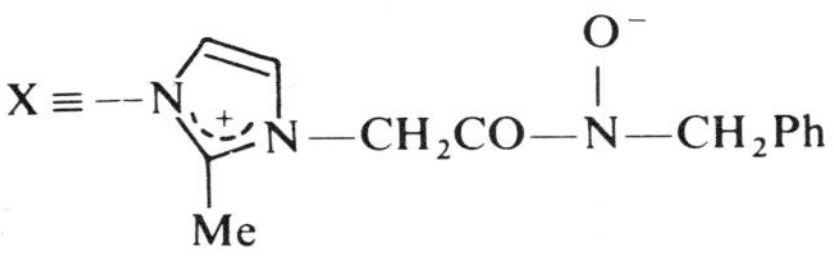

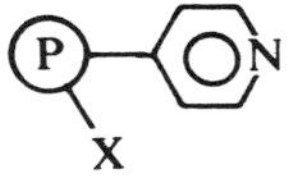

(a) X ≡ H   Hydrolysis of phenyl esters   221

(b) X ≡ pyridinium—N⁺—R   Hydrolysis of phenyl esters   222–228

(c) X ≡ N⁺—CH₂C(=N—OH)—Ph Br⁻   Hydrolysis of p-nitrophenylacetate 229

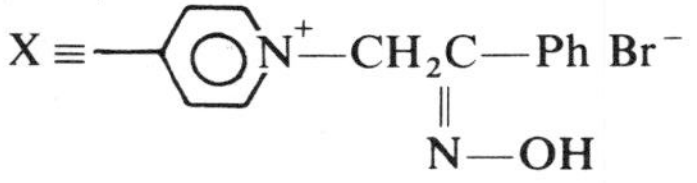

(d) X ≡ N⁺—CH₂CO—N—CH₂Ph (with O⁻)   Hydrolysis of phenyl acetate   220

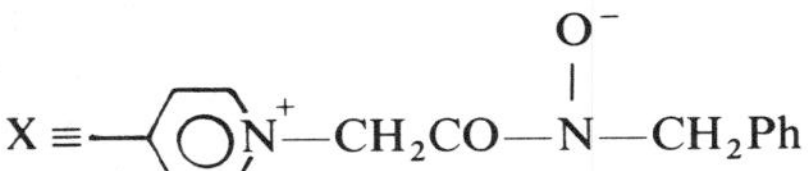

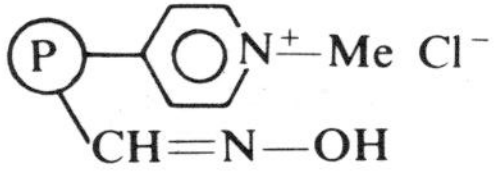

P—N⁺—Me Cl⁻
CH=N—OH   230

P—N⁺—Me X⁻
C—CH₂CHMeEt
‖
NOH   Esterolysis of O₂N—C₆H₄—OCOR   231

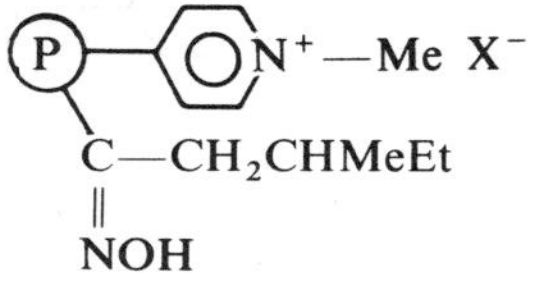

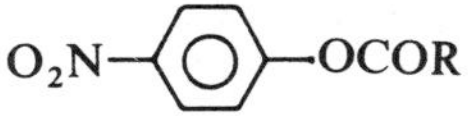

Table 4.4 (*Contd.*)

Functionalized polymer	Application	Reference
(P)—N⁺—CH₂C(=NOH)—Ph Br⁻ (with attached benzene ring)	Esterolysis of O₂N—⟨⟩—OCOR	232
(P)—N⁺—CH₂CHMeEt Br⁻		
(P)—CON(R)OH, CONH₂; R ≡ Me, Ph		193, 233
(P)—CH₂CH₂COO⁻; C₁₂H₂₅	Hydrolysis of Schiff bases	234
(P)—(CH₂)₂N⁺Me₂R; (CH₂)₂NMe₂; R ≡ C₆, C₈, C₁₂, C₁₈	Hydrolysis of *p*-nitrophenylacetate	235
—[(CH₂)₂—N(R)]ₙ—; R ≡ H	Cleavage of *p*-nitrophenylphosphate and hydrolysis of aspirin and phenyl esters	236–241
R ≡ PhCH₂—	Hydrolysis of *p*-nitrophenyl esters and cleavage of disulphide linkage of Ellman's reagent	242, 243
R ≡ C₁₂H₂₅, C₁₆H₃₃		244
(PS)—CH₂—⁺N(⟨⟩)—CONH₂ X⁻; X⁻ ≡ Cl⁻, CN⁻, SO₃²⁻		245–250

Structure	Function	Ref.
Flavin structure (R'—, Me, Me substituents; P—R, R'); $R = H, CH_2N^+Me_2C_{12}H_{25}$	Oxidation of NADPH to NADP$^+$ and thiophenol	251–253
P—N$^+$—C$_{12}$H$_{25}$ (imidazolium), Et	Proton abstraction from benzoin to give benzil	254
P—NH (imidazole)	Hydrolysis of phenyl esters	255
$\{SO_2CH_2\}_n$ ... N$^+$ R R^1 Cl$^-$; $R \equiv R^1 \equiv Me, Et$; $R \equiv Me$; $R^1 \equiv PhCH_2—, p\text{-}O_2N—C_6H_4—CH_2$	Hydrolysis of dinitrophenyl phosphate dianoin	256
P—CH=N—OH (pyridine)	Decarboxylation of O_2N—[benzisoxazole]—COO$^-$	257
P—N$^+$—CH$_2$—C(Ph)=N—OH	Decarboxylation of O_2N—[benzisoxazole]—COO$^-$	257

Table 4.4 (*Contd.*)

Functionalized polymer	Application	Reference
(P)—⬡—N^+—CH_2CO—N—OH with CH_2Ph	Decarboxylation of [benzisoxazole: O_2N—, COO^-, N—O]	257
(P)—⬡—N^+—$(CH_2)_n$—Me, $n = 1, 7, 11, 17, 21$	Decarboxylation of [benzisoxazole: O_2N—, COO^-, N—O]	258–260
(P)—⬡—CH_2—N^+Et_3 Cl^-	Decarboxylation of [benzisoxazole: O_2N—, COO^-, N—O]	261
(P)—⬡—CH_2CO—N—CH_2Ph with O^- and $(CH_2)_{21}$—Me	Decarboxylation of [benzisoxazole: O_2N—, COO^-, N—O]	262
$+(CH_2CH_2$—$N^+)_n$ with R and $C_{12}H_{25}$	Decarboxylation of 6-nitro-benzisoxazole-3-carboxylate and cyanophenyl-acetic acid	262
	Decarboxylation of oxalacetic acid	263
(Si)—$(CH_2)_3(NHCO(CH_2)_{10})_n$—$P^+Bu_3$ Br^-, $n = 1, 2$	Decarboxylation of [benzisoxazole: O_2N—, COO^-, N—O]	264

Decarboxylation of [structure: O_2N benzisoxazole-COO^-]	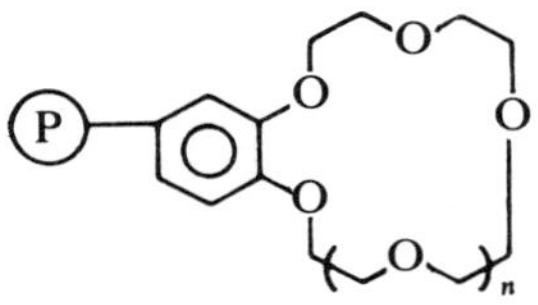 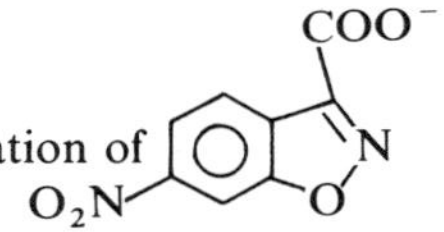 265–275
Binding of 4-[(1-pyrenyl)-butyl]trimethyl ammonium bromide	276
Binding of fluorescent molecules	277
Solubilization of pyrenes	278
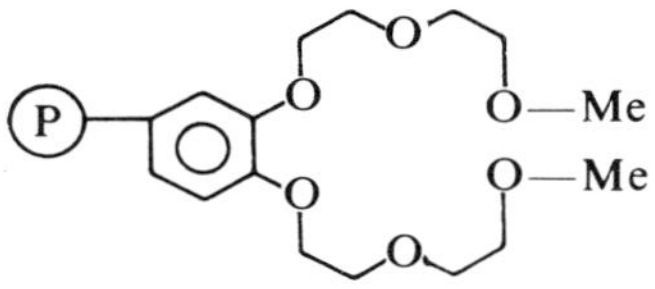 Binding of fluorescent molecules	277
Solubilization of pyrenes	278
Binding of picrates	279–281

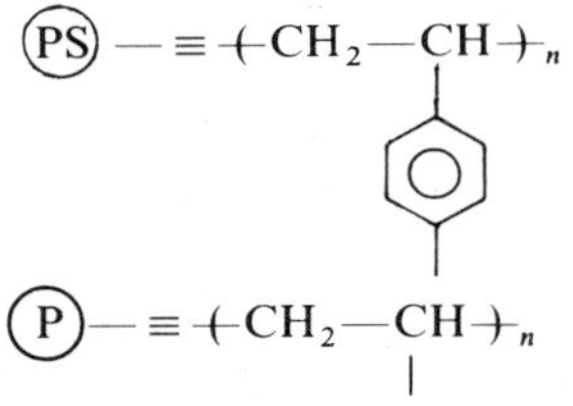

$n = 1,2$

$Z \equiv -O-, -O(CH_2)_{10}O-$
$n = 1,2$

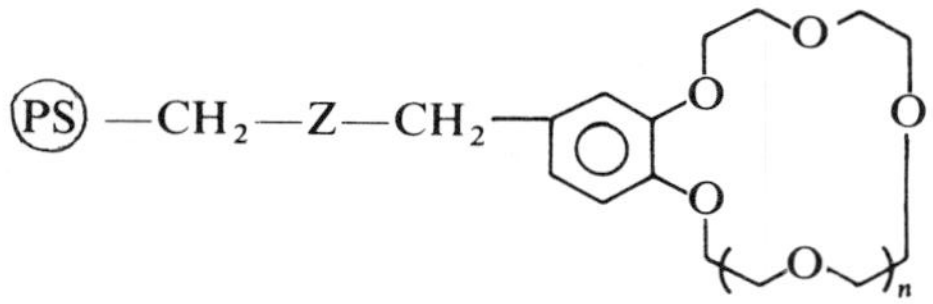

$(PS)- \equiv -(CH_2-CH)_n-$

$(C)- \equiv$ cellulose

$(Si)- \equiv$ silica

$(P)- \equiv -(CH_2-CH)_n-$

$Nafion-H \equiv -(CF_2-CF_2)_m-(OCF_2-C)_n-$
CF_3
$O-(CF_2)_2-SO_3H$

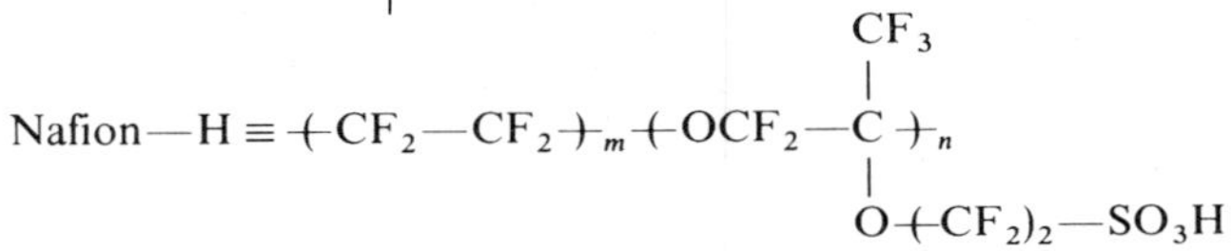

substituents (4) is larger than that of the spontaneous reaction in water without the polymer [282].

Other decarboxylation reactions investigated in the presence of quaternized polyethylenimines include those of cyanophenylacetic acid [282] and oxaloacetic acid [263]. The catalytic property of polysoaps is related to the hydrophobic microenvironment of the polymer, whereas the binding property with a variety of neutral and ionic organic solutes may arise from hydrogen bonding, charge transfer interactions or electrostatic forces.

Poly(crown ethers) and polyglymes behave as typical polycations in salt solutions of alkali and alkaline earth cations and strongly interact with neutral and anion solutes such as alkali picrates, methyl orange, p-nitrophenolate, phenolphthalein [283]. They also act in water media as efficient catalysts for decarboxylation reactions [284].

Other polysoaps include poly(vinyl pyrrolidone), which is a polymer with apolar side chains that is especially effective for interacting with dodecyl sulphate, azodyes and many aromatic species.

4.5 POLYMERIC PHASE TRANSFER CATALYSTS

The reaction between two reagents in essentially separate phases, two immiscible liquids or a liquid and a solid, is generally very slow and undetectable. The very low rates of such reactions are due to the low concentration of at least one of the compounds in each phase so that the collisions which are the fundamental requirement for a bimolecular reaction to occur are infrequent. Rapid stirring has been shown, in certain cases, to have an accelerating effect by increasing the surface contact between the layers and thereby increasing the interfacial component of the reaction [285–287]. In some other instances, solvents such as alcohols, acetone, tetrahydrofuran and dioxane which exhibit both lipophilic and hydrophilic properties can solve the problem. However, the difficulty in using these solvents is that the salts are less soluble in them than in hydrocarbons. This problem has been solved by the utilization of dipolar aprotic solvents which are more strongly cation-solvating media, e.g. dimethyl sulphoxide (DMSO), di-methyl formamide (DMF), MeCN and hexamethylphosphoramide (HMPA) are required. Unfortunately, these solvents are relatively expensive, difficult to purify and dry and often present problems in their complete removal from the products. They are also difficult to recover once the reaction is complete.

Phase transfer catalysts have been developed as a convenient solution to this problem. Phase transfer catalysis [288–289] is a relatively recent method in which liquid–liquid or solid–liquid phase-separated reactions are accelerated by the addition of a specific catalyst. In general, substances which can form ion pairs with anions or complex with the cation of a salt may function as phase transfer catalysts. Typically they are charged molecules such as organophilic onium salts (quaternary ammonium and phosphonium salts) or uncharged species such as crown ethers and cryptands.

The generally proposed mechanism for catalysis, typified by the reaction of a nucleophile in an aqueous solution with a substrate R—X in an organic phase involves two steps: (a) the transfer of anion from the aqueous phase to the organic phase via the catalyst; (b) nucleophilic displacement between the substrate and the transferred nucleophile with product formation:

$$\text{organic phase } Q^+Y^- + R - X \longrightarrow R - Y + Q^+X^-$$
$$\text{aqueous phase } Q^+Y^- + M^+X^- \rightleftharpoons M^+Y^- + Q^+X^-$$

where Q^+X^- is the onium salt or uncharged catalyst. Onium salts undergo ion exchange with the anion of the reagent in the aqueous phase or at the solid phase interface in an ion-pair equilibrium. Uncharged catalysts function by surrounding a cation with the oxygen or nitrogen electron lone pairs while at the same time providing an outer hydrocarbon coating which enhances solubility in a wide range of organic solvents. Tight complexation of the cation promotes ion separation and enhances anion activity.

The concept of heterogeneous catalysis has recently been introduced to (aqueous phase)–(organic phase) reactions by employing a polymeric catalyst in which both the catalyst and each of a pair of reactants are located in separate phases [290]. Such reactions are called 'triphase catalysed reactions'. When the catalyst species are supported on a polymer their catalytic action is essentially maintained while problems associated with the use of conventional soluble catalysts such as emulsification, which makes the experimental work-up more difficult, are eliminated. Costly catalysts such as optically active anium salts, crown ethers and cryptands are also effectively retained in this manner and can readily be reused. Furthermore, the toxicological problems associated with crown ethers are overcome.

During the last few years much interest has been focused on the application of functionalized polymers as phase transfer catalysts, as shown in Table 4.5. However, catalysis by polymeric onium salts can be considered as polymer-bound anion formation *in situ*, and the success of such processes depends on the compatibility of the polymer with both the aqueous and organic phases [379]. Although commercially anion exchange resins have been applied successfully as basic catalysts in reactions which require the transfer of hydroxide anion (section 4.1), their application as phase transfer catalysts in nucleophilic displacement reactions is restricted. Their lower catalytic activity as phase transfer catalysts may be due to (a) the reduced compatibility of the ionic groups with lipophilic substrates which limits the diffusion of the reacting species across the polymer bulk and (b) the unfavourable short inter-ionic distance which does not allow the formation of suitably loose ion pairs and imparts steric hindrance. These drawbacks can be overcome and the catalytic efficiency can be improved by increasing the flexibility of the linkage between the active centre and the polymer backbone through introduction of a spacer arm, by reducing the frequency of ionic groups, or by using more lipophilic amine groups which make the catalytic sites of the system more suited to hydrophobic interactions and less prone to hydrophilic interactions. A significant number of quaternary ammonium salts

Table 4.5 Polymeric phase transfer catalysts

Functional polymer	Application $R—X + Y^- \longrightarrow R—Z$	Reference
Ammonium salts		
ⓟ—◯—$(CH_2)_n$—$N^+R_2R'X^-$		
(a) $n = 0$; $R \equiv R' \equiv$ Me; $X^- \equiv Cl^-$, Br^-, F^-, I^-	$Y^- \equiv Cl^-$, Br^-, F^-, I^-	291, 292
(b) $n = 0$; $R \equiv R' \equiv$ Me; $X^- \equiv OH^-$		293
(c) $n = 1$; $R \equiv R' \equiv$ Me, Et, n-Bu, n-Oct; $X^- \equiv Cl^-$	$Y^- \equiv CN^-$, I^-, $R—\overset{-}{C}HCN$, Asymmetric Darzen's reaction	294–308, 295, 296, 301–303, 309–311
(d) $n = 2, 3$; $R \equiv R^1 \equiv$ n-Bu; $X^- \equiv Br^-$	$Y^- \equiv I^-$, PhS^-, Br^-, Cl^-, N_3^-, ArO^-, AcO^-	305, 309
ⓟ—◯—Z—N^+R_3 X^-		
$R \equiv$ Me; $Z \equiv -CH_2OCO(CH_2)_n-$; $n = 5, 11$	$Y^- \equiv CN^-$, I^-	300, 312–314
$R \equiv$ Me, n-Bu; $Z \equiv CH_2NHCO(CH_2)_{10}$		
Ⓢⓘ—Z—N^+R_3 X^-	$Y^- \equiv AcO^-$	305, 306, 315, 316
$Z \equiv -(CH_2)_n-$; $n = 2, 3, 6$		
—CH_2— ◯		
$R \equiv$ H, Me, Et, n-Bu, n-Oct, n-$C_{16}H_{33}$		
ⓟ—◯—CH_2—[⁺N ring] Cl^-	$Y^- \equiv Ph—\overset{-}{C}H—CN$	317
ⓟ—◯—N^+—R Cl^- or Br^-	$Y^- \equiv PhO^-$	318, 319
$R \equiv$ H, n-Bu, $PhCH_2$, $CH_2CHMeEt$		

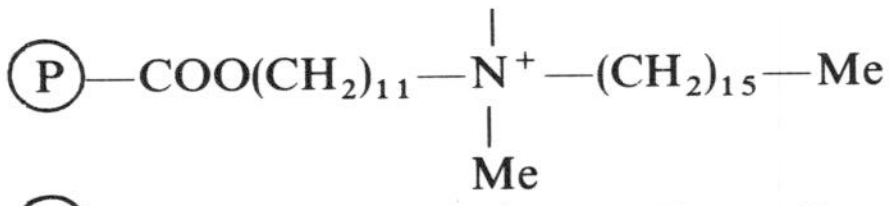

(P)—COO(CH$_2$)$_{11}$—$\overset{\displaystyle |}{\underset{\displaystyle Me}{N^+}}$—(CH$_2$)$_{15}$—Me

314

(C)—OCO(CH$_2$)$_n$—N$^+$R$_2$BuCl$^-$ or Br$^-$, with OMe

$n = 1, 2$; R ≡ Et, n-Bu

Y$^-$ ≡ PhO$^-$, AcO$^-$; reduction

320

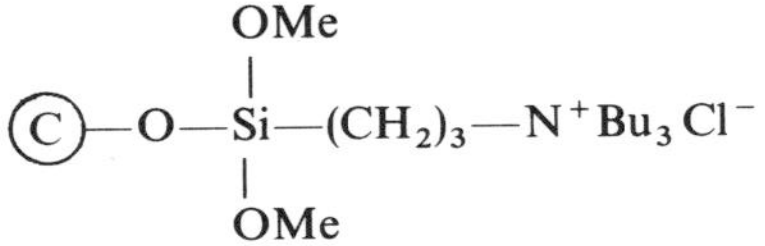

(C)—O—Si—(CH$_2$)$_3$—N$^+$Bu$_3$Cl$^-$ (OMe, Et, OMe)

Y$^-$ ≡ I$^-$, CN$^-$; reduction

321

(C)—O(CH$_2$)$_2$—$\overset{\displaystyle Et}{\underset{\displaystyle Et}{N^+}}$—CH$_2$CHOR Cl$^-$ (Me)

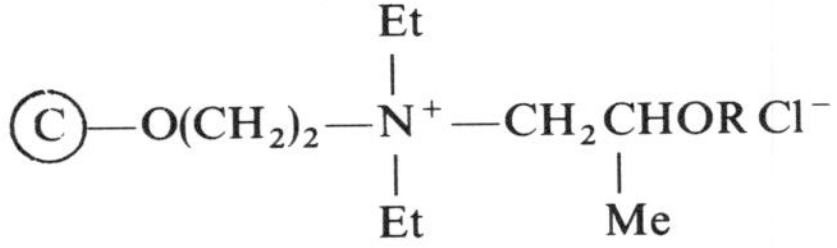

R ≡ H, SiMe$_3$, CH$_2$Ph

Y$^-$ ≡ I$^-$, CN$^-$; reduction

321

(P)—CH$_2$O(CH$_2$)$_n$—CR$_2$—NMe$_2$

$n = 1, 2$; R ≡ H, Me

Y$^-$ ≡ CN$^-$

322

Phosphonium salts

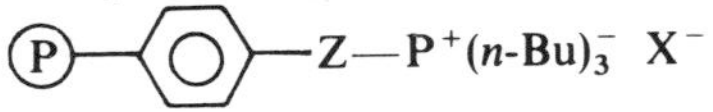

(P)—$\langle \bigcirc \rangle$—Z—P$^+$(n-Bu)$_3$ X$^-$

(a) Z ≡ —(CH$_2$)$_n$; $n = 1, 2, 3, 6, 7$
(b) Z ≡ —(CH$_2$)$_m$[NHCO(CH$_2$)$_{10}$]$_m$; $n = 1, 2, 3$; $m = 1$
(c) Z ≡ CH$_2$O(CH$_2$)$_n$

Y$^-$ ≡ CN$^-$, ArO$^-$, Cl$^-$, I$^-$, AcO$^-$,
ArS$^-$, ArCHCOMe$^-$, N$_3^-$,
SCN$^-$, S^{2-}; reduction

297–300
302, 304,
309
313, 315,
323–335,
307

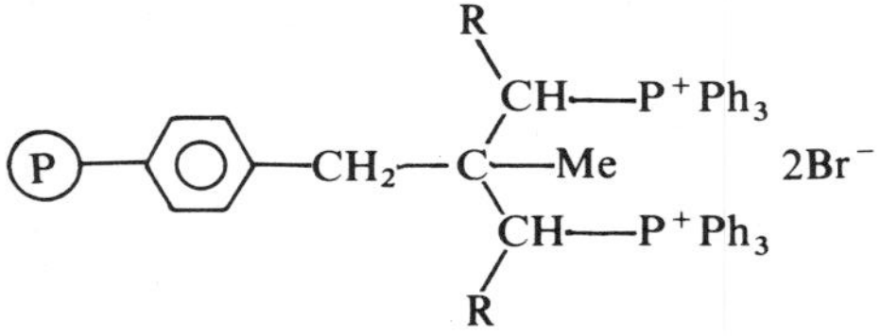

(P)—$\langle \bigcirc \rangle$—CH$_2$—$\overset{\displaystyle CH—P^+Ph_3}{\underset{\displaystyle CH—P^+Ph_3}{\overset{\displaystyle R}{\underset{\displaystyle R}{C—Me}}}}$ 2Br$^-$

R ≡ H, Me

Y$^-$ ≡ PhS$^-$, PhO$^-$

327

Table 4.5 (*Contd.*)

Functional polymer	Application $R-X + Y^- \longrightarrow R-Z$	Reference
Ⓒ—OCOCH$_2$P$^+$(Bu-n)$_3$ Cl$^-$	$Y^- \equiv AcO^-$, PhO$^-$; reduction	320
ⓈⒾ—Z—P$^+$(n-Bu)$_3$X$^-$	$Y^- \equiv I^-$, PhO$^-$; reduction	300, 315, 316, 331, 336
(a) $Z \equiv -(CH_2)_3-$		
(b) $Z \equiv (CH_2)_3[NHCO(CH_2)_m]_n$; $m = 2, 10$; $n = 1, 2$		315, 316, 332, 336
(c) $Z \equiv$ 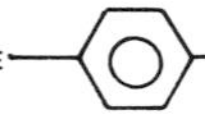 ⟨◯⟩—CH$_2$—		305

Crown ethers and cryptands

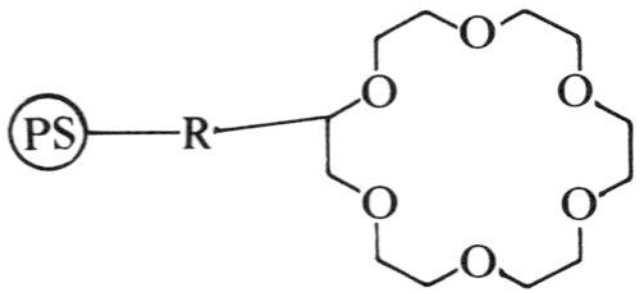

PS—R

(a) $R \equiv CH_2NH(CH_2)_2-$ $Y^- \equiv CN^-$ 297

(b) $R \equiv CH_2[NHCO(CH_2)_{10}]_n-N-(CH_2)_9-$ $Y^- \equiv I^-$, PhS$^-$, CN$^-$ 313, 334
          |
          Et

$n = 0, 1, 2$

(c) $R \equiv -CH_2-$ $Y^- \equiv I^-$, CN$^-$, PhO$^-$ 337

(d) $R \equiv -(CH_2)_n-OCH_2-$; $n = 1, 3$ $Y \equiv F^-$, CN$^-$, AcO$^-$ 334, 338

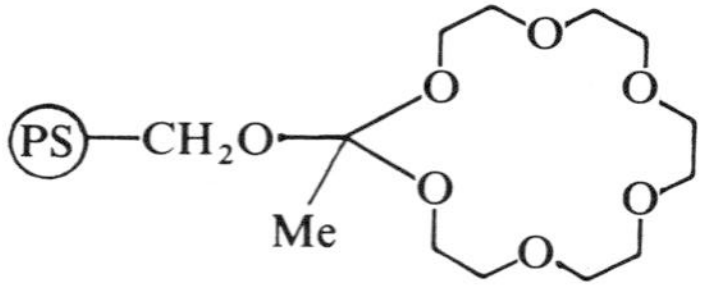

PS—CH$_2$O   Me   339

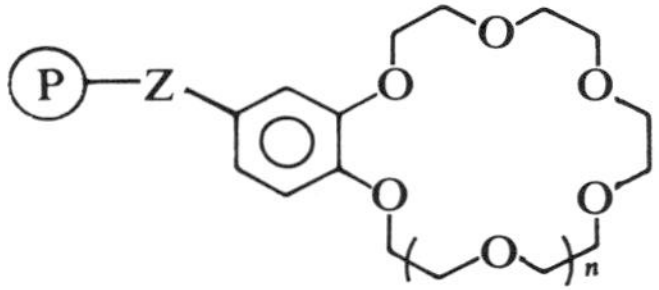

(a) $Z \equiv$ nothing; $n = 0, 1$; $R \equiv$ ⌁ $Y^- \equiv AcO^-, CN^-, PhO^-$ 340, 341

(b) $Z \equiv CH_2OCH_2$; $n = 1$; $R \equiv$ ⌁ , ⌁OMe , ⌁O⌁O⌁OMe $Y^- \equiv CN^-$; dehydrohalogenation; alkylation   342–344

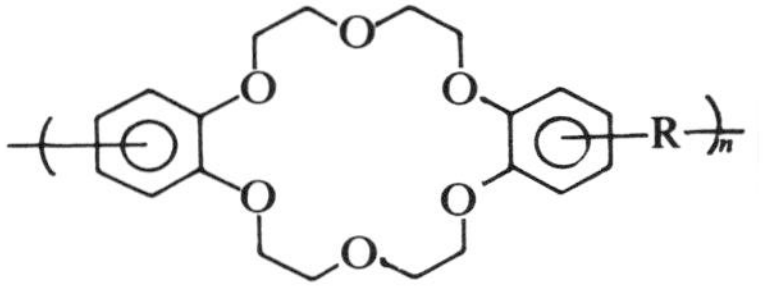

$Y^- \equiv I^-, CN^-, PhO^-$ 337

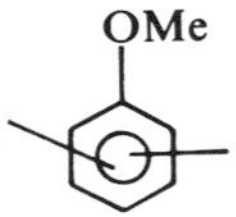

$R \equiv CH_2, -CH_2 -$

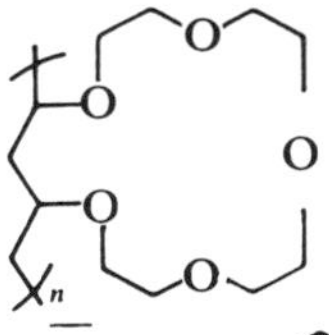

$Y^- \equiv CN^-$ 345

$Y^- \equiv I^-, CN^-, PhS^-$; reduction   297, 334

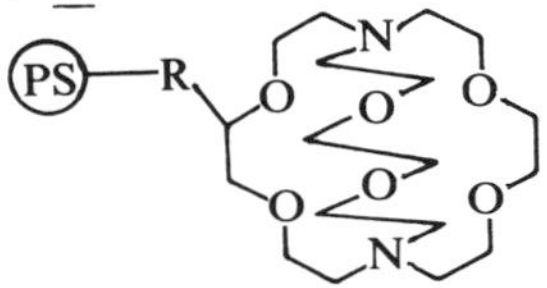

(a) $R \equiv CH_2NH(CH_2)_a$
(b) $R \equiv CH_2OCH_2$

Table 4.5 (*Contd.*)

Functional polymer	Application $R-X + Y^- \longrightarrow R-Z$	Reference
(PS)—CH$_2$O—R		346
(PS)—CH$_2$OCH$_2$	$Y^- \equiv I^-, CN^-$	347
Cosolvents		
(P)—O—CH$_2$(O—CH$_2$CH$_2$)$_n$—OR		
(a) R ≡ Me; n = 1, 2, 3, 13, 16, 7	$Y^- \equiv ArO^-, Cl^-, CN^-, AcO^-$	298, 299, 337, 348–352
(b) R ≡ H, Me; n = 5, 6, 7	Saponification of esters	349
(c) R ≡ H; n = 2–5, 13	Dehydrohalogenation	342
(d) R ≡ Me; n = 7, 16	Alkylation of nitriles, alcohols and tertiary ketones	343
(e) R ≡ H; n = 1, 4, 13	Nucleophilic substitution	299, 350
(f) R ≡ Ph	$Y^- \equiv PhO^-$; substitution	319, 352, 353
(g) R ≡ , Tetrahydrofuran-2-yl, Tosylate	$Y^- \equiv PhO^-$	319, 353, 354

(h) R $\equiv$ C$_9$H$_{19}$—C$_{11}$H$_{23}$; $n = 9$

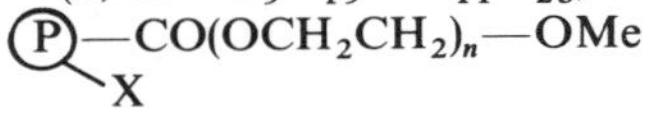

$\boxed{P}$—CO(OCH$_2$CH$_2$)$_n$—OMe

$\boxed{P}$—X

$n = 4, 8, \approx 22$; X $\equiv$ Ph, COOMe, CN, —pyridyl

$\boxed{P}$—(OCH$_2$CH$_2$)$_n$—OR / —(OCH$_2$CH$_2$)$_n$—OR

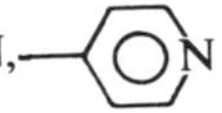

$n = 2, 3, 4$; R $\equiv$ Me, Et, quinolyl

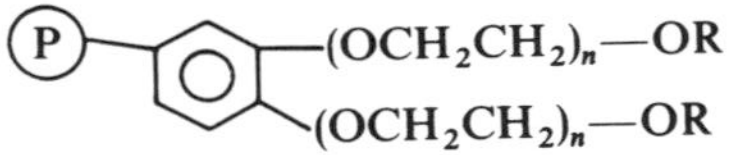

$\boxed{P}$—C$_6$H$_4$—OMe / O(CH$_2$CH$_2$O)$_3$

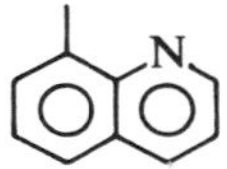

$\boxed{P}$—C$_6$H$_4$—CH$_2$—N=P(NMe$_2$)$_2$; R

R = H, Me

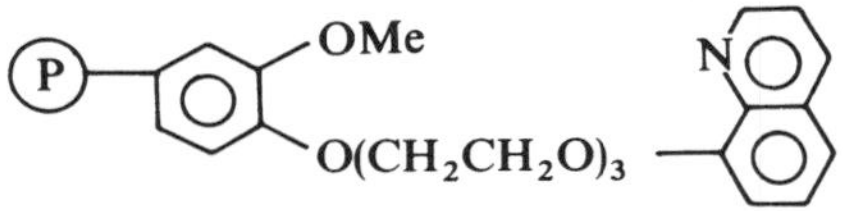

$\boxed{P}$—C$_6$H$_4$—CH$_2$—N(Me)—CHO

$\boxed{P}$—C$_6$H$_4$—(CH$_2$)$_n$—S(=O)—Me

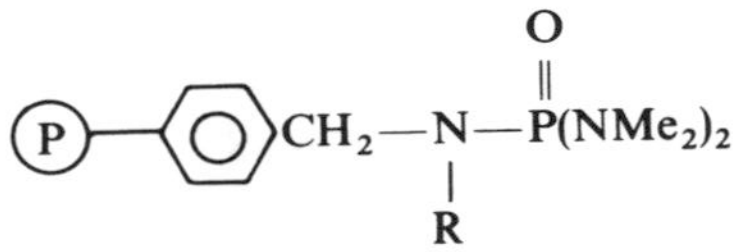

$n = 0, 1$

Y$^-$ $\equiv$ I$^-$, CN$^-$, PhO$^-$
Y$^-$ $\equiv$ ArO$^-$, AcO$^-$

Y$^-$ $\equiv$ AcO$^-$

Y$^-$ $\equiv$ AcO$^-$

Y$^-$ $\equiv$ ArO$^-$, Cl$^-$, CN$^-$, AcO$^-$

Y$^-$ $\equiv$ SCN$^-$, I$^-$, PhO$^-$, PhS$^-$

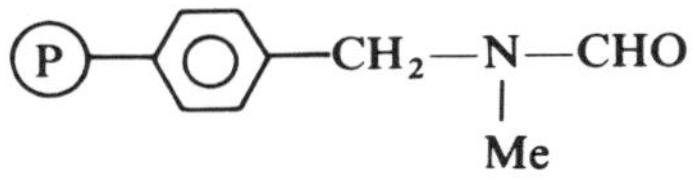

Y$^-$ $\equiv$ SCN$^-$, CN$^-$, I$^-$, PhO$^-$, PhS$^-$

353
355

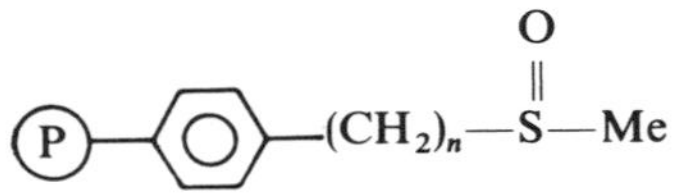

349, 356, 357

356, 357

298, 299,
358–361

362

363, 364

Table 4.5 (*Contd.*)

Functional polymer	Application $R-X + Y^- \longrightarrow R-Z$	Reference
![P]—⟨⟩—(CH₂S(=O)—CH₂)ₙ—R $n = 1, 2, 3;\ R = H, Me$	$Y^- \equiv I^-, SCN^-, CN^-, PhO^-$	365
Other catalysts		
![P]—⟨⟩—X		
(a) $X \equiv COONa, COOH, SO_3Na$	$Y^- \equiv Cl^-, CN^-, AcO^-$	299
(b) $X \equiv S^+ Me_2 FSO_3^-$	Epoxidation of ketones	366
(c) $X \equiv -CH_2-$ (pyrrolidinone)	$Y^- \equiv ArO^-, SCN^-$	367
![P]—⟨⟩—(CH₂)ₘ—SO₃H $m = 1, 2$	Hydrolysis of methyl acetate	368
$-(OCH_2CH_2)_n-$ $-(CH_2CH_2O)_n-$ (triazine)—(OCH₂CH₂)ₙ—		369–375 376
R₆-benzene ring; $R = CH_2(OCH_2CH_2)_n-$		377

P—OH $Y^- \equiv PhO^-$ 374

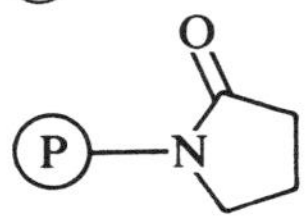

P—N(pyrrolidinone) $Y^- \equiv PhO^-$ 374

P—CHO $Y^- \equiv PhO^-$ 374

P—OMe $Y^- \equiv PhO^-$ 374

P—COMe $Y^- \equiv PhO^-$ 374

P—$CONH_2$ $Y^- \equiv PhO^-$ 374

P—$CONR_2$ $Y^- \equiv PhO^-$ 374, 378

R ≡ Me, Et, n-Pr, n-Bu, n-Oct

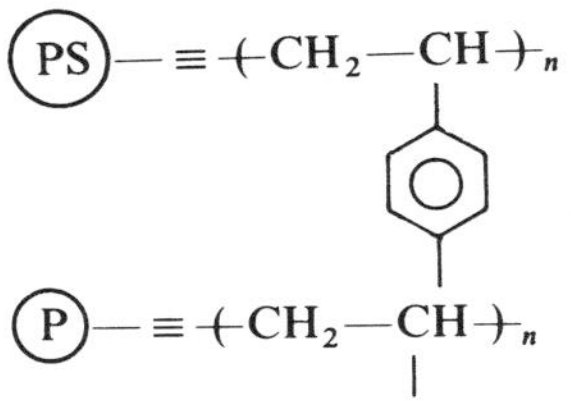

PS $\equiv \left(CH_2 - CH \right)_n$

C — ≡ cellulose

P $\equiv \left(CH_2 - CH \right)_n$

Si — ≡ silica

Nafion—H ≡ $\left(CF_2 - CF_2 \right)_m \left(OCF_2 - C \right)_n$ CF_3 $O \left(CF_2 \right)_2 - SO_3H$

bound to resins have been used in this context [380], and these are listed in Table 4.5.

The use of polymeric phosphonium salts instead of the ammonium salts as phase transfer catalysts leads to greater chemical and thermal stability as well as a higher catalytic activity; those which have been used are given in Table 4.5 together with the supported crown ethers and cryptands [297] exploited as phase transfer catalysts. With all these species there is evidence for an increase in catalytic activity when the catalyst is separated from the polymer backbone by a spacer arm [380, 313]. In general, the catalytic activity of polymer-bound crown ethers or cryptands, which is due to complexing of the cation and hence the increased nucleophilicity of the anions, is better than that achieved with polymeric onium salts.

Problems associated with the phase separation of an inorganic reagent and an organic substrate can also be overcome by the use of an appropriate solvent or cosolvent, such as the dipolar aprotic species DMSO or HMPA supported on a resin [297]. These supported species catalyse the nucleophilic displacement with higher activity than their model compounds, and overcome the difficulties resulting from conventional solvents or cosolvents. Polymer-supported linear oligoethylene oxides have also been used in this context [361, 348, 381] as well as a number of other linear polymers.

The triphase catalysis technique has been used successfully in many synthetic reactions such as the nucleophilic displacement by cyanide ion on organic halides for the preparation of nitriles. Other significant reactions accelerated by insoluble polymeric catalysts include halogen exchange in organic halides, the preparation of alkyl and aryl ethers, the generation of dichlorocarbene from aqueous sodium hydroxide solution and chloroform, dehalogenation of vic-dihalides with sodium iodide and sodium thiosulphate, and the oxidation of alcohols with aqueous sodium hypochlorite. Examples illustrating the utilization of triphase catalysts in such reactions and in other synthetic reactions are provided in Table 4.5.

4.6 POLYMER-SUPPORTED TRANSITION METAL CATALYSTS

The catalytic activity of a transition metal complex catalyst arises from the binding of the reactants to the open coordination site of the metal complex and their subsequent transfer to the product, thereby regenerating the free catalyst. Although a number of metal complexes are more effective homogeneous catalysts in that they allow the use of milder conditions in terms of both lower temperatures and lower pressures, because of their high cost and the difficulty of recovering them, they are rarely used in industrial synthesis. In addition, a loss in activity of a catalyst takes place when the complex dimerizes during the opening of the active site. Hence the basic requirement for metal complexes to function as catalysts is the availability of open coordination sites and their isolation to achieve separation of the reactive binding sites.

Thus the preparation of polymer-supported transition metal catalysts and their applications are of a great importance for industrial processes. The aim of this chapter is not to compile all the work published in this field but to illustrate the applications of polymer-supported transition metal complexes as one of the functionalized polymers used as catalysts in organic chemistry. For a fuller coverage of typical examples of polymer-supported metal catalysts of various types, see the tables summarized in earlier reviews [382–404].

Although unfunctionalized polystyrene can bind certain transition metal complexes via π-bonding through the benzene ring, functionalized polymers containing ligands with oxygen or nitrogen donors are usually used to bind transition metals. The desired metal complex is generally attached to the functionalized polymer by equilibration of the polymer with a metal complex having similar or weaker ligands. The attachment may involve either the displacement of a low molecular weight ligand or the addition of the polymeric ligand. The mode of attachment depends on the nature of the transition metal complex and the amount of crosslinking in the polymer. Most attempts to prepare polymer-supported metal complexes by first binding the metal to a monomer which is then copolymerized with unfunctionalized monomers have failed because of strong interactions between the polymerization initiator and the metal that result in a mixture of products. The presence of the metal complex on a monomer may also dramatically alter the reactivity of the monomer in the polymerization.

The optimum polymeric catalyst for a particular reaction can be determined by variation of a number of parameters: (a) the nature of the support; (b) the nature of the metal complex; (c) the degree of crosslinking; and (d) the nature of the solvent. The properties of the catalyst, made by chemically binding a soluble transition metal complex to a polymer, lie between those of homogeneous and those of heterogeneous catalysts. The polymer-bound transition metal complex catalysts offer the following advantages:

1. The catalytic sites can be separated by binding to a rigid support and hence a greater catalytic activity is gained by avoiding the formation of ligand-bridged complexes through the isolation of one reaction catalytic side from another. Hence, the highly crosslinked polymer reduces the possibility of chelation by double bonding.
2. They can be employed under conditions comparable with those of conventional homogeneous catalysts.
3. Two active catalysts anchored to the same polymer backbone can be used to conduct sequential multistep organic synthesis.
4. Organometallic complexes used as catalysts are very expensive, but losses during the use of polymeric catalysts are minimized.
5. Attaching the catalyst to the polymer backbone results in an increase in the steric environment of the catalyst. As result of this restriction around the catalyst centre, substrate selectivity is increased. The selectivity of the polymer towards substrates of different sizes can be controlled by (a) the loading of the

catalyst on the polymer support, (b) controlling the degree of resin swelling and (c) introducing optically active groups in the polymer around the active site.

One of the advantages of incorporating metals into polymer support is exemplified by a set of titanocene complexes. Soluble titanocene complexes have poor hydrogenation catalytic activity because of the formation of dimers. Attachment of titanocene to a crosslinked polymeric ligand produced a catalyst **5** which is more active than the homogeneous catalyst because the rigid matrices isolate these highly reactive species and prevent dimerization:

5

Polymer-supported metal complexes have been used as catalysts for a wide range of reactions, principally of olefinic substrates, including hydrogenation, hydroformylation, hydrosilyation, carbonylation, acetoxylation and polymerization. The majority of polymeric catalysts used in the hydrogenation of a variety of olefins involve mainly metal complexes of rhodium(I), palladium(II) and platinum(II). The rates of hydrogenation are strongly dependent on the solvent used to swell and suspend the polymer support. The rate decreases with decreasing swelling ratio of the polymer in the solvent, while increased swelling lowers the diffusion barrier to the substrate.

Hydroformylation is the reaction sequence which involves the addition of a formyl group to the terminal or internal carbon atom of an olefinic double bond and a hydrogen atom to the other.

$$R-CH=CH_2 + CO + H_2 \xrightarrow{cat} R-CH_2-CH_2CHO \text{ or } R-\underset{\underset{CHO}{|}}{C}HCH_3$$

Molecular hydrogen and CO are used as the source of these two groups. The ratio of the two aldehydes formed and the side-products produced by aldehyde hydrogenation are dependent on the catalyst used. In general, cobalt and rhodium carbonyl complexes supported on polymers are normally used as active catalysts for this reaction.

Hydrosilyation refers to the addition reaction of an organosilicon hydride to an unsaturated compound such as olefin, acetylene or ketone in the presence of a catalyst.

$$R{-}CH{=}CH_2 + HSiR'_3 \xrightarrow{\text{cat}} R{-}CH_2CH_2SiR'_3$$

This reaction is of considerable interest and potential in the organosilicon industry. The principal catalysts that have been used include polymer-supported rhodium and platinum complexes. However, the rate of hydrosilyation decreases as electron-withdrawing substituent groups are substituted in the olefin molecule.

4.7 IMMOBILIZED ENZYMES

Enzymes are natural polymeric species that are well known to catalyse biochemical transformations. They are usually characterized by high catalytic activity and high specificity, and the catalysis involves specific binding of the substrate to the active site. Enzymes can also bring about various reactions at ambient temperature and pressure and afford high reaction velocities. Compounds that are very difficult to prepare by purely chemical methods may be obtained quite readily and economically with the help of enzymes. However, the extraction and purification of enzyme catalysts are often complex and expensive processes. In addition, enzymes are invariably lost after each batch operation and lose their catalytic activity which makes their use very difficult. With the rapidly increasing knowledge of the structures and mode of action of natural enzymes, several attempts have been made to use fully synthetic polymeric catalysts which incorporate all the functional features of the specificity and the activity of the natural enzyme. Although polymeric enzymes were among the first functionalized polymers used as catalysts, they fail in most cases to attain the specificity and the activity of enzymes.

Attempts have also been made to prepare immobilized enzymes, as semisynthetic catalysts, by combination of an enzyme with a polymeric carrier. A considerable number of immobilized enzyme systems are described in the literature and a number of reviews have been published on this subject [405–414].

The immobilization of enzymes can be achieved by various means. Physical adsorption of the enzyme onto a polymer matrix without covalent bonding, using electrostatic or other non-covalent bonding mechanisms, is the oldest method of enzyme immobilization. Unfortunately, the reversible nature of the bonding of the enzyme to the support may lead to desorption of the enzyme at a critical time.

Alternatively, physical entrapment of enzyme within the polymer can be achieved by its inclusion within the pores of a highly crosslinked polymer in which the polymer forms a net-like matrix around the enzyme [415, 416]. The enzyme in the microcapsules is not actually attached. The pores are too small to allow escape of the enzyme but large enough to permit the entry of low molecular weight substances. From a practical standpoint, physically adsorbed enzyme on supports are unsatisfactory since the enzymes dissociate and readily leach into the surrounding substrate solution. They are therefore unsuitable for column or cyclic applications.

Immobilization of enzymes by chemical covalent bonding of the enzyme to a reactive polymer matrix through a group other than the active site of the enzyme is probably the most extensively used method. Such systems have many advantages: (a) they can be recovered and reused repeatedly in continuous enzymic processes; (b) they show greater environmental stability toward changes in pH or temperature and also show improved storage properties for long periods; (c) in some cases they show enhanced activity toward a substrate and the reactions involving them can be stopped at any desired stage by simple filtration; (d) the isolation and purification of the products is quite easy.

A variety of support materials that can swell in water and whose reactive groups can react with the functional groups of the enzyme under mild reaction conditions without interfering with the biologically active centre of the enzymes are used. Most of the earlier studies were directed towards the attachment of enzymes to water-insoluble polymeric supports such as polysaccharide derivatives, polyacrylamides and porous glass. The non-compatibility of styrene with relatively polar biomaterials and aqueous solvents is the most important limitation of such supports. In general, the activity of the immobilized enzyme depends on the composition of the carrier, the size of the carrier particles, and the degree of hydration of the polymer matrix. In addition, a polymeric support should have the following properties:

1. It should be chemically stable and completely insoluble in solvents and reagents and under the environmental conditions employed.
2. It should be capable of undergoing functionalization reactions.
3. It should not be susceptible to bacterial attack and degradation.
4. It should show minimal interaction with the substrates both before and after coupling of the specific ligand.
5. The matrix should form a loose porous network that permits the easy passage of large molecules and retains a good flow rate.

Since a number of major problems are encountered in using soluble enzymes as industrial catalysts, e.g. their stability, recovery and use in batch processes, the possibility of using immobilized enzymes as stable reusable catalysts shows interesting potential for continuous industrial applications. Large-scale applications of immobilized enzymes [417] include the following.

1. In the dairy industry the use of immobilized enzymes in the treatment of milk falls into three major areas: (a) immobilized rennin is used for the continuous coagulation of milk in the production of cheese; (b) since whey, the liquid waste from cheese production, contains protein and large quantities of lactose, immobilized lactase has been used to remove lactose from cheese whey; (c) stabilization of milk to extend shelf life without change in flavour is one other

area where immobilized enzyme technology may be of value, e.g. in the treatment of milk with immobilized trypsin.

2. Isomerization of glucose to fructose: the potential for using immobilized glucose isomerase in the food industry for the commercial production of fructose syrups from glucose, obtained from corn starch or from any form of cheaply available starch, e.g. potatoes, is one of the most successful industrial processes.

3. Hydrolysis of starch to glucose: the increasing requirement for glucose syrups as a substrate for the production of high fructose syrups has led to the use of immobilized amyloglucosidase for the industrial production of glucose syrups from starch.

4. Resolution of DL-amino acids: the chemical synthesis of amino acids leads to the racemic mixture but the L-form is the only isomer required on a large scale for use in both medicines and foodstuffs. Immobilized enzymes have been used for the large-scale industrial separation of enantiomers of various amino acids, e.g. amino acid acylase immobilized onto DEAE-Sephadex is used for the industrial production of L-amino acids.

5. Immobilized pectinase has been used for clarifying fruit juices.

4.8 POLYMERIC pH INDICATORS

Recently, pH indicators covalently bound to polymers such as 6 and 7 have been reported to be stable acid–base indicators [418, 419]:

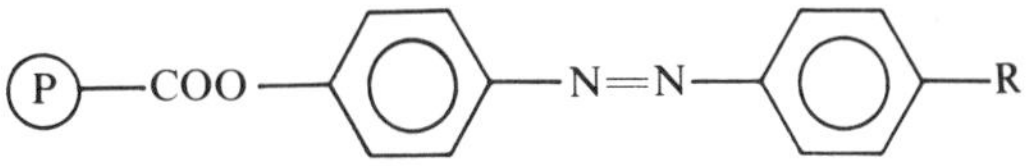

6

$R \equiv NO_2NH_2, NMe_2$

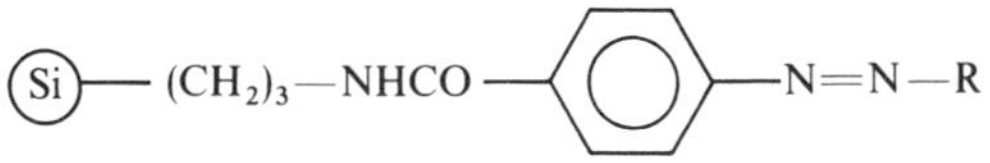

7

R = bromocresol purple, crystal violet, methyl red, bromocresol green, phenolphthalein

Polymeric pH indicators have several advantages over soluble indicators: (a) they can be used for a long time with quantitative recovery of the indicator; (b) they are not susceptible to microbial attack; (c) they are insoluble and hence do not contaminate the tested systems; and (d) they are superior in the determination of the pH values of weakly buffered or non-buffered solutions.

4.9 POLYMERIC INITIATORS

A number of polymeric initiators have been prepared and used to initiate the polymerization of vinylic monomers (Table 4.6). However, one of the main

Table 4.6 Polymeric initiators

Functionalized polymer	Application	Reference
(P)—⟨O⟩—CO—⟨O⟩—CO—O—O—Bu-t	Photoinitiator	420
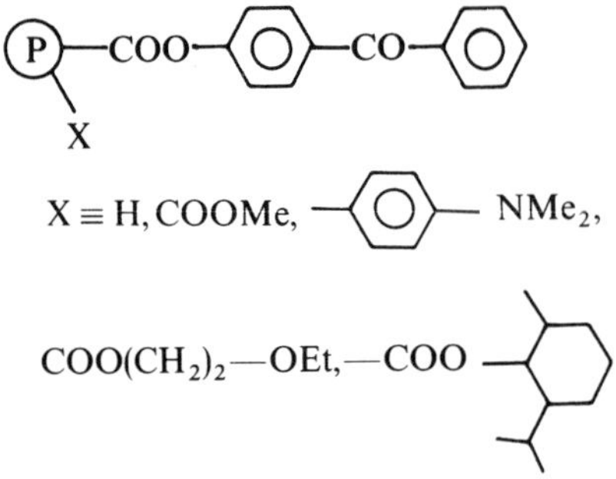 (P)—COO—⟨O⟩—CO—⟨O⟩ with X	Photoinitiator	421, 422

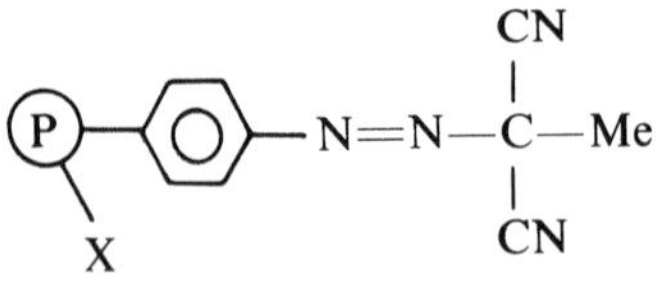

X ≡ H, COOMe, —⟨O⟩— NMe$_2$,

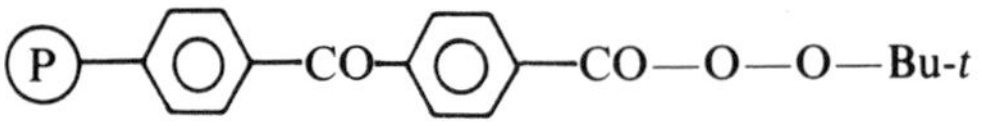

COO(CH$_2$)$_2$—OEt, —COO—⟨cyclohexyl⟩

Functionalized polymer	Application	Reference
(P)—⟨O⟩—N=N—C(CN)(CN)—Me with X	Azo-initiator	423

X ≡ H, Ph

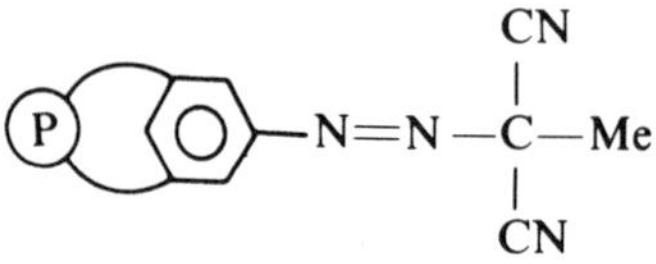 (P)⟨O⟩—N=N—C(CN)(CN)—Me	Azo-initiator	424
(P)—⟨O⟩—CH$_2$CH$_2$OC(=O)—N=N—C(=O)—OMe		425, 426
(P)—⟨O⟩—N=N—C(CN)(CN)—Me		423
(P)—⟨O⟩—N=N—C(R)(R)—OCOH$_3$		427

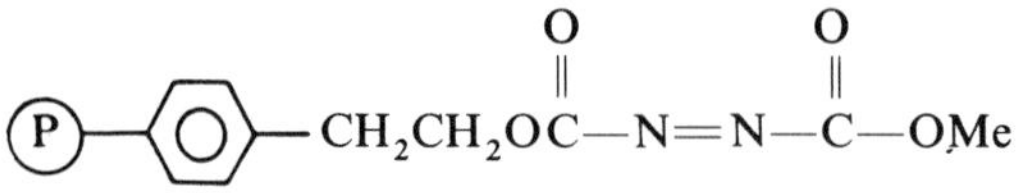

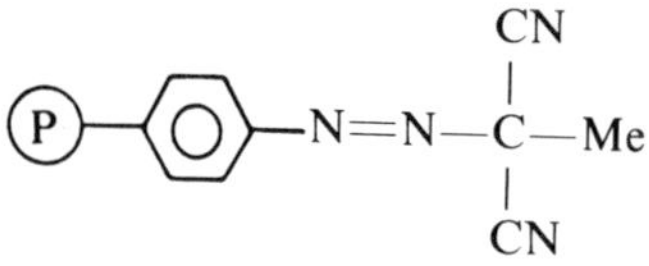

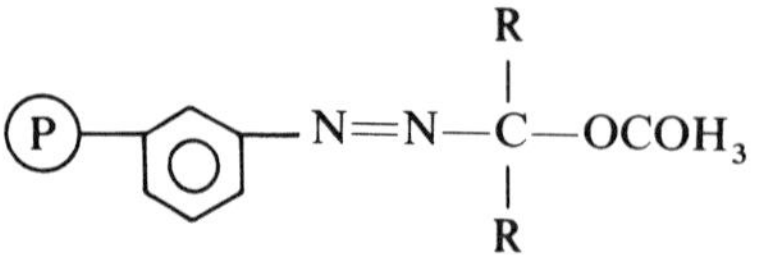

R ≡ Me, Ph

Table 4.6 (*Contd.*)

Functionalized polymer	Application	Reference

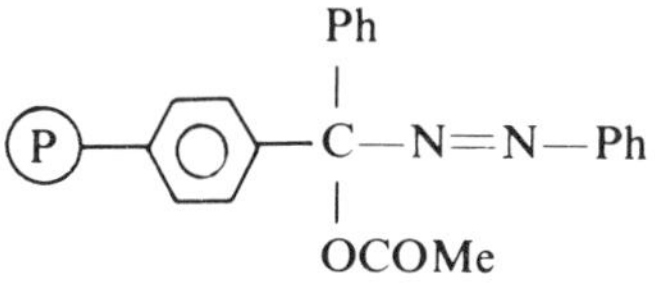

$$\text{(P)}-\text{O}-\underset{\underset{\text{OCOMe}}{|}}{\overset{\overset{\text{Ph}}{|}}{\text{C}}}-\text{N}=\text{N}-\text{Ph}$$

427

$$\text{(P)}-\text{OCO(CH}_2)_2-\underset{\underset{\text{CN}}{|}}{\overset{\overset{\text{Me}}{|}}{\text{C}}}-\text{N}=\text{N}-\underset{\underset{\text{CN}}{|}}{\overset{\overset{\text{Me}}{|}}{\text{C}}}-(\text{CH}_2)_2\text{COCl}$$

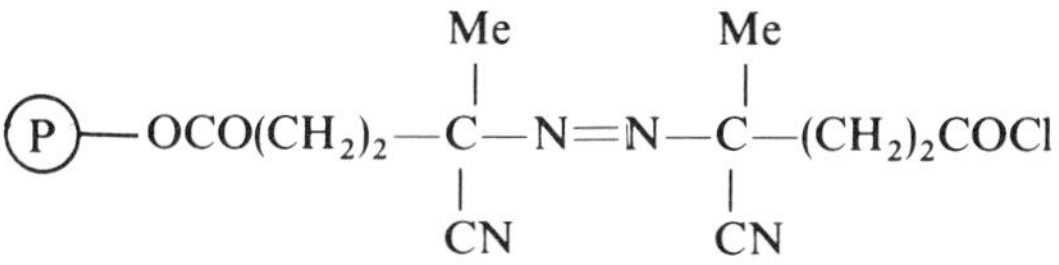

428

$$\text{(P)}-\text{O}-\underset{\underset{\text{Ph}}{|}}{\overset{\cdot}{\text{C}}}-\text{ONa}$$

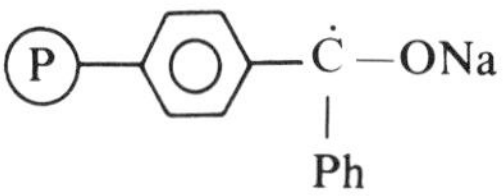

Anionic initiator   429

$$\text{(P)}-\text{O} \cdot \text{Na}$$

Anionic initiator   430

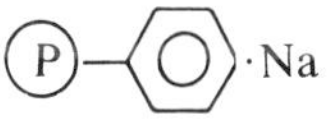

$$\text{(P)}-\text{O}-\overset{+}{\diagdown}\underset{\text{O}}{\overset{\text{O}}{\diagup}}\;\; \text{ClO}_4^{\,-}$$

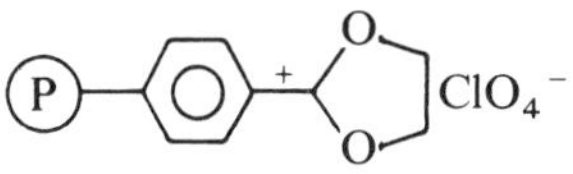

Cationic initiator   431

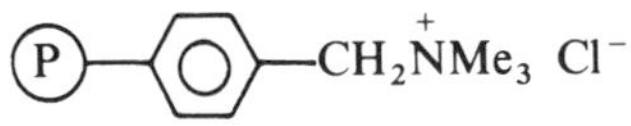

$$\text{(P)}-\text{O}-\text{CH}_2\overset{+}{\text{N}}\text{Me}_3 \;\; \text{Cl}^-$$

Cationic initiator   432

$$\text{(P)}-\text{O} \cdot \text{AlCl}_3$$

Cationic initiator   433

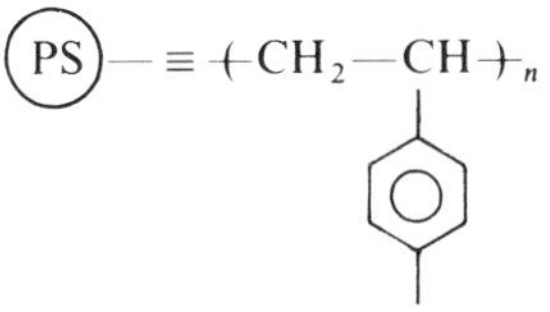

$$\text{(PS)}- \equiv +\text{CH}_2-\text{CH}\!+_n \qquad \text{(C)}- \equiv \text{cellulose}$$

$$\text{(P)}- \equiv +\text{CH}_2-\text{CH}\!+_n \qquad \text{(Si)}- \equiv \text{silica}$$

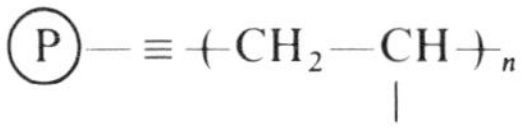

$$\text{Nafion}-\text{H} \equiv +\text{CF}_2-\text{CF}_2\!+_m +\text{OCF}_2-\underset{\underset{\text{O}+\text{CF}_2)_2-\text{SO}_3\text{H}}{|}}{\overset{\overset{\text{CF}_3}{|}}{\text{C}}}+_n$$

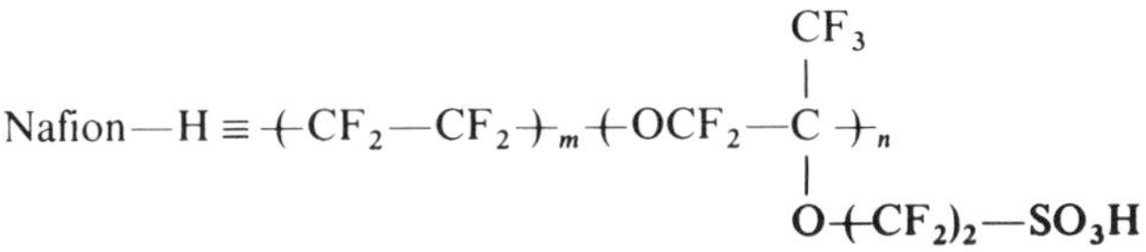

advantages of the functionalized polymer, namely its ease of separation after reaction, is probably lost because the polymeric product itself will also be insoluble if a high molecular weight is achieved.

Polymeric initiators are used for the preparation of graft copolymers in which the reactive free radical centre can be converted into grafting sites in the presence. of a polymerizable monomer. For example, heating or photolysis of the polymer initiator 9 prepared by copolymerization of styrene with 2-(4-vinyl-phenylazo)-2-methylmalonitrile, 8, in the presence of another vinyl monomer results in graft polymer 10 [423].

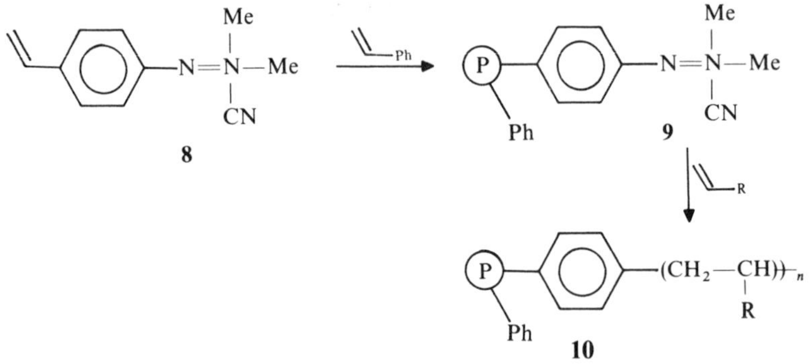

The metallated polymers prepared either by decomposition of a derivative of a transition metal and an organometallic compound on the surface of an inert support or by chemically bonding the transition metal species to a polymer which has reactive electron donor groups have been employed as polymeric Ziegler–Natta initiators of polymerization [434]. For example, organoaluminium-containing polymers 11 [435], prepared by treatment of unsaturated polymers with transition metal halides, are capable of initiating the polymerization of ethylene and α-olefins to yield graft or block copolymers [436]:

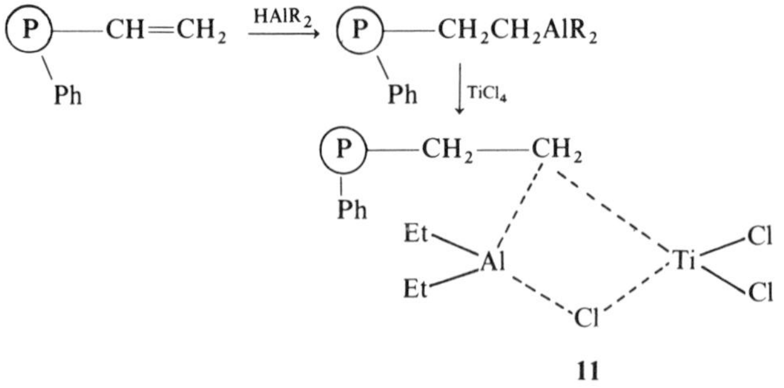

REFERENCES

1. Astle, M.J. (1957) In *Ion Exchangers in Organic and Biochemistry* (eds C. Calman and T.R.E. Kressman), Interscience, New York, Ch. 36, p. 658.
2. Helfferich, F. (ed.) (1962) *Ion Exchange*, McGraw-Hill, New York.

3. Polyanskii, N.G. (1970) *Russ. Chem. Rev.*, **31**, 496; **39**, 244.
4. Sugihara, M. (1963) *Sci. Ind.*, **37**, 334.
5. Polyanskii, N.G. and Sapozhnikov, V.K. (1977) *Russ. Chem. Rev.*, **46**, 226.
6. Klein, J. and Widdecke, H. (1979) *Chem. Ing. Tech.*, **51**, 560.
7. Klein, J. (1981) *Makromol. Chem. Suppl.*, **5**, 155.
8. Davies, C.W. and Thomas, G.G. (1952) *J. Chem. Soc.*, 1607.
9. Sakurada, I., Sakaguchi, Y., Ono, T. and Ueda, T. (1966) *Makromol. Chem.*, **91**, 243.
10. Yashikawa, S. and Kim, O.K. (1966) *Bull. Chem. Soc. Jpn*, **39**, 1515, 1729.
11. Hasek, R.H., Gott, P.G., Meen, R.H. and Martin, J.C. (1963). *J. Org. Chem.*, **28**, 2496.
12. Lawrence, L. and Moore, W.J. (1951) *J. Am. Chem. Soc.*, **73**, 3973.
13. Underwood, G.E. and Deathrage, F.E. (1952) *Science*, **115**, 95.
14. Paulson, J.C. and Deathrage, F.E. (1953) *J. Biol. Chem.*, **205**, 909.
15. Paulson, J.C., Deathrage, F.E. and Almy, E.F. (1953) *J. Am. Chem. Soc.*, **75**, 2039.
16. Whitaker, J.R. and Deathrage, F.E. (1955) *J. Am. Chem. Soc.*, **77**, 3360, 5293.
17. Kern, W., Herold, H. and Scherlag, B. (1956) *Makromol. Chem.*, **17**, 231.
18. Collins, R.F. (1957) *Chem. Ind.*, 736.
19. Kunin, R. (1958) *Ion-Exchange Resins*, Wiley, New York.
20. Kern, W. and Scherhag, B. (1958) *Makromol. Chem.*, **20**, 209.
21. Yamashita, S. (1971) *Biochem. Biophys. Acta*, **229**, 301.
22. Bora, J.M., Saund, A.K., Sharma, I.K. and Mathur, N.K. (1976) *Indian J. Chem.*, **14-B**, 722.
23. Hodge, J.E. and Rist, C.E. (1952) *J. Am. Chem. Soc.*, **74**, 1498.
24. Painter, T.J. and Morgan, W.T.J. (1961) *Chem. Ind.*, 437.
25. Denel, H. (1955) *Mitt. Geb. Lebensmittelunters. Hyg.*, **46**, 12.
26. Steinhardt, J.S. and Fugitt, C.H. (1942) *J. Res. Nat. Bur. Stand.*, **29**, 315.
27. Hartler, N. and Hyllengren, K. (1961) *J. Polym. Sci.*, **55**, 779.
28. Mill, P.J. and Crimmin, R. (1957) *Biochim. Biophys. Acta*, **23**, 432.
29. Jain, J.C., Sharma, I.K., Sahni, M.K., Gupta, K.C. and Mathur, N.K. (1977) *Indian J. Chem.*, **15-B**, 766.
30. Vesley, G.F. and Stenberg, V.I. (1971) *J. Org. Chem.*, **36**, 2543.
31. Mowery, D.F. (1961) *J. Org. Chem.*, **26**, 3484.
32. Sussman, S. (1945) *Ind. Eng. Chem.*, **38**, 1223.
33. Moffatt, J.G. (1963) *J. Am. Chem. Soc.*, **85**, 118.
34. Stenberg, V.I., Vesley, G.F. and Kubic, D. (1971) *J. Org. Chem.*, **36**, 2550.
35. Stenberg, V.I., Vesley, G.F. and Kubic, D. (1974) *J. Org. Chem.*, **39**, 2815.
36. Patwardham, S.A. and Dev, S. (1974) *Synthesis*, 348.
37. John, E.V.O. and Israelstam, S.S. (1961) *J. Org. Chem.*, **26**, 240.
38. Harms, W.M. and Eisenbraun, E.J. (1971) *Org. Prep. Proced. Int.*, **3**, 239.
39. Scott, L.T. and Naples, J.O. (1973) *Synthesis*, 209.
40. Moriyama, H., Sugihara, Y. and Nakanishi, K. (1968) *Tetrahedron Lett.*, 2351.
41. Mastagli, P., Floc'h, A. and Durr, G. (1952) *Compt. Rend.*, **235**, 1402.
42. Wasman, W.H. (1952) *J. Chem. Soc.*, 3051.
43. Frilette, V.J., Mower, E.B. and Rubin, M.K. (1964) *J. Catal.*, **3**, 25.
44. Heatha, H.W. and Gates, B.C. (1972) *Am. Inst. Chem. Eng. J.*, **18**, 321.
45. Gates, B.C. and Johanson, L.N. (1969) *J. Catal.*, **14**, 69.
46. Gates, B.C. and Schwab, G.M. (1969) *J. Catal.*, **15**, 430.
47. Gates, B.C., Wisnouskas, J.S. and Heath, H.W. (1972) *J. Catal.*, **24**, 320.
48. Pines, H. and Manassen, J. (1966) *Adv. Catal.*, **16**, 49.
49. Price, P. and Israelstam, S.S. (1964) *J. Org. Chem.*, **29**, 2800.
50. Loev, B. and Massengale, J.T. (1957) *J. Org. Chem.*, **22**, 988.
51. Wesley, R.B. and Gates, B.C. (1974) *J. Catal.*, **34**, 288.
52. Hasegawa, H. and Higashimura, T. (1981) *Polym. J.*, **13**, 915.
53. Newman, M.S. (1953) *J. Am. Chem. Soc.*, **75**, 4740.

54. Arcus, C.L., Howard, T.L. and South, D.S. (1964) *Chem. Ind.*, 1756.
55. Bodamer, G. and Kunin, R. (1951) *Ind. Eng. Chem.*, **43**, 1082.
56. Astle, M.J. and Oscar, J.A. (1961) *J. Org. Chem.*, **26**, 1713.
57. Walter, M., Besendorf, H. and Schmider, O. (1961) *Helv. Chim. Acta*, **44**, 1546.
58. Neier, W. and Woellner, J. (1973) *Chem. Tech.*, 95.
59. Billimoria, J.D. and Maclagan, N.F. (1954) *J. Chem. Soc.*, 3257.
60. Delmas, M. and Gaset, A. (1980) *Synthesis*, 871.
61. Misic-Vikovic, M.M., Vukoric, D.V. and Hadzismajlovic, D.E. (1980) *J. Chem. Res.*, (s), 258.
62. Affrossman, S. and Murray, J.P. (1966) *J. Chem. Soc. B*, 1015.
63. Haskell, V.C. and Hammett, L.P. (1949) *J. Am. Chem. Soc.*, **71**, 1234.
64. Rioesz, P. and Hammett, L.P. (1954) *J. Am. Chem. Soc.*, **76**, 992.
65. Chen, C.H. and Hammett, L.P. (1958) *J. Am. Chem. Soc.*, **80**, 1329.
66. Arai, K. and Ise, N. (1975) *Makromol. Chem.*, **176**, 37.
67. Arai, K., Hagiwara, N. and Ise, N. (1975) *Nipon Kagaku Kaishi*, 201.
68. Legator, M.C. (1971) US Patent 3 567 420.
69. Morawetz, M. and Shafer, J.A. (1963) *J. Phys. Chem.*, **67**, 1293.
70. Svec, F., Bares, M., Zajic, J. and Kalal, J. (1977) *Chem. Ind.*, 159.
71. Fayed, S., Delmas, M. and Gaset, A. (1982) *Synth. Commun.*, **12**, 1121.
72. Mariani, E. and Baldass, F.V. (1950) *Ric. Sci.*, **20**, 324.
73. Deuel, H., Solms, J., Anyas-Weisz, L. and Huber, G. (1951) *Helv. Chim. Acta*, **34**, 1849.
74. Samelson, H. and Hammett, L.P. (1956) *J. Am. Chem. Soc.*, **78**, 524.
75. Kosower, E.M. and Patton, P.W. (1961) *J. Org. Chem.*, **26**, 1318.
76. Kirby, A.J. and Varvoglis, A.G. (1967) *J. Am. Chem. Soc.*, **89**, 415.
77. Bobbitt, J.M. and Doolittle, R.E. (1964) *J. Org. Chem.*, **29**, 2293.
78. Galat, A. (1948) *J. Am. Chem. Soc.*, **70**, 3945.
79. Arcus, C.L., Gonzales, C.G. and Linnecar, D.F.C. (1969) *J. Chem. Soc. Chem. Commun.*, 1377.
80. Prini, R.F. and Baumgartner, E. (1974) *J. Am. Chem. Soc.*, **96**, 4489.
81. Prini, R.F. and Turyn, D. (1973) *J. Chem. Soc., Faraday Trans. 2*, **69**, 1326.
82. Hinman, R.L. and Lang, J. (1964) *J. Org. Chem.*, **29**, 1449.
83. Astle, M.J. and Zaslowsky, J.A. (1952) *Ind. Eng. Chem.*, **44**, 2867.
84. Bergmann, E.D. and Corett, R. (1956) *J. Org. Chem.*, **21**, 107; (1958) *J. Org. Chem.*, **23**, 1507.
85. Austerweil, G.V. and Palland, R. (1953) *Bull. Soc. Chim. Fr.*, 678.
86. Bocheme, W.R. and Koo, J. (1961) *J. Org. Chem.*, **26**, 3589.
87. Schmidle, C.J. and Mansfield, R.C. (1952) *Ind. Eng. Chem.*, **44**, 1388.
88. Mastagli, P., Zafiriadis, Z., Durr, G., Floc'h, M.A. and Lagrange, G. (1953) *Bull. Soc. Chim. Fr.*, 639.
89. Durr, G. (1956) *Compt. Rend.*, **242**, 1630.
90. Astle, M.J. and Abbot, F.P. (1956) *J. Org. Chem.*, **21**, 1228.
91. Astle, M.J. and Gergel, W.C. (1956) *J. Org. Chem.*, **21**, 493.
92. Hein, R.W., Astle, M.J. and Shelton, J.R. (1961) *J. Org. Chem.*, **26**, 4874.
93. Loretta, N.B. (1957) *J. Org. Chem.*, **22**, 346.
94. McCain, G.H. (1958) *J. Org. Chem.*, **23**, 632.
95. Astle, M.J. and Zaslowsky, J.A. (1952) *Ind. Eng. Chem.*, **44**, 2871.
96. Rowe, E.J., Kaufman, K.L. and Piantadosi, C. (1958) *J. Org. Chem.*, **23**, 1622.
97. Bobbitt, J.M. and Scola, D.A. (1960) *J. Org. Chem.*, **25**, 560.
98. Chatterjee, B.G., Rao, V.V. and Mazumder, B.N.G. (1965) *J. Org. Chem.*, **80**, 4101.
99. Cainelli, G. and Manescalchi, F. (1975) *Synthesis*, 723.
100. Shimo, K. and Wakamatsu, S. (1963) *J. Org. Chem.*, **28**, 504.
101. Galat, A. (1952) *J. Am. Chem. Soc.*, **74**, 3890.

102. Weinstock, J. and Boekelheide, V. (1953) *J. Am. Chem. Soc.*, **75**, 2546.
103. LeGoffic, F., Sicsic, S. and Vincent, C. (1976) *Tetrahedron Lett.*, 2845.
104. Yoshida, J.I., Hashimoto, J. and Kawabata, N. (1981) *Bull. Chem. Soc. Jpn*, **54**, 309.
105. Shinkai, S., Tsuji, H., Hara, Y. and Manabe, O. (1981) *Bull. Chem. Soc. Jpn*, **54**, 631.
106. Blossey, E.O., Turner, L.M. and Neckers, D.C. (1973) *Tetrahedron Lett.*, 1823; (1975) *J. Org. Chem.*, **40**, 959.
107. Scott, L.T. and Naples, J.O. (1976) *Synthesis*, 738.
108. Sket, B. and Zupan, M. (1983) *J. Macromol. Chem.*, **A-19**, 643.
109. Neckers, D.C., Kooistra, D.A. and Green, G.W. (1972) *J. Am. Chem. Soc.*, **94**, 9284.
110. Kelly, J.T. (1958) US Patent 2 843 642.
111. Huang, T.J. and Yurchak, S. (1974) US Patent 3 855 343.
112. Magnotta, V.L., Gates, B.C. and Schuit, G.C.A. (1976) *J. Chem. Soc. Chem. Commun.*, 342.
113. Magnotta, V.L. and Gates, B.C. (1977) *J. Polym. Sci., Polym. Chem. Ed.*, **15**, 1341; (1977) *J. Catal.*, **46**, 266.
114. Dooley, K.M. and Gates, B.C. (1984) *J. Polym. Sci., Polym. Chem. Ed.*, **22**, 2859.
115. Whang, K.J., Lee, K.I. and Lee, Y.K. (1984) *Bull. Chem. Soc. Jpn*, **57**, 2341.
116. Kinstle, J.F. and Quinlan, G.L. (1981) *Polym. Prepr., Am. Chem. Soc., Div. Polym. Chem.*, **22**, 166 (1).
117. Olah, G.A., Kaspi, J. and Bubala, J. (1977) *J. Org. Chem.*, **42**, 4187.
118. Olah, G.A. and Mehrotra, S.K. (1982) *Synthesis*, 962.
119. Olah, G.A., Narang, S.C., Meidar, D. and Salem, G.F. (1981) *Synthesis*, 282.
120. Olah, G.A. and Meidar, M. (1978) *Synthesis*, 358.
121. Olah, G.A., Fung, A.P. and Malhotra, R. (1981) *Synthesis*, 474.
122. Kaspi, J. and Olah, G.A. (1978) *J. Org. Chem.*, **43**, 3142.
123. Konishi, H., Suetsugu, K., Okano, T. and Kiji, J. (1982) *Bull. Chem. Soc. Jpn*, **55**, 957.
124. Olah, G.A., Malhotra, R., Narang, S.C. and Olah, J.A. (1978) *Synthesis*, 672.
125. Kaspi, J., Montgomery, D.D. and Olah, G.A. (1978) *J. Org. Chem.*, **43**, 3147.
126. Olah, G.A., Meidar, D., Malhotra, R., Olah, J.A. and Narang, S.C. (1980) *J. Catal.*, **61**, 96.
127. Olah, G.A. and Meidar, D. (1978) *Synthesis*, 671.
128. Olah, G.A., Husain, A., Gupta, B.G.B. and Narang, S.C. (1981) *Synthesis*, 471.
129. Olah, G.A., Keumi, T. and Meidar, D. (1978) *Synthesis*, 929.
130. Olah, G.A. and Narang, S.C. (1978) *Synthesis*, 690.
131. Olah, G.A., Fung, A.P. and Meidar, D. (1981) *Synthesis*, 280.
132. Olah, G.A., Husein, A. and Singh, B.P. (1983) *Synthesis*, 892.
133. Petrakis, K.S. and Fried, J. (1983) *Synthesis*, 891.
134. Olah, G.A., Yamato, T., Iyer, P.S. and Prakash, G.K.S. (1986) *J. Org. Chem.*, **51**, 2826.
135. Olah, G.A., Arvanaghi, M. and Krishnamurthy, V.V. (1983) *J. Org. Chem.*, **48**, 3359.
136. Olah, G.A. and Fung, A.P. (1981) *Synthesis*, 473.
137. Olah, G.A. (1984) US Patents 4 465 893, 4 467 130.
138. Olah, G.A., Krishnamurthy, V.V. and Narang, S.C. (1982) *J. Org. Chem.*, **47**, 596.
139. Kanemoto, S., Saimoto, H., Oshima, S. and Nozaki, H. (1984) *Tetrahedron Lett.*, 3317.
140. Waller, F.J. (1982) US Patent 4 356 318; (1983) US Patents 4 414 409, 4 416 801; (1984) US Patents 443 3839, 4 444 329, 4 463 103.
141. Cainelli, G. Contento, M. and Panunzio, M. (1983) *Synthesis*, 306.
142. Grot, W. (1975) *Chem. Ing. Tech.*, **47**, 617.
143. Olah, G.A., Prakash, G.K.S. and Sommer, J. (1979) *Science*, **206**, 4414.
144. Olah, G.A., Iyer, P.S. and Prakash, G.K.S. (1986) *Synthesis*, 513.
145. Cassidy, H.G. and Kuhn, K.A. (eds) (1965) *Oxidation–Reduction Polymers*, Wiley Interscience, New York.

146. Lindsey, A.S. (1970) *Rev. Macromol. Chem.*, **41**, 1.
147. Bailey, D. and Vogl, O. (1976) *J. Macromol. Sci. Rev.*, **C-14**, 267.
148. Dawson, D.J., Otteson, K.M., Wang, P.C. and Wingard, R.E. (1978) *Macromolecules*, **11**, 320.
149. Dawson, D.J. (1981) *Aldrichimica Acta*, **14**, 23.
150. Dawson, D.J., Glass, R.D. and Wingard, R.E. (1976) *Chem. Technol.*, (6), 724.
151. Weinshenker, N.M. (1977) *Polym. Prepr., Am. Chem. Soc., Div. Polym. Chem.*, **18** (1), 531.
152. Furia, T. (1977) *Food Technol.*, **31** (5), 34.
153. Leonard, W.J. (1978) *Polymeric Delivery Systems* (ed. R. Kostelnik), Gordon and Breach, New York.
154. Neckers, D.C. (1978) *Chem. Technol.*, (8), 108.
155. Blossey, E.C., Neckers, D.C., Thayer, A.L. and Schaap, A.P. (1975) *J. Am. Chem. Soc.*, **95**, 5820.
156. Paczkowski, J. and Neckers, D.C. (1985) *Macromolecules*, **18**, 1245.
157. Blossey, E.C. and Neckers, D.C. (1974) *Tetrahedron Lett.*, 323.
158. Schaap, A.P., Thayer, A.L., Blossey, E.C. and Neckers, D.C. (1974) *J. Am. Chem. Soc.*, **97**, 3241.
159. Nakahira, T., Shinomiya, E., Fukumoto, T., Iwabuchi, S. and Kojima, K. (1978) *Eur. Polym. J.*, **14**, 317.
160. Moser, R.E. and Cassidy, H.G. (1964) *J. Polym. Sci.*, **B-2**, 545.
161. Sanchez, G., Weill, G. and Knoesel, R. (1978) *Makromol. Chem.*, **179**, 131.
162. Asai, N. and Neckers, D.C. (1980) *J. Org. Chem.*, **45**, 2903.
163. Neckers, D.C., Gupta, S., Gupta, I., Thijs, L. and Damen, J. (1981) *Polym. Prepr., Am. Chem. Soc., Div. Polym. Chem.*, **22** (1), 167.
164. Bourdelande, J.L., Font, J. and Sanchezferrando, F. (1980) *Tetrahedron Lett.*, 3805.
165. Gupta, S.N., Thijs, L. and Neckers, D.C. (1980) *Macromolecules*, **13**, 1037.
166. Nishikubo, T., Iizawa, T. and Yamada, M. (1981) *J. Polym. Sci., Polym. Lett. Ed.*, **19**, 177.
167. Bailey, D., Tirrell, D., Pinazzi, C. and Vogl, O. (1978) *Macromolecules*, **11**, 312.
168. Kamogawa, H., Takayanagi, Y. and Nanasawa, M. (1981) *J. Polym. Sci., Polym. Chem. Ed.*, **19**, 2947.
169. Yoshida, S. and Vogl, O. (1980) *Polym. Prepr., Am. Chem. Soc., Div. Polym. Chem.*, **21** (1), 203.
170. Sumida, Y., Yoshida, S. and Vogl, O. (1980) *Polym. Prepr., Am. Chem. Soc., Div. Polym. Chem.*, **21** (1), 201.
171. Negishi, N., Tsunemitsu, K. and Shinohara, I. (1981) *Polym. J.*, **13**, 411.
172. Morawetz, H. (1969) *Adv. Catal.*, **20**, 341.
173. Overberger, C.G., Guteri, A.C., Kawakami, Y., Mathias, L.J., Meenakshi, A. and Tomono, T. (1978) *Pure Appl. Chem.*, **50**, 309.
174. Overberger, C.G. and Salamone, J.C. (1969) *Acc. Chem. Res.*, **2**, 217.
175. Kunitake, T. and Okahata, Y. (1976) *Adv. Polym. Sci.*, **20**, 159.
176. Overberger, C.G., Morimoto, M. and Cho, I. (1969) *Macromolecules*, **2**, 553; (1971) *J. Am. Chem. Soc.*, **93**, 3228.
177. Overberger, C.G. and Morimoto, M. (1971) *J. Am. Chem. Soc.*, **93**, 3222.
178. Overberger, C.G., Glowaky, R.C. and Vandewyer, P.H. (1973) *J. Am. Chem. Soc.*, **95**, 6008.
179. Overberger, C.G. and Sannes, K.N. (1974) *Angew. Chem., Int. Ed. Engl.*, **13**, 99.
180. Overberger, C.G., Pierre, T.S., Vorchheimer, N., Lee, J. and Yaroslavsky, S. (1965) *J. Am. Chem. Soc.*, **87**, 296.
181. Bruice, T.C. and Herz, J.L. (1964) *J. Am. Chem. Soc.*, **86**, 4109.
182. Overberger, C.G. and Shen, C.M. (1971) *Bioorg. Chem.*, **1**, 1.
183. Overberger, C.G. and Okamoto, Y. (1972) *Macromolecules*, **5**, 363.

184. Overberger, C.G., Corett, R., Salamone, J.C. and Yaroslavsky, S. (1968) *Macromolecules*, **1**, 331.
185. Overberger, C.G., Pierre, T.S., Yaroslavsky, C. and Yaroslavsky, S. (1966) *J. Am. Chem. Soc.*, **88**, 1184.
186. Letsinger, R.L. and Klaus, I.S. (1964) *J. Am. Chem. Soc.*, **86**, 3884.
187. Overberger, C.G., Sitaramaiah, R., Pierre, T.S. and Yaroslavsky, S. (1965) *J. Am. Chem. Soc.*, **87**, 3270.
188. Overberger, C.G. and Maki, H. (1970) *Macromolecules*, **3**, 214, 220.
189. Shimidzu, T., Furuta, A. and Nakamoto, Y. (1974) *Macromolecules*, **7**, 160.
190. Kunitake, T. and Okahata, Y. (1975) *Bioorg. Chem.*, **4**, 136.
191. Overberger, C.G., Pacansky, T.J., Pierre, T.S. and Yaroslavsky, S. (1974) *J. Polym. Sci. Symp.*, **46**, 209.
192. Overberger, C.G., Salamone, J.C. and Yaroslavsky, S. (1967) *J. Am. Chem. Soc.*, **89**, 6231.
193. Kunitake, T. and Okahata Y. (1976) *J. Am. Chem. Soc.*, **98**, 7793.
194. Kunitake, T., Okahata, Y. and Ando, R. (1974) *Macromolecules*, **7**, 140.
195. Kunitake, T. and Okahata, Y. (1974) *Chem. Lett.*, 1057; (1976) *Macromolecules*, **9**, 15.
196. Overberger, C.G. and Pacansky, J. (1974) *J. Polym. Sci. Symp.*, **45**, 39.
197. Shimidzu, G., Furuta, A., Watanabe, T. and Kato, S. (1974) *Macromol. Chem.*, **175**, 119.
198. Photaki, I. and Dailsiotou, M.S. (1976) *J. Chem. Soc., Perkin Trans. 1*, 589.
199. Overberger, C.G. and Dixon, K.W. (1977) *J. Polym. Sci., Polym. Chem. Ed.*, **15**, 1863.
200. Okamoto, Y. (1978) *Nippon Kagaku Kaishi*, 870.
201. Kunitake, T. and Shinkai, S. (1971) *J. Am. Chem. Soc.*, **93**, 4247, 4256.
202. Kunitake, T. and Shinkai, S. (1972) *Makromol. Chem.*, **151**, 127.
203. Imanishi, Y., Amimoto, Y., Sugiwara, T. and Higashimura, T. (1976) *Makromol. Chem.*, **177**, 1401.
204. Klotz, I.M., Roger, G.P. and Scarpa, I.S. (1971) *Proc. Natl. Acad. Sci. USA*, **68**, 263.
205. Johnson, T.W. and Klotz, I.M. (1973) *Macromolecules*, **6**, 788.
206. Kiefer, H.C., Congdon, W.I., Scarpa, I.S. and Klotz, I.M. (1972) *Proc. Natl. Acad. Sci. USA*, **69**, 2155.
207. Saegusa, T., Ikeda, H. and Fujii, H. (1972) *Macromolecules*, **5**, 108.
208. Spetnagel, W.J. and Klotz, I.M. (1977) *J. Polym. Sci., Polym. Chem. Ed.*, **15**, 621.
209. Klotz, I.M., Royer, G.P. and Sloniewsky, A.R. (1969) *Biochemistry.*, **8**, 4752.
210. Suh, J. and Klotz, I.M. (1977) *Bioorg. Chem.*, **1**, 165.
211. Pshezhetskii, V.S., Lukjanova, A.P. and Kabanov, V.A. (1975) *Bioorg. Khim. (USSR)*, **1**, 1458.
212. Pshezhetskii, V.S., Lukjanova, A.P. and Kabanov, V.A. (1977) *Eur. Polym. J.*, **13**, 423.
213. Meyers, W.E. and Royer, G.P. (1977) *J. Am. Chem. Soc.*, **99**, 614.
214. Overberger, C.G., Pierre, T.S., Vorchheimer, N. and Yaroslavsky, S. (1963) *J. Am. Chem. Soc.*, **85**, 3513.
215. Overberger, C.G. and Podsiodly, C.J. (1974) *Bioorg. Chem.*, **3**, 16.
216. Overberger, C.G., Pierre, T.S. and Yaroslavsky, S. (1965) *J. Am. Chem. Soc.*, **87**, 4310.
217. Overberger, C.G. and Podsiodly, J.C. (1974) *Bioorg. Chem.*, **3**, 35.
218. Kunitake, T., Shimada, F. and Aso, C. (1969) *J. Am. Chem. Soc.*, **91**, 2716.
219. Shinkai, S. and Kunitake, T. (1973) *Polym. J.*, **4**, 253.
220. Okahata, Y. and Kunitake, T. (1977) *J. Polym. Sci., Polym. Chem. Ed.*, **15**, 2571.
221. Letsinger, R.L. and Savereide, T.J. (1962) *J. Am. Chem. Soc.*, **84**, 114, 3122.
222. Kirsh, Y.E., Kabanov, V.A. and Kargin, V.A. (1968) *Polym. Sci. USSR*, **10**, 407.
223. Okubo, T. and Ise, N. (1973) *J. Org. Chem.*, **38**, 3120.

224. Kabanov, V.A., Kirsh, Y.E., Papisov, I.M. and Torchilin, V.P. (1972) *Vysokomol. Soedin. Ser. B* , **4**, 405; (1973) *Chem. Abstr.*, **78**, 102304-j.
225. Kunitake, T., Shinkai, S. and Hirotsu, S. (1975) *J. Polym. Sci., Polym. Lett. Ed.*, **18**, 377.
226. Starodubtrev, S.G., Kirsh, Y.E. and Kabanov, V.A. (1974) *Eur. Polym. J.*, **10**, 739.
227. Kirsh, Y.E. and Kabanov, V.A. (1970) *Dokl. Akad. Nauk USSR*, **195**, 109; (1971) *Chem. Abstr.*, **74**, 54331-e.
228. Osada, Y. and Chiba, T. (1979) *Makromol. Chem.*, **180**, 1617.
229. Kirsh, Y.E., Rahnanskaya, A.A., Lukovkin, G.M. and Kabanov, V.A. (1974) *Eur. Polym. J.*, **10**, 393.
230. Kirsh, Y.E., Lebedeva, T.A. and Kabanov, V.A. (1975) *J. Polym. Sci., Polym. Lett. Ed.*, **18**, 207.
231. Harrison, C.R. and Hodge, P. (1974) *J. Chem. Soc. Chem. Commun.*, 1009; (1976) *J. Chem. Soc., Perkin Trans. 1*, 605.
232. Aglietto, M. Ruggeri, G., Tarquini, B. and Ciardelli, F. (1980) *Polymer*, **21**, 541.
233. Kunitake, T., Okahata, Y. and Ando, R. (1974) *Bull. Chem. Soc. Jpn*, **47**, 1509.
234. Suh, J. and Klotz, I.M. (1978) *J. Polym. Sci., Polym. Chem. Ed.*, **16**, 1943.
235. Okahata, Y. and Kunitake, T. (1978) *J. Polym. Sci., Polym. Chem. Ed.*, **16**, 1865.
236. Prini, R.F. and Turyn, D. (1972) J. *Chem. Soc. Chem. Commun.*, 1013.
237. Rudolfo, T., Hamilton, J.A. and Cordes, E.H. (1974) *J. Org. Chem.*, **39**, 2281.
238. Turyn, D., Baumgartner, E. and Prini, R.F. (1974) *Biophys. Chem.*, **2**, 269.
239. Klotz, I.M. and Stryker, V.H. (1968) *J. Am. Chem. Soc.*, **90**, 2717.
240. Royer, G.P. and Klotz, I.M. (1969) *J. Am. Chem. Soc.*, **91**, 5885.
241. Birk, Y. and Klotz, I.M. (1971) *Bioorg. Chem.*, **1**, 275.
242. Pshezhetskii, V.S., Murtazaeva, G.A. and Kabanov, V.A. (1974) *Eur. Polym. J.*, **10**, 571.
243. Weatherhead, R.H., Stacey, K.A. and Williams, A. (1978) *J. Chem. Soc., Perkin Trans 2*, 800.
244. Pshezhetskii, V.S., Lukjanova, A.P. and Kabanov, V.A. (1977) *J. Mol. Catal.*, **2**, 49.
245. Kurusu, Y., Nakashima, K., Nakashima, M. and Okawara, M. (1968) *Kogyo Kagaku Zasshi*, **71**, 934.
246. Shinkai, S., Tamaki, K. and Kunitake, T. (1975) *Bull. Chem. Soc. Jpn*, **48**, 1918.
247. Shinkai, S., Tamaki, T. and Kunitake, T. (1976) *J. Polym. Sci., Polym. Lett. Ed.*, **14**, 1.
248. Shinkai, S., Tamaki, T. and Kunitake, T. (1977) *Makromol. Chem.*, **178**, 133.
249. Shinkai, S. and Kunitake, T. (1976) *Biopolymer.*, **15**, 1129.
250. Shinkai, S. and Kunitake, (1977) *Makromol. Chem.*, **178**, 145.
251. Spetnagela, W.J. and Klotz, I.M. (1978) *Biopolymers*, **17**, 1657.
252. Shinkai, S., Yamada, S. and Kunitake, T. (1978) *J. Polym. Sci., Polym. Lett. Ed.*, **16**, 137.
253. Shinkai, S., Yamada, S. and Kunitake, T. (1978) *Macromolecules*, **11**, 65.
254. Shinkai, S. and Kunitake, T. (1977). *Polym. J.*, **9**, 423.
255. Overberger, C.G. and Yuen, P.S. (1970) *J. Am. Chem. Soc.*, **92**, 1667.
256. Ueda, T., Harada, S. and Ise, N. (1972) *Polym. J.*, **3**, 476.
257. Morawetz, H. (1978) *Isr. J. Chem.*, **17**, 287.
258. Kunitake, T., Shinkai, S. and Hirotsu, S. (1977) *J. Org. Chem.*, **42**, 306.
259. Shinkai, S., Hirakawa, S., Shimomura, M. and Kunitake, T. (1981) *J. Org. Chem.*, **46**, 868.
260. Kunitake, T., Shinkai, S. and Hirotsu, S. (1976) *Biopolymers*, **15**, 1143.
261. Yamazaki, N., Nakahama, S., Hirao, A. and Kawabata, J. (1980) *Polym. J.*, **12**, 231.
262. Suh, J., Scarpa, I.S. and Klotz, I.M. (1976) *J. Am. Chem. Soc.*, **98**, 7060.
263. Spetnagel, W.J. and Klotz, I.M. (1976) *J. Am. Chem. Soc.*, **98**, 8199.
264. Tundo, P. and Venturello, P. (1976) *Tetrahedron Lett.*, 2581.
265. Smid, J., Kimura, K., Shirai, M. and Sinta, R. (1981) *Polym. Prepr., Am. Chem. Soc., Div. Polym. Chem.*, **22** (1), 163.

266. Shah, S.C. and Smid, J. (1976) *J. Am. Chem. Soc.*, **98**, 5198.
267. Smid, J., Shah, S., Wong, L. and Hurley, J. (1975) *J. Am. Chem. Soc.*, **97**, 5932.
268. Wong, L. and Smid, J. (1977) *J. Am. Chem. Soc.*, **99**, 5637.
269. Smid, J. (1976) *Pure Appl. Chem.*, **48**, 343.
270. Shah, S., Wong, L. and Smid, J. (1977) *Polym. Prepr., Am. Chem. Soc., Div. Polym. Chem.*, **18** (1), 766.
271. Smid, J., Wong, L., Varma, A.J. and Shah, S.C. (1979) *Polym. Prepr., Am. Chem. Soc. Div. Polym. Chem.*, **20** (1) 1063.
272. Shirai, M. and Smid, J. (1980) *J. Polym. Sci., Polym. Lett. Ed.*, **18**, 659.
273. Kemp, D.S. and Paul, K.G. (1970) *J. Am. Chem. Soc.*, **92**, 2553.
274. Kemp, D.S. and Paul, K.G. (1975) *J. Am. Chem. Soc.*, **97**, 7305.
275. Kemp, D.S., Cox, D.D. and Paul, K.G. (1975) *J. Am. Chem. Soc.*, **97**, 7312.
276. Roland, B. and Smid, J. (1983) *J. Am. Chem. Soc.*, **105**, 5269.
277. Kimura, K. and Smid, J. (1982) *Macromolecules*, **15**, 966.
278. Roland, B., Kimuira, K. and Smid, J. (1984) *J. Colloid Interface Sci.*, **97**, 392.
279. Sinta, R., Lamb, B. and Smid, J. (1983) *Macromolecules*, **16**, 1382.
280. Sinta, R., Rose, P.S. and Smid, J. (1983) *J. Am. Chem. Soc.*, **105**, 4337.
281. Smid, J. and Sinta, R. (1980) In *Crown-Ethers and Phase-Transfer Catalysis in Polymer Science* (eds L.J. Mathias and C.E. Carraher), Plenum, New York, pp. 329–43.
282. Suh, D., Scarpa, I.S. and Klotz, I.M. (1976) *J. Am. Chem. Soc.*, **98**, 7060.
283. Smid, J. (1981) *Makromol. Chem. Suppl.*, **5**, 203; (1982) *Pure Appl. Chem.*, **54**, 2129.
284. Shah, S.C. and Smid, J. (1978) *J. Am. Chem. Soc.*, **100**, 1426.
285. Menger, F.M. (1970) *J. Am. Chem. Soc.*, **92**, 5965.
286. Tomita, A., Ebina, N. and Tamai, Y. (1977) *J. Am. Chem. Soc.*, **99**, 5725.
287. Makosza, M. and Bialecka, E. (1977) *Tetrahedron Lett.*, 183.
288. Dehmlow, E.V. (1974) *Angew. Chem. Int. Ed. Engl.*, **13**, 170; (1977) *Angew. Chem., Int. Ed. Engl.*, **16**, 493.
289. Dehmlow, E.V. (1975) *Chem. Technol.*, **5**, 210.
290. Regen, S.L. (1979) *Angew. Chem., Int. Ed. Engl.*, **18**, 421.
291. Cainelli, G., Manescalchi, F. and Panunzio, M. (1976) *Synthesis*, 472.
292. Horiki, K. (1978) *Synth. Commun.*, **8**, 515.
293. Copelin, H.B., Falls, N. and Crane, G.B. (1957) US Patent 2 779 781.
294. Cainelli, G. and Manescalchi, F. (1979) *Synthesis*, 141.
295. Regen, S.L. (1977) *J. Org. Chem.*, **42**, 875.
296. Regen, S.L. (1975) *J. Am. Chem. Soc.*, **97**, 5956.
297. Cinquini, M., Colonna, S., Molinari, H., Montanari, F. and Tundo, P. (1976) *J. Chem. Soc. Chem. Commun.*, 394.
298. Regen, S.L. and Nigam, A. (1978) *J. Am. Chem. Soc.*, **100**, 7773.
299. Regen, S.L., Heh, J.C.K. and McLick, J. (1979) *J. Org. Chem.*, **44**, 1961.
300. Molinari, H., Montanari, F., Quici, S. and Tundo, P. (1979) *J. Am. Chem. Soc.*, **101**, 3920.
301. Zadeh, H.K., Dou, H.J.M. and Metzger, J. (1978) *J. Org. Chem.*, **43**, 156.
302. Regen, S.L. (1976) *J. Am. Chem. Soc.*, **98**, 6270.
303. Noguchi, H., Sugawara, M. and Uchida, Y. (1980) *Polymer*, **21**, 861.
304. Chiles, M.S. and Reeves, P.C. (1979) *Tetrahedron Lett.*, 3367.
305. Rolla, F., Roth, W. and Horner, L. (1977) *Naturwissenschaften*, **64**, 377.
306. Bram, G. and Decodts, G. (1980) *Tetrahedron Lett.*, 5011.
307. Tomoi, M. and Ford, W.T. (1980) *J. Am. Chem. Soc.*, **102**, 7140.
308. Tomoi, M. and Ford, W.T. (1981) *J. Am. Chem. Soc.*, **103**, 3821, 3828.
309. Chiles, M.S., Jackson, D.D. and Reeves, P.C. (1980) *J. Org. Chem.*, **45**, 2915.
310. Serita, H., Ohtani, N., Matsunage, T. and Kimura, C. (1979) *Kobunshi Ronbunsu*, **36**, 527.
311. Colonna, S., Fornasier, R. and Pfeiffer, U. (1978) *J. Chem. Soc., Perkin Trans. 1*, 8.

312.   Brown, J.M. and Jenkins, J.A. (1976) *J. Chem. Soc. Chem. Commun.*, 453.
313.   Molinari, H., Montanari, F. and Tundo, P. (1977) *J. Chem. Soc. Chem., Commun.*, 639.
314.   Regen, S.L., Czech, B. and Singh, A. (1980) *J. Am. Chem. Soc.*, **102**, 6638.
315.   Tundo, P. and Venturello, P. (1979) *J. Am. Chem. Soc.*, **101**, 6606.
316.   Tundo, P. (1977) *J. Chem. Soc. Chem. Commun.*, 641.
317.   Balakrishnan, T. and Ford, W.T. (1981) *Tetrahedron Lett.*, 4377.
318.   Chiellini, E., Solaro, R. and D'Antone, S. (1977) *Makromol. Chem.*, **178**, 3165.
319.   Mackenzie, W.M. and Sherrington, D.C. (1981) *Polymer*, **22**, 431.
320.   Akelah, A. and Sherrington, D.C. (1982) *Eur. Polym. J.*, **18**, 301.
321.   Kise, H., Araki, K. and Seno, M. (1981) *Tetrahedron Lett.*, 1017.
322.   Meada, H., Hayashi, Y. and Teramura, K. (1980) *Chem. Lett.*, 677.
323.   Tomoi, M., Ogawa, E., Hosokawa, Y. and Kakiuchi, H. (1982) *J. Poly. Sci., Polym. Chem. Ed.*, **20**, 3015.
324.   Tomoi, M., Ogawa, E. and Hosokawa, Y. (1982) *J. Polym. Sci., Polym. Chem. Ed.*, **20**, 3421.
325.   Tomoi, M., Hosokawa, Y. and Kakiuchi, H. (1984) *J. Polym. Sci., Polym. Chem. Ed.*, **22**, 1243.
326.   Tomoi, M., Hosokawa, Y. and Kakiuchi, H. (1983) *Makromol. Chem., Rapid Commun.*, **4**, 227.
327.   Idoux, J.P., Wysocki, R., Young, S., Tureot, J., Ohlman, C. and Leonard, R. (1983) *Synth. Commun.*, **13**, 139.
328.   Ohtani, N., Wilkie, C.A., Nigam, A. and Regen, S.L. (1980) *Macromolecules*, **14**, 516.
329.   Regen, S.L., Bolikal, D. and Barcelon, C. (1981) *J. Org. Chem.*, **46**, 2511.
330.   Regen, S.L. and Besse, J.J. (1979) *J. Am. Chem. Soc.*, **101**, 4059.
331.   Tundo, P. (1978) *Synthesis*, 315.
332.   Montanari, F., Quici, S. and Tundo, P. (1983) *J. Org. Chem.*, **48**, 199.
333.   Anelli, P.L., Montanari, F. and Quici, S. (1983) *J. Chem. Soc., Perkin Trans. 2*, 1827.
334.   Montanari, F. and Tundo, P. (1981) *J. Org. Chem.*, **46**, 2125.
335.   Tomoi, T., Shiiki, S. and Kakiuchi, H. (1986) *Makromol. Chem.*, **187**, 357.
336.   Tundo, P. and Venturello, P. (1981) *J. Am. Chem. Soc.*, **103**, 856.
337.   Funkunishi, K., Czech, B. and Regen, S.L. (1981) *J. Org. Chem.*, **46**, 1218.
338.   Manecke, G. and Kramer, A. (1981) *Makromol. Chem.*, **182**, 3017.
339.   Tomoi, M., Abe, O., Ikeda, M., Kihara, K. and Kakiuchi, H. (1978) *Tetrahedron Lett.*, 3031.
340.   Cook, F.L., Robertson, J.R. and Ernst, W.A. (1981) *Polym. Prepr., Am. Chem. Soc., Div. Polym. Chem.*, **22**(1), 161.
341.   Akabori, S., Miyamoto, S. and Tanabe, H. (1979) *J. Polym. Sci., Polym. Chem. Ed.*, **17**, 3933.
342.   Kimura, Y. and Regen, S.L. (1983) *J. Org. Chem.*, **48**, 195.
343.   Kimura, Y., Kirszensztejn, P. and Regen, S.L. (1983) *J. Org. Chem.*, **48**, 385.
344.   Rugia, M.J., Czeck, A., Czech, B.P. and Bartsch, R. A. (1986) *J. Org. Chem.*, **51**, 2945.
345.   Mathias, L.J. and Canterbarry, J.B. (1981) *Polym. Prepr., Am. Chem. Soc., Div. Polym. Chem.*, **22**(1), 38.
346.   Merck Co. (1977) *Kontakte* (1), 30.
347.   Tomoi, M., Kihara, K. and Kakiuchi, H. (1979) *Tetrahedron Lett.*, 3485.
348.   Regen, S.L. and Dulak, L. (1977) *J. Am. Chem. Soc.*, **99**, 623.
349.   Sinta, R. and Smid, J. (1980) *Macromolecules*, **13**, 339.
350.   Regen, S.L., Besse, J.J. and McLick, J. (1979) *J. Am. Chem. Soc.*, **101**, 116.
351.   Regen, S.L. (1977) *J. Am. Chem. Soc.*, **99**, 3838.
352.   MacKenzie, W.M. and Sherrington, D.C. (1978) *J. Chem. Soc. Chem. Commun.*, 541.
353.   Heffernan, J.G., MacKenzie, W.M. and Sherrington, D.C. (1981) *J. Chem. Soc., Perkin Trans. 2*, 514.
354.   MacKenzie, W.M. and Sherrington, D.C. (1980) *Polymer*, **21**, 791.

355. Wakui, T., Xu, W.Y., Shen, C.S. and Smid, J. (1986) *Makromol. Chem.*, **187**, 533.
356. Hiratani, K., Renter, P. and Manecke, G. (1979) *Isr. J. Chem.*, **18**, 208.
357. Hiratani, K., Renter, P. and Manecke, G. (1979) *J. Mol. Catal.*, **5**, 241.
358. Regen, S.L., Nigam, A. and Besse, J.J. (1978) *Tetrahedron Lett.*, 2757.
359. Tomoi, M., Ideda, M. and Kakiuchi, H. (1978) *Tetrahedron Lett.*, 3757.
360. Kobler, H., Schuster, K.H. and Simchen, G. (1978) *Liebigs Ann. Chem.*, 1946.
361. Tomoi, M., Takubo, T., Ikeda, M. and Kakiuchi, H. (1976) *Chem. Lett.*, 473.
362. Kondo, S., Inagaki, Y. and Tsuda, K. (1984) *J. Polym. Sci., Polym. Lett. Ed.*, **22**, 249.
363. Kondo, S., Ohta, K. and Tsuda, K. (1983) *Makromol. Chem. Rapid Commun.*, **4**, 145.
364. Kondo, S., Ohta, K., Ojika, R., Yasui, H. and Tsuda, K. (1985) *Makromol. Chem.*, **186**, 1.
365. Janout, V., Hrudkova, H. and Cefelin, P. (1984) *Coll. Czech. Chem. Commun.*, **49**, 2096.
366. Farrall, M.J., Durst, T. and Frechet, J.M.J. (1979) *Tetrahedron Lett.*, 203.
367. Janout, V. and Cefelin, P. (1986) *Tetrahedron Lett.*, 3525.
368. Doscher, F., Klein, J., Pohl, F. and Widdecke, H. (1980) *Makromol. Chem. Rapid Commun.*, **1**, 297.
369. Lehmkuhi, H., Rabet, F. and Hauschild, K. (1977) *Synthesis*, 184.
370. Yanagida, S., Noji, Y. and Okahara, M. (1977) *Tetrahedron Lett.*, 2893.
371. Lee, D.G. and Chang, V.S. (1977) *J. Org. Chem.*, **43**, 1532.
372. Kitazume, T. and Ishikawa, N. (1978) *Chem. Lett.*, 283.
373. Hirao, A., Nakahoma, S., Takahashi, M. and Yamazaki, N. (1978) *Makromol. Chem.*, **179**, 915.
374. Kelly, J., MacKenzie, W.M. and Sherrington, D.C. (1979) *Polymer*, **20**, 1048.
375. Santaniello, E., Ferraboschi, P. and Sozzani, P. (1979) *Tetrahedron Lett.*, 4581; (1981) *J. Org. Chem.*, **46**, 4584.
376. Fornasier, R. and Montanari, F. (1976) *Tetrahedron Lett.*, 1381.
377. Ohtomi, M., Yoneyama, S., Arai, K. and Akabori, S. (1976) *Nippon Kagaku Kaishi*, 1878.
378. Regen, S.L., Mehrotra, A. and Singh, A. (1981) *J. Org. Chem.*, **46**, 2182.
379. Chiellini, E., Solaro, R. and D'Antone, S. (1981) *Makromol. Chem. Suppl.*, **5**, 82.
380. Brown, J.M. and Jenkins, J.A. (1976) *J. Chem. Soc. Chem. Commun.*, 458.
381. MacKenzie, W.M. and Sherrington, D.C. (1978) *J. Chem. Soc. Chem. Commun.*, 541.
382. Pittman, C.U. and Evans, G.O. (1971) *Chem. Technol.*, (1), 416.
383. Manassen, J. (1971) *Platinum Met. Rev.*, **15**, 142.
384. Kohler, N. and Dawans, F. (1972) *Rev. Inst. Fr. Pet.*, **27**, 105.
385. Pittman, C.U. and Evans, G.O. (1973) *Chem. Technol.*, **3**, 560.
386. Bailar, J.C. (1974) *Catal. Rev. Sci. Eng.*, **10**, 17.
387. Cernia, E.M. and Graziani, M. (1974) *J. Appl. Polym. Sci.*, **18**, 2725.
388. Michalska, Z.M. and Webster, D.E. (1974) *Platinum Met. Rev.*, **18**, 65; (1975) *Chem. Technol.*, **5**, 117.
389. Davydova, S.L. and Plate, N.A. (1975) *Coord. Chem. Rev.*, **16**, 195.
390. Yermakov, Y.I. (1976) *Catal. Rev. Sci. Eng.*, **13**, 77.
391. Robinson, A.L. (1976) *Science*, **194**, 1261.
392. Grubbs, R.H., Sweet, E.M. and Phisanbut, S. (1976) In *Catalysis in Organic Synthesis* (eds R.N. Rylander and H. Greenfield), Academic Press, New York, p. 153.
393. Pittman, C.U., Jacobson, S., Smith, L.R., Clements, W. and Hiramoto, H. (1976) In *Catalysis in Organic Synthesis* (eds R.N. Rylander and H. Greenfield), Academic Press, New York, p. 161.
394. Hartley, F.R. and Vezey, P.N. (1977) *Adv. Organomet. Chem.*, **15**, 189.
395. Grubbs, R.H. (1977) *Chem. Technol.*, **7**, 512.
396. Chauvin, Y., Commereuc, D. and Dawans, F. (1977) *Prog. Polym. Sci.*, **5**, 95.
397. Tsuchida, E. and Nishide, H. (1977) *Adv. Polym. Sci.*, **24**, 1.
398. Yuffa, A.Y. and Lisichkin, G.V. (1978) *Russ. Chem. Rev.*, **47**, 751.

138 Polymeric catalysts

399. Villadsen, J. and Livbjerg, H. (1978) *Catal. Rev. Sci. Eng.*, **17**, 203.
400. Davydov, V. (1979) *Acta Polym.*, **30**, 119.
401. Pittman, C.U. (1980) In *Polymer Supported Reactions in Organic Synthesis* (eds P. Hodge and D.C. Sherrington), Wiley, New York, p. 249.
402. Whitehurst, D.D. (1980) *Chem. Technol.*, **1**, 44.
403. Bailey, D.C. and Langer, S.H. (1981) *Chem. Rev.*, **81**, 109.
404. Hagihara, H., Sonogashira, K. and Takahashi, S. (1981) *Adv. Polym. Sci.*, **41**, 149.
405. Stark, G.R. (1969) *Immobilized Enzymes*, Academic Press, New York.
406. Zaborsky, O. (1973) *Immobilized Enzymes*, CRC Press, Cleveland, OH.
407. Goldman, R., Goldstein, L. and Katchalski, E. (1974) In *Biochemical Aspects of Reactions on Solid Supports*, (ed. G.R. Stark), Academic Press, New York, 1974, Ch. 1, p. 1.
408. Salmona, M., Saronia, C. and Garattini, S. (eds) (1974) *Immobilized Enzymes*, Raven Press, New York.
409. Messing, R.A. (ed.) (1975) *Immobilized Enzymes for Industrial Reactors*, Academic Press, New York.
410. Weetal, H.H. (ed.) (1975) *Immobilized Enzymes, Antigens, Antibodies and Peptides*, Marcel Dekker, New York.
411. Suckling, C.J. (1977) *Chem. Soc. Rev.*, **6**, 215.
412. Chibata, I. (ed.) (1978) *Immobilized Enzymes*, Halstead Press, New York.
413. Manecke, G. and Vogl, H.G. (1978) *Pure Appl. Chem.*, **50**, 655.
414. Manecke, G. and Storck, W. (1978) *Angew. Chem., Int. Ed. Engl.*, **17**, 657.
415. Vandegaer, J.E. (ed.) (1974) *Microencapsulation*, Plenum Press, New York.
416. Chang, T.M.S. (1976) *J. Macromol. Sci. Chem.*, **A-10**, 245.
417. Olson, A.C. and Cooney, C.L. (eds) (1973) *Immobilized Enzymes in Food and Microbial Processes*, Plenum Press, New York.
418. Hatanaka, H., Sugiyama, K., Nakaya, T. and Imoto, M. (1974) *Makromol. Chem.*, **175**, 1855.
419. Harper, G.B. (1975) *Anal. Chem.*, **47**, 349.
420. Gupta, S.N., Thijs, L. and Neckers, D.C. (1981) *J. Polym. Sci., Polym. Chem. Ed.*, **19**, 855.
421. Carlini, C., Ciardelli, F., Donati, D. and Gurzoni, F. (1983) *Polymer*, **24**, 599.
422. Flamigni, L., Barigelletti, F. and Bortolus, P. (1984) *Eur. Polym. J.*, **20**, 171.
423. Kerber, R., Nuyken, O. and Steinhausen, R. (1976) *Makromol. Chem.*, **177**, 1357; (1977) *Makromol. Chem.*, **178**, 1833.
424. Kerber, R., Gerum, J. and Nuyken, O. (1979) *Makromol. Chem.*, **180**, 609.
425. Campbell, D.S., Loeber, D.E. and Tinker, A.J. (1978) *Polymer*, **19**, 1106; (1979) *Polymer*, **20**, 393; (1984) *Polymer*, **25**, 1141.
426. Campbell, D.S. and Tinker, A.J. (1984) *Polymer*, **25**, 1146.
427. Nuyken, O., Schuster, H. and Kerbor, R. (1983) *Makromol. Chem.*, **184**, 2285.
428. Samal, R.K., Iwata, H. and Ikada, Y. (1983) In *Physicochemical Aspects of Polymer Surfaces*, vol. 2 (ed. K.L. Mittal), Plenum Press, New York, p. 801.
429. Braun, D. and Loflund, J. (1962) *Makromol. Chem.*, **53**, 219.
430. Greber, G. and Toelle, J. (1962) *Makromol. Chem.*, **52**, 208.
431. Okada, M. and Sumitomo, H. (1972) *Makromol. Chem.*, **162**, 285.
432. Ouchi, T., Izumi, T., Inaba, M. and Imoto, M. (1983) *J. Polym. Sci., Polym. Chem. Ed.*, **21**, 2101.
433. Sivaram, S. and Naik, K.M. (1978) *J. Appl. Polym. Sci.*, **22**, 3293.
434. Davydova, S.L., Plate, N.A. and Kargin, V.A. (1968) *Russ. Chem. Rev.*, **37**, 984.
435. Greber, G. and Egle, G. (1962) *Makromol. Chem.*, **53**, 206.
436. Greber, G. and Egle, G. (1963) *Makromol. Chem.*, **64**, 68; (1967) *Makromol. Chem.*, **101**, 104.

5
Separations on functionalized polymers

Functionalized polymers can be used to bind one or more species out of a complex mixture selectively, either by ionic binding, e.g. ion exchange chromatography, or through electrostatic forces by neutral polymers such as polymeric crown ethers, cryptands or coordination polymers. It is also possible to induce chemical bonding between the species to be separated and functional groups of a polymer. After separation of the desired species with the functional group of the polymer from the other components of the mixture, the separated species is recovered by a release from the polymer by a simple means such as altering the pH:

$$\text{(P)}-X + (S_1 + S_2) \longrightarrow \text{(P)}-X...S_1 + S_2 \longrightarrow \text{(P)}-X...S_1 \longrightarrow \text{(P)}-X + S_1$$

Generally, the functionalized polymer used in the separation must fulfil certain criteria. The functional bond (ligand, chelating) must be stable under the conditions of separation and not interfere with the polymer backbone. Furthermore, the functional group–substrate bond must be reversible and specific. Polymeric materials have been used widely as chromatographic column stationary phases and have proved useful in both gas and liquid phase separations. The use of polymeric carriers in metal ion separations, the resolution of racemic mixtures and other separations will be discussed in this chapter.

5.1 METAL ION SEPARATIONS

Chelating polymers have received considerable attention owing to their inherent advantages over simple ion exchange resins, e.g. their greater selectivity to bind metal ions [1–5]. The selective adsorption of metal ions depends on a difference in the stability constant of the complex between a polymer ligand and a metal ion and can be accomplished by operating the adsorption at a definite pH. However, the selectivity of most chelating groups for metals resides predominantly in their ability to form chelates with cations. Although chelating resins are used to remove transition metal ions from aqueous solutions in high yield, they do not meet the requirements of an important industrial process, i.e. to remove a particular metal ion from aqueous solutions containing several metal ions, by the method of controlling the pH of the bulk aqueous phase and that of drainage in a large quantity.

Polymers containing coordination groups have been used for a selective separation of metal ions depending on the coordination number and the geometry of the ions. Polymeric ligands 1 with a maximum of four donor atoms and a structure which restricts the donor lone pairs to a tetrahedral arrangement were found to favour chelation of ions such as Cu^{2+} over Fe^{3+} [6] and hence have high ability to discriminate between Cu^{2+} and Fe^{3+}.

1

The bond length and angle between the central metal ion and the coordination ligands in a metal complex are strictly determined by the kind of metal ion and ligand. When a coordinated stereostructure of a polymer–metal complex is fixed, then the central metal ions are removed while the stereostructure is maintained fixed, and the remaining polymer ligand may have pockets especially fitted for the same metal ion which was removed from the polymer matrix. Such a selective adsorption of metal ions has been realized by fixing a stereostructure of a polymer chain by crosslinking its metal complex. Such a crosslinking procedure corresponds to a kind of template formation. For example, when a metal complex of polyvinylpyridine is crosslinked with α-dibromobutane and the metal ions are then removed by washing with dilute acid, the remaining metal-free resin shows selectivity to adsorb the same type of metal ions that were removed [7, 8].

In general, the selectivity coefficient of chelating polymers for specific separation is influenced by (a) the nature and kind of ions to be separated, (b) the nature and structure of the chelating group, (c) the molecular structure surrounding the chelating group, (d) the degree of crosslinking (a higher degree of crosslinking results in a lower degree of metal ion adsorption and a lower stability of the resulting metal complex), (e) the solvent and (f) the temperature of separation.

In addition to chelating polymers, polymers attached to macrocyclic ethers have been shown to complex metal ions. Recently, three different types of polymeric ethers, crown ethers, cryptands and podands, have been introduced as a new class of neutral ligands to replace the conventional ion exchangers. The ion-chelating properties of these neutral ligands have led to their selective applications in extraction or separation of trace metals or isotopes and organic solutes. Their selectivity depends on the number and chemical structure of the ring member as well as on the diameter and steric factors of the ring [9, 10]. A number of neutral ligands with considerably enhanced selectivity to bind metal ions have therefore been developed over recent years. Although at present liquid–liquid ion exchange is used widely in metallurgical extractions polymer-supported systems will have a significant role to play in the future. For example, polymeric macrocyclic hexaketone **2** which is a very strong host of uranyl (UO_2^{2-}) ion, has recently been synthesized and utilized successfully for the extraction of uranium directly from sea water [11]. The characteristics of the hexaketone are that (a) six oxygen atoms can form hexadentate coordination in near coplanarity which corresponds to the crystalline structure of various uranyl salts, (b) β-diketone can easily be dissociated in sea water to form a strong ligand, ketoenolate anion, and (c) the bound uranyl ion is readily released by treatment with dilute aqueous acid. Table 5.1 lists some of the selective separations that have been achieved with functionalized polymers.

2

5.2 RESOLUTION OF RACEMIC MIXTURES

The need for optically pure materials in both the pharmaceutical and the agrochemical industry continues to grow. In general, racemic mixtures consist of enantiomers which are optically active isomers with completely identical physical and chemical properties. The difference in their stereochemical structures can be identified only by measuring the rotation value or by interaction with chiral reagents to form diastereomers which have different properties. However, the conventional methods of resolution, such as fractional crystallization, require the presence of specific functional groups in the enantiomers and in addition it is necessary to prepare and then to hydrolyse the diastereomers [142–145]. For these reasons, these methods are essentially restricted to acid and base racemic mixtures, often involve high losses and do

Table 5.1 Metal ion separations

Functionalized polymer	Application	Reference
(PS)—N=N— (8-hydroxyquinoline, OH, N)	Extraction of Ni, Cu and Co / Separation of Cu, Zn and Al	12–14 / 15, 16
(P)—O—CO— —N=N— (8-hydroxyquinoline, OH, N)	Separation of Cu^{II} and Fe^{III}	17
(P)— imidazole, $^+$NH, COO$^-$, COOH	Selective for heavy metal ions	18
(PS)— pyrazole, COO$^-$, COOH, $^+$N—H, H—N	Selective for heavy metal ions	18
(PS)—N—N=N triazole, N—H, S	Chelating agent	19
(PS)— thiazole, S, N, C—NHR	Selective separation of Hg^{2+}	20
(PS)— thiazole, S, N, C=NMe		20

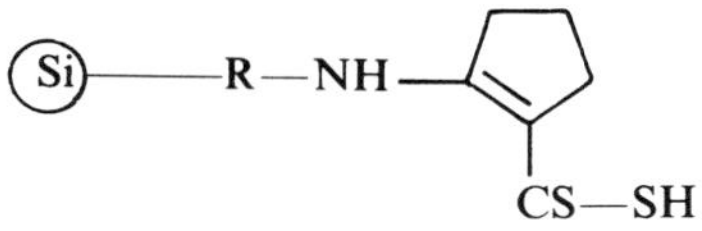

21

Extraction of metal ions

(Si)——R—NH—⟨cyclopentene⟩CS—SH

R ≡ —(CH$_2$)$_3$—, (CH$_2$)$_3$,NH(CH$_2$)$_2$—

22

Separation of various metal ions

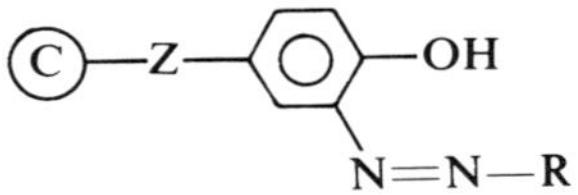

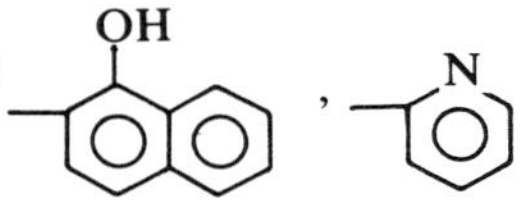

(C)—Z—⟨benzene⟩ OH, N=N—R

R ≡ ⟨pyridine⟩ N , ⟨naphthalene⟩ OH

Z ≡ —OCH$_2$—, —O(CH$_2$)$_2$—SO$_2$—

22, 23

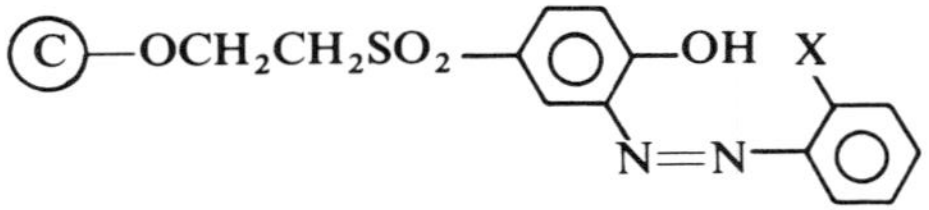

Selective for UO$_2$$^{2+}$ ions

(C)—OCH$_2$CH$_2$SO$_2$—⟨benzene⟩ OH X, N=N—⟨benzene⟩

22

(C)—Z—⟨structure⟩ OH HO, OH, N=N, HO

24

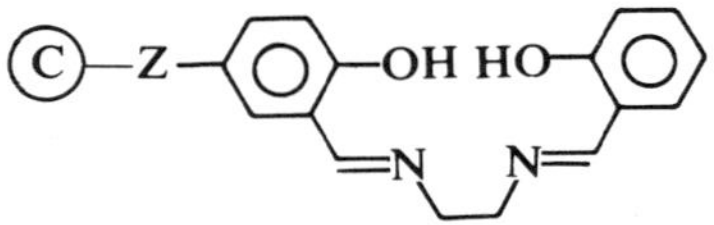

Binding of metal ions

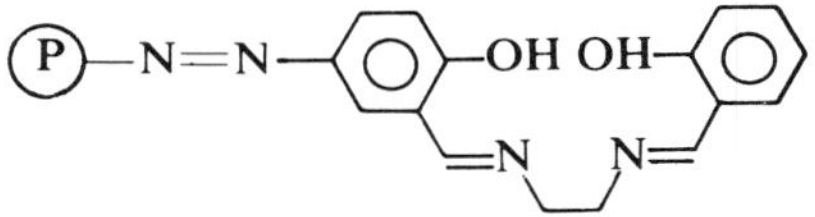

(P)—N=N—⟨macrocyclic structure⟩ OH OH

Table 5.1 (*Contd.*)

Functionalized polymer	Application	Reference
(PS)— naphthol with OH and NO substituents	Binding of metal ions	25, 26
(PS)—CH$_2$— aromatic ring with OH, ON substituents	Selective chelating agent	27
(P)—O—CO— phenyl —N=N—C=N—NHR with R'	Binding of Co^{2+}, Pb^{2+} and Hg^{2+}	28
(PS)—CH$_2$—N with R and NH—R' biphenyl	Selective separation of Cu, Ag and Ni	6

(a) R ≡ H; R' ≡ CH$_2$— pyridine (N)

(b) R ≡ R' ≡ CH$_2$— pyridine (N)

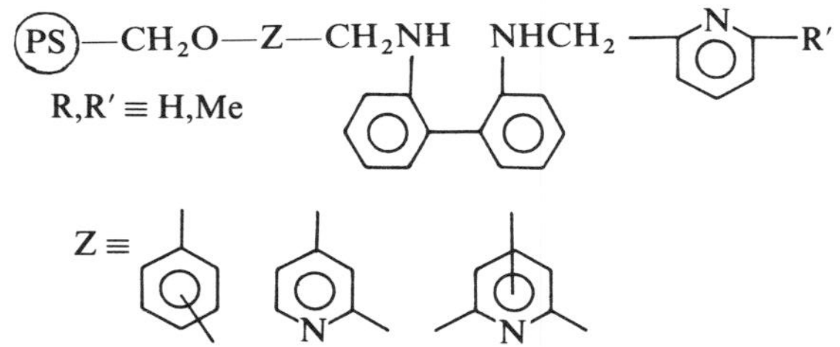

(PS)—CH$_2$O—Z—CH$_2$NH   NHCH$_2$ N—R'

R,R' ≡ H,Me

Z ≡ (structures)

Selective separation of Cu, Ag and Ni — 6

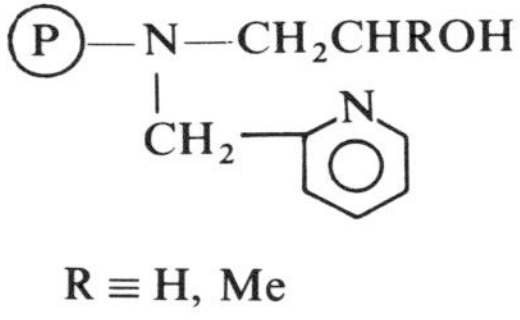

(P)—N—CH$_2$CHROH
        |
       CH$_2$

R ≡ H, Me

Selective for Cu — 16, 29–31

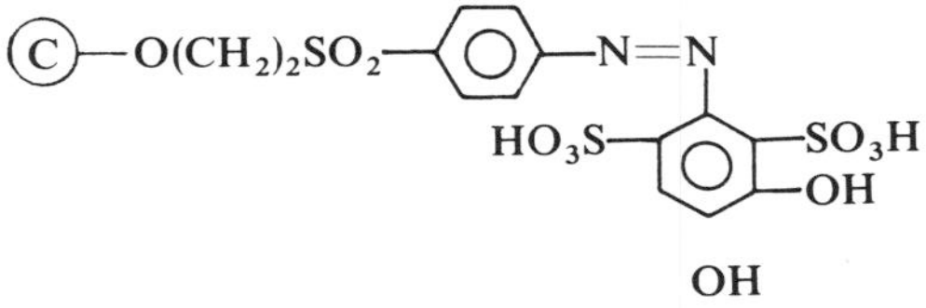

(C)—O(CH$_2$)$_2$SO$_2$—   —N=N—   SO$_3$H, OH
HO$_3$S

Adsorption of Fe^{3+}, Cu^{2+} and Hg^{2+} — 32

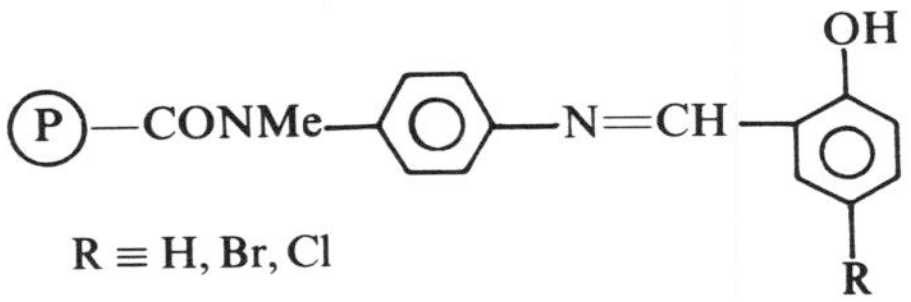

OH   OH

Adsorption of various metal ions — 33

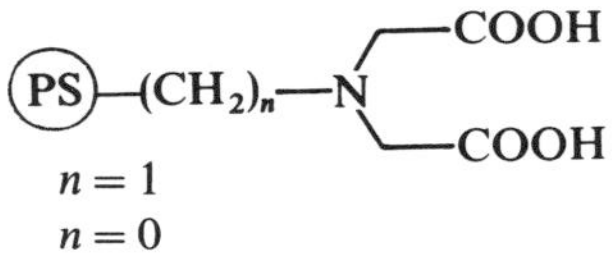

(P)—CONMe—   —N=CH—   R

R ≡ H, Br, Cl

Separation of transition metals — 34–39
Separation of UO$_2^{2+}$ from water — 40, 41

(PS)—(CH$_2$)$_n$—N   COOH
                        COOH

n = 1
n = 0

Table 5.1 (*Contd.*)

Functionalized polymer	Application	Reference
PS—CH$_2$—N—(CH$_2$)$_3$—N$\begin{smallmatrix}\text{COOH}\\\text{COOH}\end{smallmatrix}$ $\quad\quad$ COOH		35
C—N(CH$_2$COOH)$_2$	Chelating agent	42
PS—CH$_2$—(N—CH$_2$CH$_2$)$_n$ $\quad\quad\quad$ R	Adsorption of various metal ions	
R ≡ H; n = 1, 2, 3		43, 44
R ≡ —(CH$_2$)$_m$COOH; m = 1, 2		45
R ≡ —(CH$_2$)$_2$SH, —CS—SNa		46
P—COOCH$_2$CHCH$_2$NRR' $\quad\quad\quad\quad\quad$ OH	Adsorption of Cu^{2+}	47–49
R ≡ H, Et R' ≡ Et, (CH$_2$)$_2$NH$_2$, (CH$_2$)$_2$OH	Separation of Ag$^+$ and Ag^{2+}	50, 51
P—CONHCH$_2$NHCOCH$_2$SH	Extraction of metal ions	52
Si—(CH$_2$)$_3$—NHR R ≡ H, Me, (CH$_2$)$_2$NH$_2$	Selective extraction of U	53, 54

Structure	Application	Ref.
P—CO—N—R with OH; (a) R ≡ H, (b) R ≡ Me	Separation of Ag, Au and various transition metals	55, 56
	Determination of trace metals	57
	Chelating agent	58
PS—SO$_2$—NH—C(=NH)—NH$_2$	Selective separation of various metals	59
P— (pyridyl-oxy)	Selective separation of Cu and Ni	60
P—N$^+$—(CH$_2$)$_4$—$^+$N—⬡—P, 2Br$^-$	Selective adsorption of metal ions	7
PS—(CH$_2$)$_n$SH, n = 1	Selective ion exchanger	12, 61
n = 0	Selective for Hg^{2+}	62
		63–68
PS—CH$_2$—N(Me)—R; R ≡ (CH$_2$)$_n$COOH; n = 1, 2; R ≡ COMe	Separation of Ag, Hg, Au, Bi and Cu	69
PS—COO(CH$_2$)$_6$OCOCH$_2$SH	Separation of U, Th, Zr, Au and Pd	70
PS—(CH$_2$)$_3$CON(n-Bu)$_2$		70
PS—(CH$_2$)$_4$—NBuAc		70

Table 5.1 (*Contd.*)

Functionalized polymer	Application	Reference
(PS)—S—C(=S)—N(R')—R	Adsorption of various metal ions	
(a) R ≡ H; R' ≡ Me		71
(b) R ≡ Me; R' ≡ CH_2COOH		72
(c) R ≡ R' ≡ CH_2CH_2OH		73
(PS)—$CH_2NH(CH_2)_2$C(=S)—S—	Adsorption of Hg^{2+}	74
(PS)—CONH$(CH_2)_n$C(=O)—N(R)—OH; $n = 1, 2, 35$ R ≡ H, Me, $CH_2CONMeOH$	Selective for Fe^{3+}	58, 75
(P)—C(X OH)=CH—CO—Me		
X ≡ Ph		76, 77
X ≡ H		78
(PS)—P(OH)$_2$; (PS)—P(=O)(OH)$_2$	Adsorption of UO_2^{2+}	79
(PS)—CH_2P(=O)(OH)$_2$		80–83
(PS)—$COCH_2CO$—C_2F_5	Binding of Cu^{2+}, Ni^{2+} and UO_2^{2+}	84

Structure	Function	Ref.
(Si)—R—NHCOCH$_2$COCH$_3$	Complexing of various metal ions	85
(PS)—CH$_2$— , (Si)—(CH$_2$)$_3$— (pentane-2,4-dione groups)	Binding of Cu^{2+}, Fe^{3+} and Cr^{3+}	40
(PS)—CHMe—C=NOH , HO (β-substituted oxime)	Selective adsorption of Cu^{2+} and Mo^{4+}	86
(PS) β-diphenylglyoxime	Selective for Pd	87
(PS) ethylenediglycoldibutyl ether	Selective for Au	87
—(N—(CH$_2$)$_2$NH(CH$_2$)$_2$—)$_n$ with C(=S)S$^-$ group	Adsorption of various metal ions	88, 89
NH NH$_2$ NH NH NHCONH$_2$, COOBu-t	Selective adsorption of metal ions	90
—(NH—R′—NH—C(=S)—R—C(=S)—)$_n$	Adsorption of various metal ions	91, 92
—(O—CF$_2$CF(CF$_3$)—)—O(CF$_2$)$_2$SO$_3^-$	Selective separation of metal ions	93

Table 5.1 (*Contd.*)

Functionalized polymer	Application	Reference
$R \equiv -(CH_2)_3$; $R' \equiv -(CH_2)_4$	Selective for Hg	94, 95
$R \equiv$ [benzene ring], $R' \equiv -CO-$[benzene ring]$-CO-$; $(CH_2)_2Z-(CH_2)_2$; $Z \equiv O, S$ $-(-S-CH-CH_2)_n-$ $R \equiv$ Me; CH_2N-CHR'; $R' \equiv$ Et, Ph (Me Me)	Adsorption of Cu	96
$Z \equiv -CH_2-$, $-SO_2-$; $R \equiv -CH_2-$[benzene ring]$-CH_2-$, $-(CH_2CH_2NH)_mCH_2CH_2-$; $m \equiv 1,2$	Adsorption of various metals	97, 98

Structure	Application	Ref.
$+NH-R-NH+_n$ (Cu complex); $R = $ ... , Me, $COOR$	Selective separation of Cu^{2+}	99
Polyurethane·polyether Polyurethane·polyester	Extraction of Sn	100
(crown ether, X); $X = H, Me$; $R = $	Complexing of alkali metal ions	101–104
(crown ether), $n = 1,2$	Complexing of alkali metal ions Binding of fluorescent molecules	101–105 106, 107
$PS-CH_2OCH_2-$ (crown)	Cation binding	108

Table 5.1 (*Contd.*)

Functionalized polymer 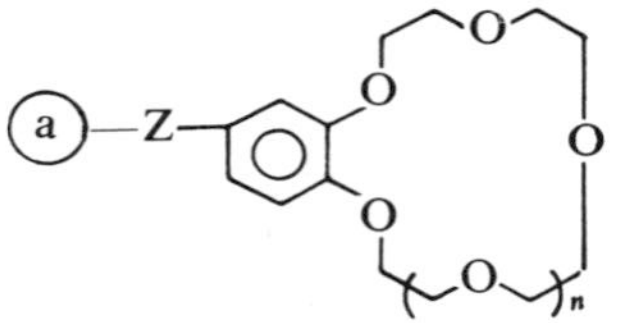	Application	Reference

(a) ≡ (PS); $n = 0, 1, 2, 4$; $Z = -CH_2-$ Binding of alkali metal ions   109,110

(a) ≡ (P); $Z = COOCH_2$, CH_2OCH_2, $CONH$, $O(CH_2)_2OCOCH=CH$; $n = 1, 2$ 110–114

(a) ≡ (Si); $n = 1$; $Z = -(CH_2)_3NHCO-$ Separation of alkali metal ions   115

$Z = -CH_2-$ 9

(a) ≡ $-(COCHNH)_m$; $n = 1$; $R = -(CH_2)_2COOCH_2-$ 116

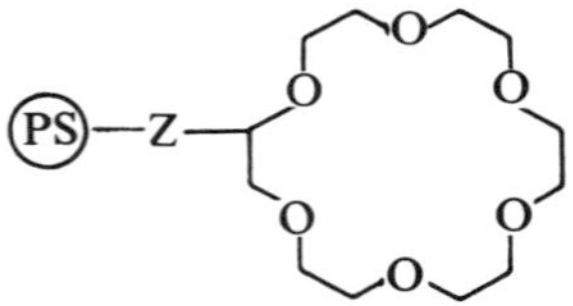

Binding of alkali metals

$Z \equiv -(CH_2)_mOCH_2-$; $m = 1, 3$ 117

$Z \equiv -CH_2NEt(CH_2)_9$ 118, 119

$Z \equiv -CH_2(NHCO(CH_2)_{10})_n-NEt(CH_2)_9$ 118

120

121, 122

123

124, 125

126

Cation binding

Complexation of transition metal anions

Binding of alkali metal ions

Selective for Cu^{2+}

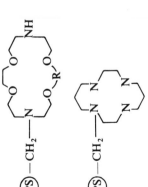

Table 5.1 (*Contd.*)

Functionalized polymer	Application	Reference
		127
$m = 1, 2$	Selective for K, Pb and Cs	128
	Binding of alkali metal ions	129–133
$R \equiv$		125

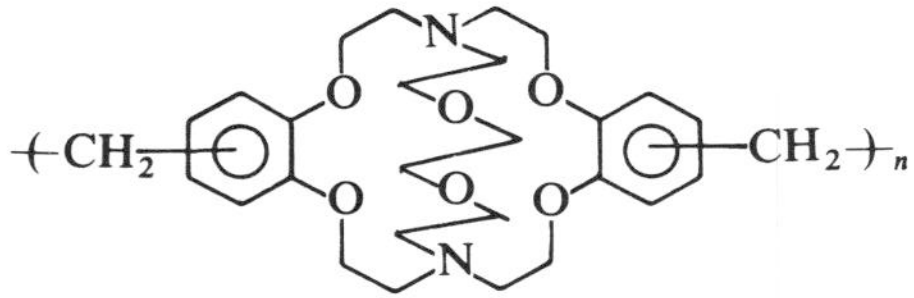

125

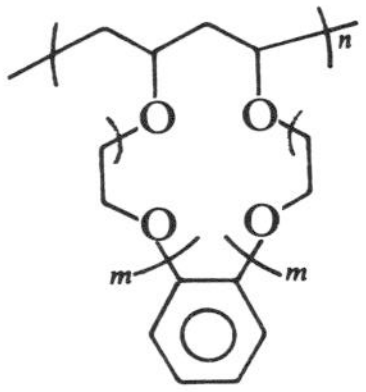

134

$(CH_2)_n$

$m = 0, 1$

(PS) —CH_2—

11

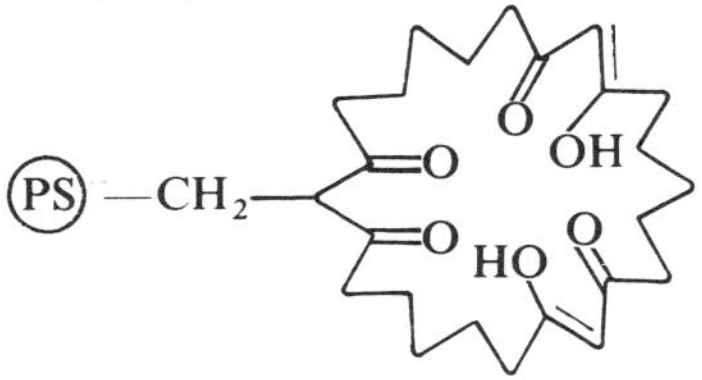

Extraction of UO_2^{2+} from seawater

119, 135, 136

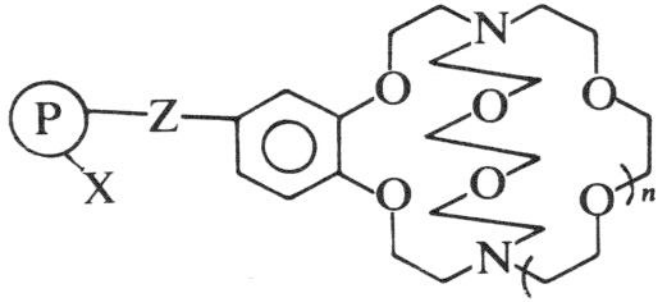

$Z \equiv -CH_2OCH_2-$, $-CH_2NH(CH_2)_9$;

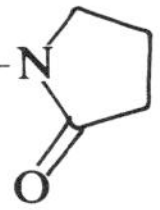

$X \equiv H$, $-N$, Ph

Table 5.1 (*Contd.*)

Functionalized polymer	Application	Reference
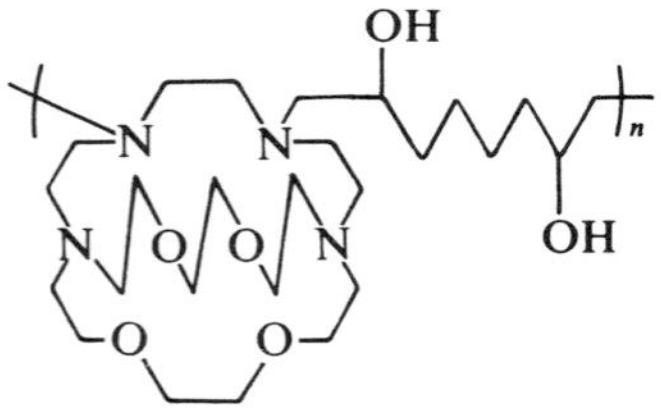	Binding of various metal ions	137
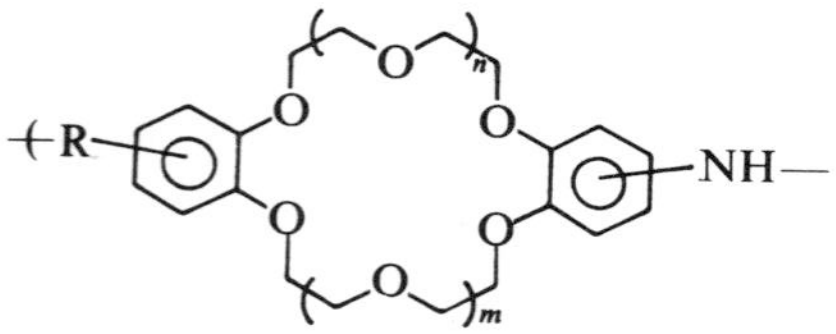		138–140
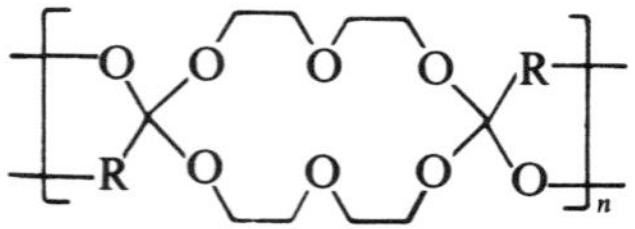		141

$n = 1, 2, 3; m = 0, 1, 2, 3$

PS — ≡ $+CH_2—CH+_n$

C — ≡ cellulose

Si — ≡ silica

a — ≡ agarose

P — ≡ $+CH_2—CH+_n$

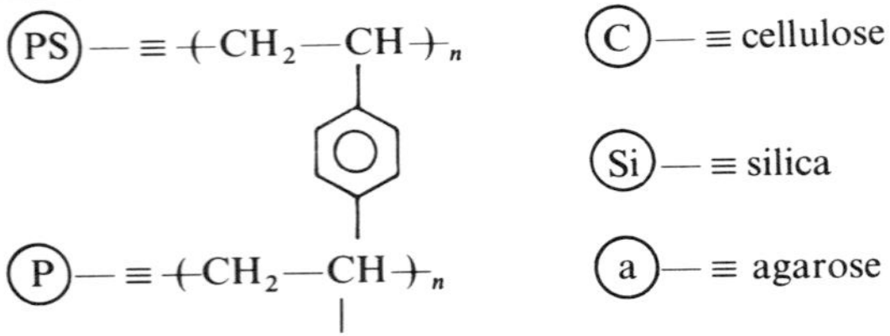

not always yield optically pure samples. In addition to the traditional methods of resolution of racemic mixtures, considerable progress has been reported on resolution by chromatographic methods using naturally occurring or synthetic polymers as stationary phases. In general, chromatographic resolutions are of two types. The first involves separation of diastereomers, formed by reaction with chiral compounds, on inactive stationary phases by gas chromatography or ion exchange chromatography. Although gas chromatography is a suitable method for the resolution of volatile diastereomers and for the quantitative analytical determination of optical purity, ion exchange chromatography can result in racemization of the enantiomers. The alternative method is the method of choice for direct resolution for preparative scales using either a chiral eluent or a chiral stationary phase. Resolution of racemic mixtures by liquid chromatography has been attempted on a wide variety of asymmetric stationary phases by various methods [146–148], as listed in Table 5.2.

5.2.1 Optically active polymers

The attachment of a large number of chiral ligands to silica gel supports has been described and used as optically active stationary phases in gas chromatography for the determination of configurations and optical purity on an analytical scale [208]. However, in recent years a number of significant advances have been made in the use of optically active polymeric materials in the resolution of racemates by liquid chromatography owing to their application in the separation of optical isomers on a preparative scale. Various chiral polar synthetic polymers are reported to achieve enhanced resolution of many racemates in non-polar eluents by the formation of temporary diastereomers through hydrogen bond formation between the enantiomers and a chiral stationary phase. Since the two adsorbates, i.e. diastereomers, are not equally stable, one enantiomer (that forming the weaker adsorbate) will pass through the column faster than the other. Chromatographic resolutions of racemic acids and bases have also been carried out on optically active basic and acidic ion exchangers respectively. For example, N-[(S)-1-phenylethyl]acrylamides and other optically active polymers (**3**) have been described as optically active adsorbents for the resolution of racemic mandelic acid and mandelamide [153, 158–161]:

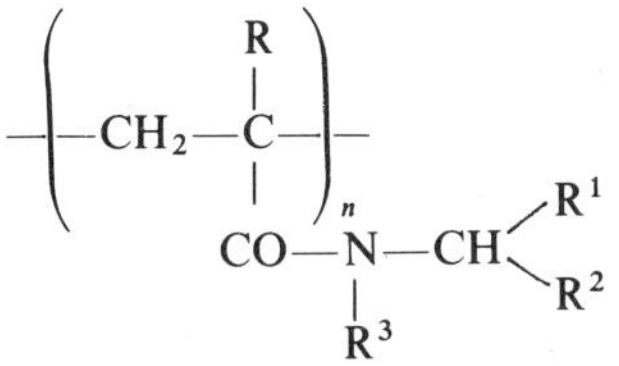

3

The optical yield was found to depend on the type and amount of crosslinking in the polymer used in the synthesis of **3**, which employs optically active ion

Table 5.2 Resolution of racemic mixtures

Functional polymer	Application	Reference
Optically active polymers		
PS—NH—(triazine ring with N≡N, Cl)—N—CHMeCHPh, Me, —OH	Separation of mandelic acid	149
PS—CH₂—N—*CH—R′ (R, Me)	Resolution of mandelic acid and mandelamide	
(a) R ≡ H; R′ ≡ Ph		150, 151, 152
(b) R ≡ Me; R′ ≡ Me, CHPhOH, CHPhOAc		153
(c) R ≡ Me; R′ ≡ Ph		150
(d) R ≡ H; R′ ≡ −CH₂NH₂		152
PS—CH₂N⁺R₂(CHMePh)Cl⁻	Resolution of mandelic acid	150
R ≡ Me	Separation of sugar anomers	154
R ≡ H, Me		
Si—(CH₂)₃NHCOCH—CHMe₂, NHAc	Resolution of amino acids	155
Si—CH₂CHMeCONHCHCONHCMe₃, CHMe₂	Separation of antipodes of NHCOCHMeOH and R—CHNH—R′, CHOH (with aromatic ring, cyclohexyl)	156

Structure	Description	Ref.
(PS)—SO$_3$—N (pyrrolidine)—COOH	Separation of α-methylbenzylamine	157
(P)—CO—N—CH*—R″ (with R, R′); R ≡ H, Me; R′ ≡ Me, COOH, COOEt, COOBu-t; R″ ≡ Me, Ph, PhCH$_2$, c-C$_6$H$_{11}$, c-C$_6$H$_{11}$CH$_2$—, p-I—C$_6$H$_4$—, p-HO—C$_6$H$_4$CH$_2$—, p-MeOCOC$_6$H$_4$CH$_2$—, CHPhOH, 1-naphthyl, (indol-3-yl)CH$_2$—	Resolution of mandelic acid, mandelamide, α-substituted 2-phenylacetamides, benzoin derivatives, N-acylamino acid esters and antimalarial drugs	153, 158–162 163
(P)—COO—CH*—CHMe—N—R′ (with R, Me); R ≡ Ph, c-C$_6$H$_{11}$; R′ ≡ H, Me, PhCH$_2$—, MeCO—, PhCO—	Resolution of mandelic acid and mandelamide	153
(P)—COO—C$_6$H$_4$—CH$_2$CH—NH—COMe (with COOEt)	Resolution of racemic mandelic acid and mandelamide	161
(P)—COO—CH*—CHMe—N$^+$Me$_3$ Cl$^-$ (with Ph)	Resolution of mandelic acid and mandelamide	153

Table 5.2 (*Contd.*)

Functional polymer	Application	Reference
(Si)—CH$_2$CH—CONHCHCONH—CMe$_3$ 　　　　|　　　| 　　　Me　　CHMe$_2$	Resolution of chiral amides	156
Ligand exchange resolution		
  R ≡ H, OH; Z ≡ —(CH$_2$)$_m$——(m = 1,3), 　—CONHCH$_2$—, —SO$_2$—; ⓐ ≡ (PS), (Si), (P)	Resolution of racemic mixtures	164–171
  Z ≡ —CH$_2$—, —SO$_2$—, —CONHCH$_2$—; R ≡ CH$_2$SH, CH$_2$SCH$_2$COOH, 　CH$_2$S(CH$_2$)$_2$NH$_2$, CH$_2$Ph, 　(CH$_2$)$_2$SMe, CH$_2$S(CH$_2$)$_2$SCH$_2$CHCOOH 　　　　　　　　　　　　　　　| 　　　　　　　　　　　　　　　NH$_2$	Resolution of racemic amino acids	168, 169 172–176

Structure	Application	Ref.
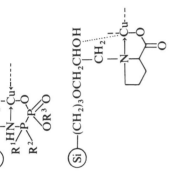	Resolution of racemic amino acids	177–179
	Separation of amino acids	180
	Resolution of tryptophan, phenyl alanine, tyrosine	181
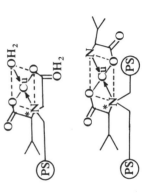	Resolution of racemic amino acids	183
	Resolution of racemic amino acids	182

Table 5.2 (*Contd.*)

Functional polymer	Application	Reference
Asymmetric cavities		
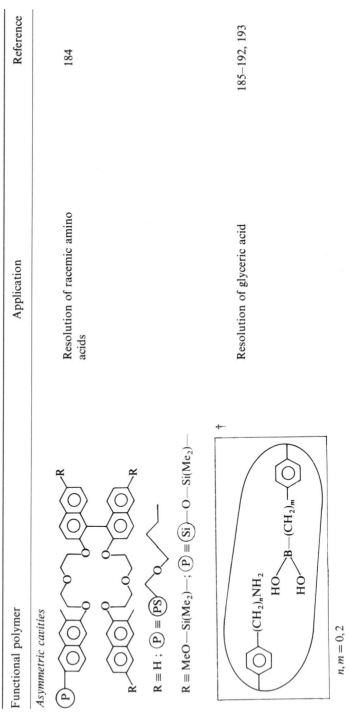	Resolution of racemic amino acids	184
	Resolution of glyceric acid	185–192, 193

Resolution of glyceric acid   194, 204

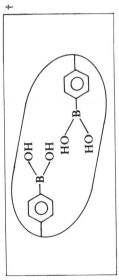

n = 0, 1

Resolution of mannopyranoside
derivatives   195, 196, 197

Resolution of 4-nitrophenyl-o-
D-mannoside   198

Resolution of mannitol   199, 197

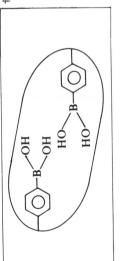

Resolution of amino acid esters
and anilides   200

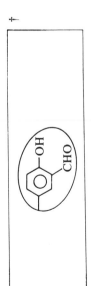

Table 5.2 (*Contd.*)

Functional polymer	Application	Reference
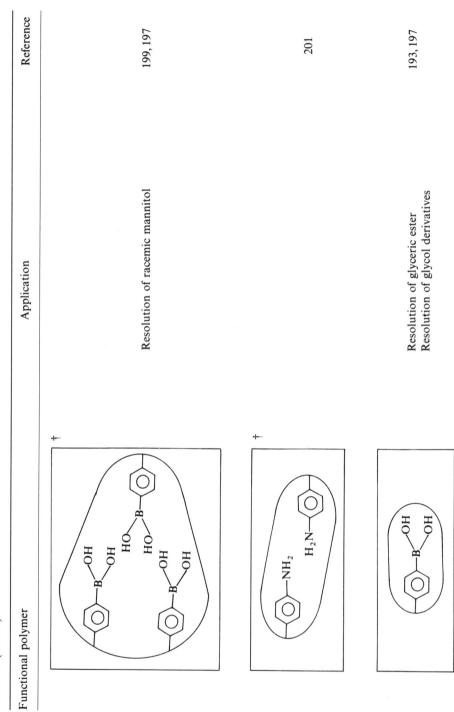	Resolution of racemic mannitol	199, 197
		201
	Resolution of glyceric ester Resolution of glycol derivatives	193, 197

Polymeric chiral charge transfer complexes

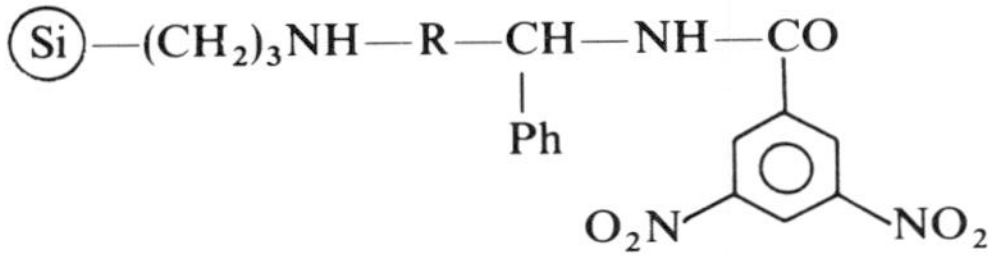

$(Si)-(CH_2)_3NH-R-CH-NH-CO-$ [Ph] dinitrophenyl (NO_2, O_2N)

$R \equiv -CO-, \ -COOCH_2-$

Separation of enantiomers of sulphoxides, lactones, alcohol derivatives, amines, amino acids, hydroxy acids and mercaptans

202

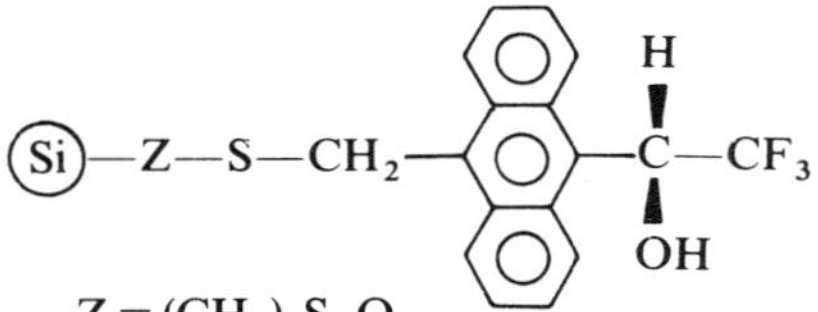

$(Si)-Z-S-CH_2-$ (anthracene with $C(CF_3)(H)(OH)$)

$Z \equiv (CH_2)_3S, \ O$

Separation of enantiomers of sulphoxides, lactones, alcohol derivatives, amines, amino acids, hydroxy acids, mercaptans

202, 203

$(P)-COOC(Ph)_3$

Resolution of many racemic mixtures such as Träger base and menthol

204

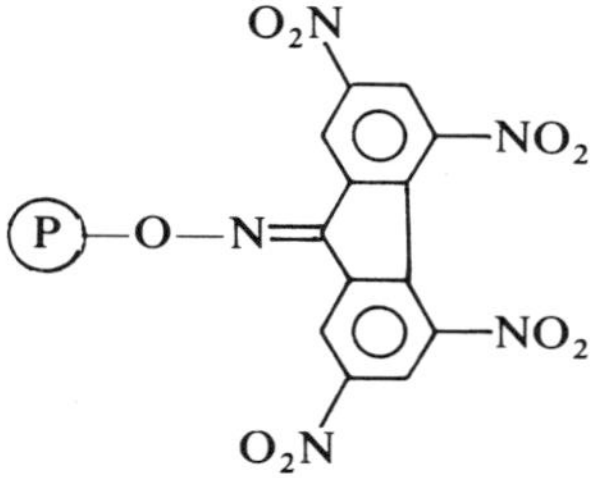

$(P)-O-N=$ (fluorenone derivative with O_2N, NO_2 groups)

Resolution of racemic helicenes

205

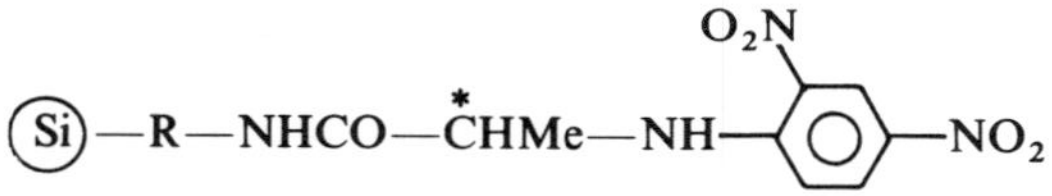

$(Si)-R-NHCO-^*CHMe-NH-$ (dinitrophenyl: O_2N, NO_2)

Resolution of racemic helicenes

206

Table 5.2 (*Contd.*)

Functional polymer	Application	Reference
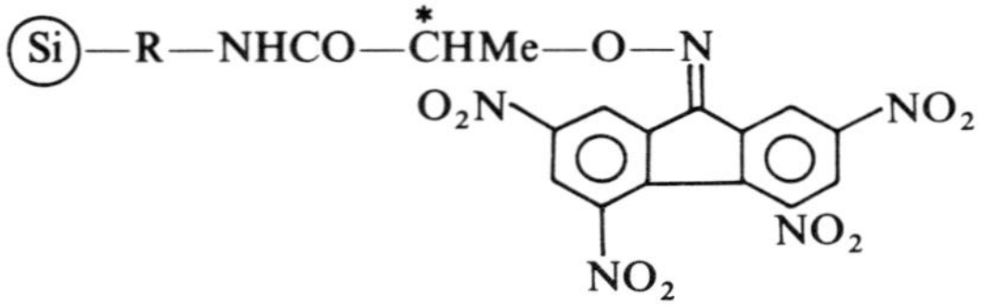 (Si)—R—NHCO—*CHMe—O—N= (structure with NO₂, NO₂, O₂N, NO₂ groups on helicene)	Resolution of racemic helicenes	207
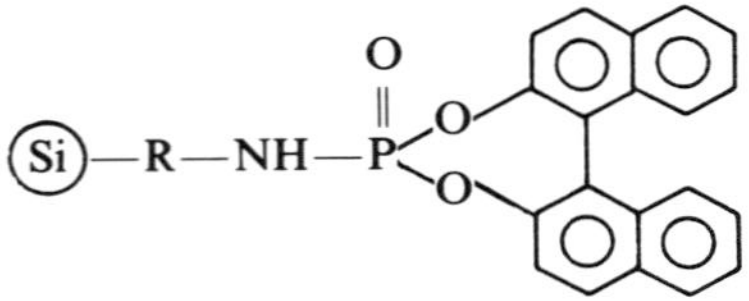 (Si)—R—NH—P(=O)(—O—)(—O—) (binaphthyl phosphate structure)	Resolution of racemic helicenes	207

$R \equiv —(CH_2)_3—, \ —(CH_2)_3NH(CH_2)_3—$

*Chiral centre
†Definitive chiral cavity with the stereoconfiguration of the template after its removal.

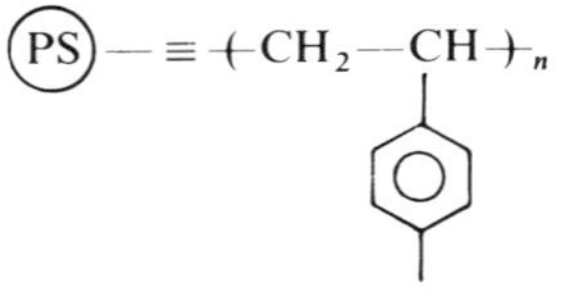

(C)— ≡ cellulose

(Si)— ≡ silica

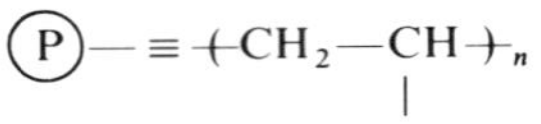

(a)— ≡ agarose

(PS)— ≡ $(CH_2—CH)_n$ —⟨benzene ring⟩—

(P)— ≡ $(CH_2—CH)_n$

exchange resins [162, 202, 204] or ion exchangers based on Sephadex [159]. Many chromatographic resolutions of racemic acids and bases have also been reported. An optically active strongly basic or weakly basic ion exchange resin such as those of L-(-)-N,N-dimethyl-1-phenylethylamine (**4b**) and of (-)-α-phenylethylamine (**4a**) respectively were found partially to resolve the optical isomers of mandelic acid [150]:

$$\text{(PS)} - \text{CH}_2\text{N} \underset{R}{\overset{\text{CHMePh}}{<}}$$

   4a R ≡ H
   4b R ≡ Me

5.2.2 Ligand exchange resolution

Other resolutions of interest are those which are carried out on asymmetric complex-forming stationary phases by means of ligand exchange chromatography [146, 147]. The principle of this type of chromatography depends on the covalent attachment of optically active ligand to an insoluble polymeric backbone, followed by complexing with a metal ion capable of undergoing cation interchange such as copper(II) or nickel(II). Interaction between the complexed stationary phase and the enantiomers to be separated occurs during the formation of coordination bonds inside the coordination sphere of the complex-forming metal ion, which thus ensures α stationary phase–substrate interaction. The generated complex must be kinetically labile and can readily decompose and reform a number of times during the chromatographic process.

Because of differences in the rates of formation and dissociation of the two diastereomeric complexes (i.e. their ability to break down and reform) and in their thermodynamic stabilities (i.e. enantioselectivity in their formation), the two enantiomers can be separated by a release from the stationary complexes at different rates of elution. Because of their ease of complexation with metal ions, ammonia or amine are usually used as eluents. The order of elution of the enantiomers depends on their nature as well as the nature of the ligand attached to the support and on the actual binding strengths of the ligand and the metal. The loading capacity of the stationary phase depends on the number of coordination sites available and can be advantageously high. Nevertheless, the method is restricted to compounds with amino and/or carboxyl groups that can also be resolved conventionally via diastereomeric salts. Most of the materials resolved by this approach have been amino acids and their derivatives.

Resolutions of racemic amino acids have been achieved successfully on columns of asymmetric resins containing L-proline [147]. After charging with Cu^{2+} (ratio of 2:1), the stationary ligands **5** from complexes **6** which contain two fixed ligands per metal ion. When the mixture of enantiomers is then added, sorption takes place and the stationary complexes **6** are converted into the mixed complexes **7**.

The two complexes of structure **7** with the D- and L-amino acids, i.e. diastereomers, differ in energy content, which affects the equilibrium between the free and complexed amino acids. The complex **7** which contains the L-amino acid is usually less stable so that it is eluted more rapidly with water, and the D-amino acid complex is only eluted with a base such as ammonia.

Amino acid derivatives have also been separated using other ligand exchange resins such as a polystyrene resin containing the (N-carboxymethyl-L-valine)-copper(II) complex **8** [183].

5.2.3   Asymmetric cavities

(a)   Polymeric crown ethers

Optically active crown ethers covalently attached to a polymeric support have been developed and used for the chromatographic separation of enantiomers of primary amine salts such as the salts of α-amino acids and their esters [184]. In this type of separation, the degree of differentiation exhibited by a host molecule for a pair of enantiomers depends on the formation of temporary diastereomeric complexes, usually held together by numerous hydrogen bonds, with the asymmetric cavity of the crown ether, in which each guest pair has a different degree of stability; this leads to their chromatographic resolution.

The chiral binaphthyl derivatives have stable configurations since the rotation around the binaphthyl axis is sterically hindered. Thus the optically pure (S, S)-dibinaphthyl-22-crown-6 covalently bonded to polystyrene resin (9a) or to silica gel (9b) shows large differences in bonding ability for enantiomers with separation factors (α) of 1.4–26.

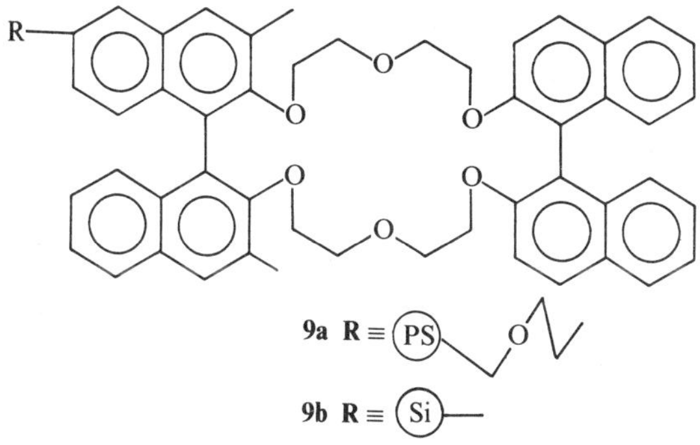

9a R ≡ (PS)

9b R ≡ (Si)—

(b)   Enzyme-analogue polymers

Highly crosslinked polymers containing chiral cavities, have recently been prepared and used for the resolution of racemic mixtures in a similar manner to enzymatic processes. The principle of this type of resolution is the preparation of enzyme-like synthetic polymers containing asymmetric cavities with functional groups fixed in exact stereospecific positions [185–192, 195, 196, 198–201]. This can be achieved by polymerizing vinyl monomers (A and B) attached to an optically active template molecule (T). Subsequent removal of T leaves behind a chiral cavity 10.

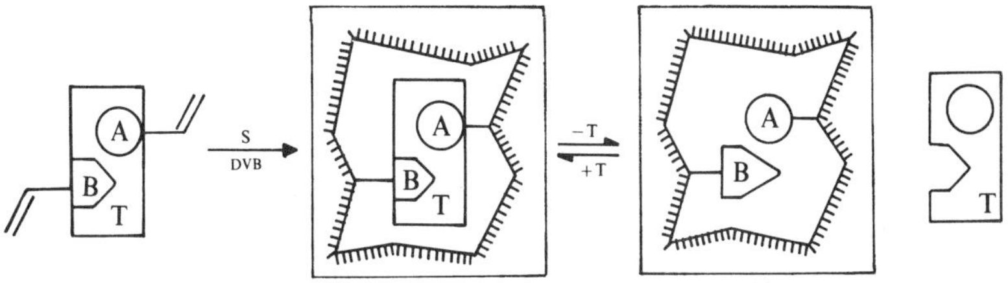

10

To stabilize the asymmetric structure of the polymer and hence to avoid the loss of the stereoregularity of the functional groups in the cavities after splitting off the template molecules, the polymer is prepared under conditions that lead to a rigid low-swellable matrix which can be achieved by using a high degree of crosslinking. In addition, the functional groups of the cavities must undergo an easily reversible interaction and cleavage reaction with the template molecules that have been used in the preparation of the cavities, analogous to the active site in natural enzymes. Asymmetric cavities prepared in this way preferentially interact with the enantiomer which has been used as a template. In this process the enantiomers are not bound adsorptively but in a reverse equilibrium reaction; thus almost complete resolution of a racemic mixture is possible at every flow rate on the column. For example [191], polymeric boronic acid with asymmetric cavities has been prepared by copolymerization of the optically active ester **11** with a very high proportion of crosslinked agent to **12** and subsequent hydrolytic cleavage of the optically active matrix **14**. From a solution of racemic methyl-α-mannopyranoside (**14a**), the polymer **13** takes up in preference the enantiomer originally polymerized-in.

Several enzyme-like polymers of this type have been used for the resolution of enantiomers of various racemic mixtures, as listed in Table 5.2.

5.2.4   Polymeric chiral charge transfer complexes

The resolution of racemic mixtures of chiral arenes, such as helicenes, which have no specific functional groups in their moities has recently been achieved by using optically active charge transfer acceptors covalently linked to polymeric supports. The different stabilities of the diastereomeric charge transfer complexes formed enable the chromatographic resolution of such racemic arenes and also the separation of mixtures of helicenes into their components. R(-)-2-(2, 4, 5, 7-tetranitro-9-fluorenylidene-aminooxy)propionic acid covalently bound to silica **15** has been used for the resolution of a series of racemic helicenes into the enantiomers [207]. Some other polymeric charge transfer acceptors used for resolution of racemic mixtures are shown in Table 5.2.

15

5.3   OTHER SEPARATIONS

A variety of other interesting separations employing specifically functionalized polymers are listed in Table 5.3. Ion exchange chromatography has been used for the separation of various ionic compounds or mixtures of ionic and non-ionic compounds such as surface-active agents [255] and has also been applied to cosmetic analysis [256]. A very interesting separation of a minor component from a complex mixture has recently been achieved in the removal of allergenic

Table 5.3 Other separations

Functionalized polymer	Application	Reference
P—OH	Selective adsorption of cystine and mercaptans	209
PS—$CH_2NMeCONH$—NH_2	Trapping aldehyde side-products in photochemical reactions	210
P—CONH—⟨ring⟩—S_xCl	Selective separation of tryptophan and tryptophan-containing peptides	211
P—$CONH(CH_2)_2NHCO$—⟨ring, S_xCl, O_2N⟩ $x = 2, 3$		
PS—$CH_2CH_2NH_2$	Separation of allergenic substituents from natural oils	212, 213
PS—$B(OH)_2$	Separation of glycolipids from natural lipids and phospholipids	214
	Separation of diol mixture	215
C—O—R—CONH—⟨ring⟩—$B(OH)_2$	Separation of nucleotides	216, 217

—R—$= -CH_2-, \ -(CH_2)_2NHCO(CH_2)_2$

Structure	Description	Page
PS—$CH_2N^+Me_3$ OH^-	Separation of N-nitrosodiethanol amine in cosmetics	218
P—$COOH$	Separation of alkylpyridines in beer	219
Si—$(CH_2)_3OCH_2$—$\begin{array}{c} X \\ OH \end{array}$ $X \equiv OH, NH_2$	Separation of polar compounds	220
Si—$(CH_2)_3NH(CH_2)_2NH_2$		220
P—pyridine–X $X \equiv H, Ph$	Separation of HCl from acid chlorides	221
P—N^+—$CH_2CONHNH_2$ Br^- or Cl^-	Separation of ketones from non-ketonic materials	222
P—OH	Separation of amine and pyridine compounds	221
PS—$CH_2N(CH_2COO^-)_2FeCl^{2+}$	Recovery of phenolic compounds from water	223
PS—CH_2—Z—CH_2CH—OH / CH_2—OH $Z \equiv O, S$	Separation of aldehyde and ketones from	224

Table 5.3 (*Contd.*)

Functionalized polymer	Application	Reference
(Si)—(CH$_2$)$_3$OCH$_2$CH—OR 　　　　　　　　｜ 　　　　　　　CH$_2$OH R ≡ H, —(CH$_2$)$_2$NMe$_2$	Separation of oligonucleotides	225
(Si)—(CH$_2$)$_3$—[uracil ring]	Separation of adenine and its derivatives	226
(Si)—CH$_2$CH$_2$NEt$_2$	Quantitative isolation of glycolipids	227
(PS)—OH	Removal of ε-caprolactam from aqueous solutions	228
(P)—COO—R—COOH R ≡ —CH$_2$—, —(CH$_2$)$_3$OCH$_2$—, —(CH$_2$)$_2$OCO(CH$_2$)$_2$—	Separation of protein	229
(PS)—SO$_3$H	Separation of amines, guanidines and hydroxycinnamic acid amides	230
	Separation of xanthines	231

Separation of gases

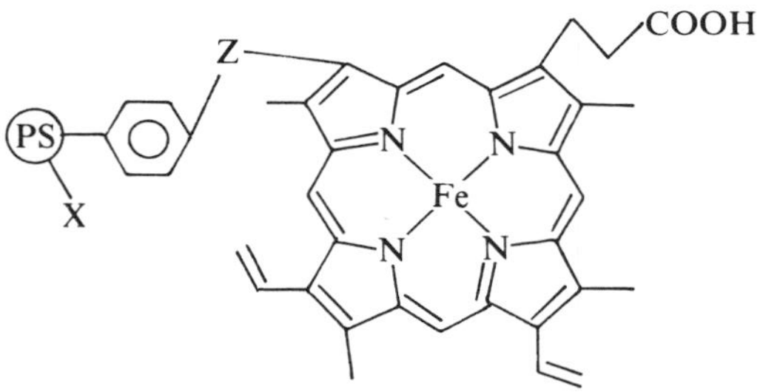

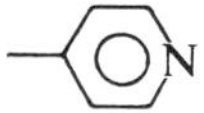

Binding of O_2

232, 233

$Z \equiv CH_2OCO(CH_2)_2 —$, $— NHCO(CH_2)_2 —$;

$X \equiv Ph$, Ph and —⟨O⟩—N

Reversible binding of O_2

234–240

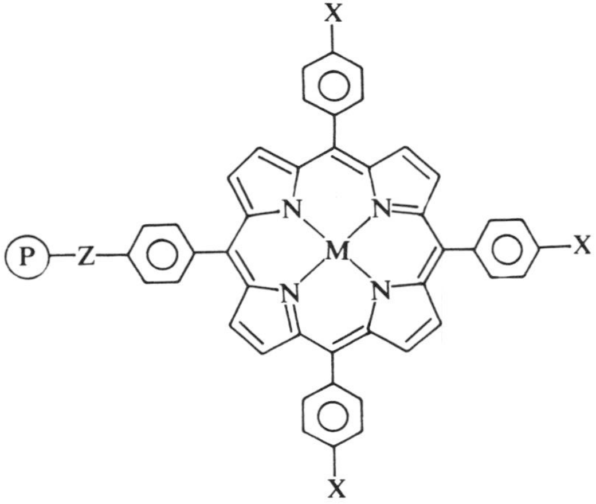

$Z \equiv — CH_2OCO—$, $—CH_2NH—$, $— NHCO—$,
$—CH_2—N=N—$; $X \equiv H$, NH_2, COOH;

$M \equiv$ Co, Fe; (P) = (PS), (Si),

Table 5.3 (*Contd.*)

Functionalized polymer	Application	Reference
	Reversible binding of O_2, CO	234, 235 239, 241
$n = 0, 1; R \equiv H, \text{(PS)}; X \equiv Ph, H$	Binding of O_2	242, 243
	Binding of O_2	244

Structure	Application	Ref.
$+CH_2-\langle\bigcirc\rangle-CH=N\diagdown\,_R\diagup N=CH-\langle\bigcirc\rangle-O)_n$ with M	Reversible binding of O_2	245
$R \equiv (CH_2)_4, \quad -\langle\bigcirc\rangle-CH_2-\langle\bigcirc\rangle-CH_2-;$		
$M \equiv V^{2+}, Mn^{2+}$		
$+CH_2C\underset{\parallel}{=}C\underset{\parallel}{\longrightarrow}_n \cdot$ (Pd or Cu), NOH NOH	Adsorption of CO, CO_2 and O_2	246
(PS)$-CH_2CN\cdot$(Pd complex)	Adsorption of olefin gases	247
(P)$-COOCH_2CH_2CN\cdot$(Pd complex)	Adsorption of olefins	247
(PS)$+PPh_2)_3\cdot CoH$	Adsorption of N_2	248, 249
(P)$-$[cyclopentadienyl]$-Mn(CO)_3$, Ph Me	Adsorption of N_2	250

Solvent purification

Structure	Application	Ref.
(PS)$-NH_2\cdot$(metal complex)	Removal of O_2 from water	251, 252
(P)$-OH$	Removal of O_2 from solutions	253

Table 5.3 (*Contd.*)

Functionalized polymer	Application	Reference
Affinity chromatography		
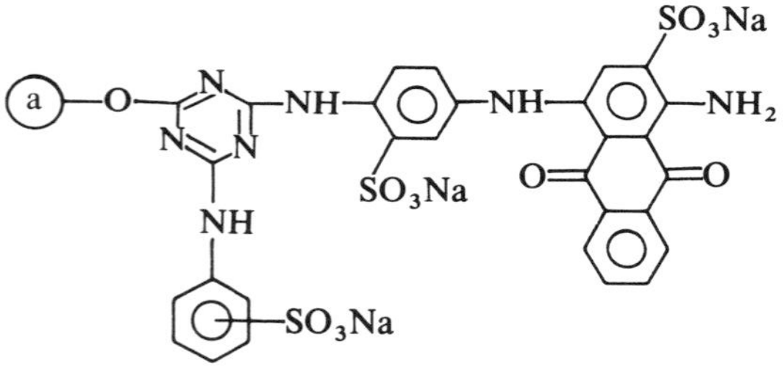	Purification of enzymes and proteins	254
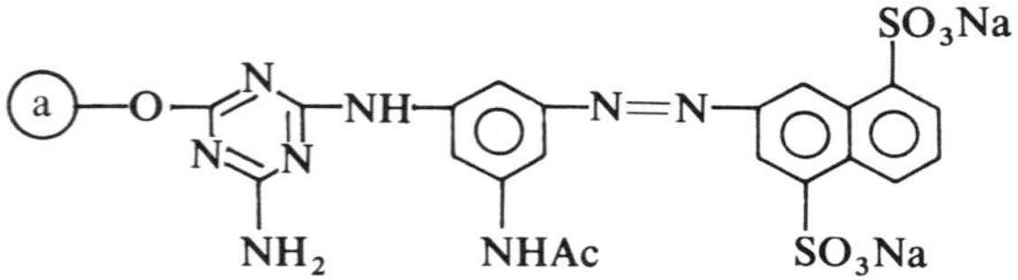		254
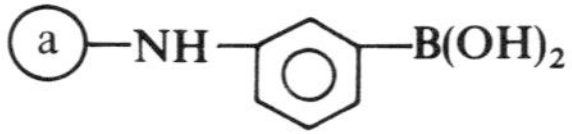	Purification of glycoproteins	254

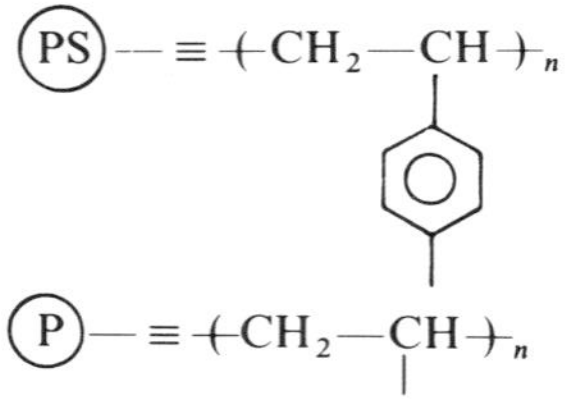

PS — $\equiv \left(\text{CH}_2\text{—CH}\right)_n$

P — $\equiv \left(\text{CH}_2\text{—CH}\right)_n$

C — $\equiv$ cellulose

Si — $\equiv$ silica

A — $\equiv$ agarose

substances from natural oils used in the cosmetic industry [212, 213]. A polystyrene resin functionalized with primary alkyl amine groups (16) reacts stoichiometrically with the allergen and traps it from the oil. Subsequent treatment of the resin with methyl iodide and sodium bicarbonate releases the free allergen 17.

The coordinating reaction of resin-bound metal ions can also be used for the adsorption of coordinating organic molecules, which can then be isolated from each other. Application of such a coordination reaction of resin-bound metal ions could lead to the effective isolation of coordinating compounds such as amino acids and amines from their mixtures or to the selective collection of a trace compound possessing coordinating ability.

5.3.1   Separation of gas molecules

Some polymer–metal complexes can absorb gas molecules such as O_2 or CO or olefins reversibly (Table 5.3). Because it is important to concentrate small molecules such as gases efficiently and to make use of them, there is increasing interest in the synthesis of stable dioxygen complexes of Fe(II)-porphyrin. It is well known that the iron(II) or cobalt(II) complexes possessing a large π-conjugate equatorial ligand, such as a tetrapyrrole ring, and a suitable axial base show an ability to bind molecular oxygen. However, these complexes cannot bind molecular oxygen reversibly because of the oxidation reaction. Much work has been directed towards avoiding this bimolecular reaction between the dioxygen complex and excess Fe(II)-porphyrin. One method is to incorporate Fe(II)-porphyrin compounds into a polymer or to bind them at the surface of a solid support. In the oxygenation of Fe(II)-porphyrin complexes covalently and separately bound to polymer matrices, the oxidation via dimerization can be minimized by the dilution effect of the macromolecular chain. For example, the Fe(II)-porphyrin and Co(II)-porphyrin complexes attached to a rigid modified polymer or silica gel, as haemoglobin models, are capable of absorbing molecular oxygen reversibly at low temperatures.

The fixation of nitrogen by polymer–metal complexes has also been studied. However, in nitrogen complexation the polymer–metal complexes exhibit some different properties and reactivities from those of low molecular weight complexes [257, 258]. The polymer system was found to be highly active due to the stability of the nitrogen complex because of the locally high concentration of the active sites, i.e. the metal complex. Other polymer–metal complexes have been used to adsorb CO and olefin reversibly as shown in Table 5.3.

5.3.2   Solvent purification

Functionalized polymers have been used in the purification of some solvents, depending on the solvent and impurity. Several analytical tests, particularly those involving low level aldehyde and peroxide analysis, require solvents which are essentially aldehyde free. For example, to offer convenient and effective removal of borohydride-reducible impurities, polymer-bound borohydride was used to purify ethanol and the distillation step was no longer necessary [259].

5.3.3 Affinity chromatography

In addition to adsorption chromatography, other chromatographic separations based on synthetic polymers as stationary phases are of interest. These include gel filtration, gel permeation and ion exchange chromatography. However, separation in such chromatographic methods is achieved on the basis of the sizes, shapes or overall charge of the molecules in a mixture, which are often not sufficiently different for separation in some cases, e.g. the separation of mixtures of proteins.

The technique of affinity chromatography [260–273] has been applied to almost every area of biochemistry for the highly specific separation, isolation or purification of biologically active materials, e.g. enzymes, antigens, antibodies, from a complex mixture. This approach is based on the unique biological property of many proteins to bind ligands specifically and reversibly. A selective adsorbent is prepared by covalently attaching to an insoluble inert support a ligand that is capable of interacting specifically with the desired biologically active molecules to be separated. In principle, only substrate with appreciable affinity for the ligand will be retained on the support, while others will pass through. Thus a single enzyme contained in a complex mixture will attach to a polymer to which a specific inhibitor is bound. The substrate–polymer complex is easily separated from the other components of the mixture and the specifically adsorbed substrate can then be released from the polymer and eluted by altering the composition of the solvent to favour dissociation.

Affinity chromatography has enormous potential utility as a separation method in many fields of biochemistry and medicine where any particular ligand interacts specifically with a biomolecule. For example, specific adsorbents can be used to purify enzymes, antibodies, nucleic acids and drug- or hormone-binding receptor proteins. Furthermore, the technique can be employed for concentrating dilute protein solutions, for separating denatured from

18

biologically active forms of proteins, and for the resolution of protein components resulting from specific chemical modifications of purified protein.

A recent trend in affinity chromatography involves the use of completely synthetic immobilized ligands instead of natural ligands. An example of such a synthetic ligand is the triazinyl dye derivative **18**, which has been employed for the purification of enzymes and proteins by ionic and apolar interactions with specific surface sites on the proteins [254]. However, because the dye-ligand is not a true biological ligand, the interaction with proteins is not usually completely specific.

The successful application of affinity chromatography depends on the following.

1. The nature of the stationary phase is important. An ideal insoluble carrier should possess the following properties: it must interact very weakly with substrate to minimize its non-specific adsorption; it should exhibit good flow properties; it must possess chemical groups to allow the chemical linkage of a variety of ligands; it must be mechanically and chemically stable. In addition to polystyrene derivatives, various hydrophilic polymers derived from polysaccharide, polyacrylamide and organic derivatives of silica have been used as insoluble carriers for affinity chromatography.
2. Another factor is the stereochemical alternation of the ligand resulting from its covalent linkage to the carrier.
3. Spacer groups are important in the determination of the accessibility of the coupled ligand to the substrate. Since the direct coupling of the ligand to the matrix backbone exerts steric hindrance for the ligand–substrate interaction, the ligand must be sufficiently distant from the backbone of the solid support, as illustrated in Fig. 5.1.
4. The steric properties and flexibility of the spacer group are relevant.
5. The chemical bonds must be stable.

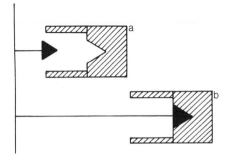

Fig. 5.1 Ligand–enzyme interaction: (a) the distance of ligand from backbone is short to permit ligand–enzyme interaction; (b) ligand attached to backbone via spacer group allow ligand–enzyme interaction.

6. The experimental conditions for washing and elution must be selected carefully.

REFERENCES

1. Bolto, R.A. (1980). *M. Macromol. Sci. Chem.*, **A-14**, 107.
2. Nicless, G. and Marshall, G.R. (1964) *Chromatogr. Rev.*, **6**, 154.
3. Blasius, E. and Brozio, B. (1967) Chelating ion-exchange resins. In *Chelates in Analytical Chemistry*, vol. 1, (eds H.A. Flaschka and A.J. Barnard), Marcel Dekker, New York, p. 49.
4. Tsuchida, E. and Nishide, H. (1977) *Adv. Polym. Sci.*, **24**, 1.
5. Kaneko, M. and Tsuchida, E. (1981) *J. Polym. Sci., Macromol. Rev.*, **16**, 397.
6. Melby, L.R. (1975) *J. Am. Chem. Soc.*, **97**, 4044.
7. Nishide, H., Deguchi, J. and Tsuchida, E. (1976) *Chem. Lett.*, 169.
8. Nishide, H. and Tsuchida, E. (1976) *Makromol. Chem.*, **177**, 2295.
9. Blasius, E., Janzen, K.P., Klein, W., Klotz, H., Nguyen, V.B., Tien, T.N., Pfeiffer, R., Scholten, G., Simon, H., Stockmemer, H. and Toussant, A. (1980) *J. Chromatogr.*, **201**, 147.
10. Smid, J. (1981) *Makromol. Chem., Suppl.*, **5**, 203.
11. Tabushi, I., Kobuke, Y., and Nishiya, T., (1979) *Nature*, **280**, 665.
12. Parrish, J.R. (1956) *Chem. Ind.*, (18), 137.
13. Pennington, L.D. and Williams, M.R. (1959) *Ind. Eng. Chem.*, **51**, 759.
14. Patchornik, A., Kalir, R., Fridkin, M. and Warshawasky, A. (1976) US Patent 3 974 110.
15. Vernon, F. and Nyo, K.M. (1978) *Sep. Sci. Technol.*, **13** (3), 263.
16. Vernon, F. (1979) *Hydrometallurgy*, **4**, 147.
17. Gruber, H. (1981) *Monatsheft. Chem.*, **112**, 445.
18. Manecke, G. and Schlegel, R. (1976) *Makromol. Chem.*, **177**, 3191; (1978) *Makromol. Chem.*, **179**, 19.
19. Grasshoff, J.M., Reid, J.L. and Taylor, L.D. (1978) *J. Polym. Sci., Polym. Chem. Ed.*, **16**, 2401.
20. Sugii, A., Ogawa, N. and Hashizume, H. (1980) *Talanta*, **27**, 627.
21. Seshadri, T. and Kettrup, A. (1982) *Z. Anal. Chem.*, **310**, 1.
22. Burba, P., Gleitsmann, B. and Lieser, K.M. (1978). *Z. Anal. Chem.*, **289**, 28.
23. Burba, P. and Lieser, K.H. (1977) *Angew. Makromol. Chem.*, **64**, 197.
24. Djamal, M.G. and Lieser, K.H. (1983) *Angew Makromol. Chem.*, **113**, 129.
25. Ghosh, J.P. and Das, H.R. (1981) *Talanta*, **28**, 274.
26. Ghosh, J.P., Pramanick, J. and Das, H.R. (1981) *Talanta*, **28**, 957.
27. Sugii, A. and Ogawa, N. (1979) *Talanta*, **26**, 970.
28. Gruber, H. (1981) *Monatsheft. Chem.*, **112**, 587.
29. Jones, K.C. and Grinstead, R.R. (1977) *Chem. Ind.*, 637.
30. Jones, K.C. and Pyper, R.A. (1979) *J. Met.*, **31**, 19.
31. Grinstead, R.R. (1979) *J. Met.*, **31**, 13.
32. Burba, P., Griesbach, M. and Lieser, K.H. (1977) *Z. Anal. Chem.*, **284**, 257.
33. Laverty, J.J. and Gardlund, Z.G. (1971) *J. Polym. Sci.*, A-1, **9**, 243.
34. Herring, R. (1967)*Chelatbildende Ionenaustauscher*, Akademic Verlag, Berlin.
35. Blasius, E. and Bock, I. (1964) *J. Chromatogr.*, **14**, 244.
36. Morris, L.R., Mock, R.A., Marshall, C.A. and Howe, J.H. (1959) *J. Am. Chem. Soc.*, **81**, 377.
37. Blasius, E. and Brozio, B. (1965) *J. Chromatogr.*, **18**, 572.
38. Morris, L.R. (1959) US Patents 2 875 162, 2 888 441.
39. Bohm, R. (1958) Ger. Patent 1 034 360.

40. Yeh, H.C., Eichinger, B.E. and Andersen, N.H. (1981) *Polym. Prepr., Am. Chem. Soc., Div. Polym. Chem.*, **22** (2), 184.
41. Gregor, H.P. (1954) *Angew. Chem.*, **66**, 143.
42. Chaikina, Y.A., Gal'braikh, L.S. and Rogovin, Z.A. (1965) *Polym. Sci. USSR*, **7**, 2212.
43. Saegusa, T., Kobayashi, S. and Yamada, A. (1975) *Macromolecules*, **8**, 390.
44. Shambhu, N.B., Theodorakis, M.C. and Digenis, G. (1977) *J. Polym. Sci.*, *A-1*, **15**, 525.
45. Saegusa, T., Kobayashi, S. and Yamada A. (1977) *J. Appl. Polym. Sci.*, **21**, 2481.
46. Saegusa, T., Kobayashi, S., Hayashi, K. and Yamada, A. (1978) *Polym. J.*, **10**, 403.
47. Kalalova, E., Radova, Z., Kalal, J. Svec, F. (1977) *Eur. Polym. J.*, **13**, 287.
48. Kalalova, E., Radova, Z. (1977) *Eur. Polym. J.*, **13**, 293.
49. Kalalova, E., Radova, Z., Ulbert, K., Kalal, J. and Svec, F. (1977) *Eur. Polym. J.*, **13**, 299.
50. Svec, F., Kalalova, E., Tlustakova, M. and Kalal, J. (1980) *Angew. Makromol. Chem.*, **92**, 133.
51. Schmidt, E.M., Inczedy, J., Laki, Z. and Szabadka, O. (1980) *J. Chromatogr.*, **201**, 73.
52. Deratani, A. and Sebille, B. (1981) *Anal. Chem.*, **53**, 1742.
53. Northcott, S.E. and Leyden, D.E. (1981) *Anal. Chim. Acta*, **126**, 117.
54. Fritz, J.S. and King, J.N. (1976) *Anal. Chem.*, **48**, 570.
55. Vernon, F. and Zin, W.M. (1981) *Anal. Chim. Acta*, **123**, 309.
56. Vernon, F. and Eccles, H. (1976) *Anal. Chim. Acta*, **82**, 369; **83**, 187.
57. Colella, M.B., Siggia, S. and Barnes, R.M. (1980) *Anal. chem.*, **52**, 967, 2347.
58. Winston, A. and McLaughlin, G.R. (1976) *J. Polym. Sci., Polym. Chem. Ed.*, **14**, 2155.
59. Goldstein, S., Gulko, A. and Schmuckler, G. (1972) *Isr. J. Chem.*, **10**, 893.
60. Sugii, A., Ogawa, N., Iinuma, Y. and Yamamura, H. (1981) *Talanta*, **28**, 551.
61. Gregor, H.P., Dolar, D. and Hoeschele, G.K. (1955) *J. Am. Chem. Soc.*, **77**, 3675.
62. Braun, D. (1960) *Chim. Aarau.*, **14**, 24.
63. Hering, R. (1963) *Z. Chem.*, **3**, 30, 69, 108, 153, 233.
64. Kuhn, G. and Hering, R. (1965) *Z. Chem.*, **5**, 316.
65. Hering, R. (1965) *Z. Chem.*, **5**, 29, 113, 149, 195, 402.
66. Hering, R. (1961) *J. Prakt. Chem.*, **14**, 285.
67. Hering, R., Kruger, W. and Kuhn, G. (1962) *Z. Chem.*, **2**, 374.
68. Pohlaudt, C. and Fritz, J.S. (1979) *J. Chromatogr.*, **176**, 189.
69. Moyers, E.M. and Fritz, J.S. (1976) *Anal. Chem.*, **48**, 117.
70. Orf, G.M. and Fritz, J.S. (1978) *Anal. Chem.*, **50**, 1328.
71. Nakagawa, T., Oishi, N. and Fujiwara, Y. (1976) *J. Appl. Polym. Sci.*, **15**, 745.
72. Nakagawa, T., Taniguchi, J. and Fujiwara, Y. (1976) *J. Appl. Polym. Sci.*, **15**, 733.
73. Hiratani, K., Matsumoto, Y. and Nakagawa, T. (1978) *J. Appl. Polym. Sci.*, **22**, 1787.
74. Kobayashi, N., Osawa, A., Shimizu, K. and Fujisawa, T. (1977) *J. Polym. Sci., Polym. Lett. Ed.*, **15**, 329.
75. Varaprasad, D.V.P.R., Rosthauser, J. and Winston, A. (1984) *J. Polym. Sci., Polym. Chem. Ed.*, **22**, 2131.
76. Hoeschele, G.K., Andelman, J.B. and Gregor, H.P. (1958) *J. Phys. Chem.*, **62**, 1239.
77. Geisa, R.C., Donaruma, L.G. and Tomie, E.A. (1963) *J. Appl. Polym. Sci.*, **7**, 1515.
78. Teyssie, P. (1963) *Makromol. Chem.*, **66**, 133.
79. Egawa, H., Nonaka, T. and Ikari, M. (1984) *J. Appl. Polym. Sci.*, **29**, 2045.
80. Kennedy, J. and Davies, R.V. (1956) *Chem. Ind.*, 378.
81. Kennedy, J. and Ficken, G.E. (1958) *J. Appl. Chem.*, **8**, 465.
82. Kennedy, J., Lane, E.S. and Robinson, B.K. (1958) *J. Appl. Chem.*, **8**, 459.
83. Kennedy, J. and Wheeler, V.J. (1959) *Anal. Chim. Acta*, **20**, 412; *Chem. Ind.*, 1577.
84. Philips, H., Kinstle, J.F. and Adcock, J.L. (1981) *J. Polym. Sci., Polym. Chem. Ed.*, **19**, 175.

85. Seshadri, T. and Kettrup, A. (1979) *Z. Anal. Chem.*, **296**, 247.
86. King, J.N. and Fritz, J.S. (1978) *J. Chromatogr.*, **153**, 507.
87. Warshawsky, A. (1974) *Talanta*, **21**, 624, 962.
88. Barnes, J.H. and Esslemont, G.F. (1976) *Makromol. Chem.*, **177**, 307.
89. Kida, S., Hirano, S., Ando, F. and Nonaka, Y. (1977) *Nippon Kagaku Kaishi*, 915.
90. Guilbault, L.J., Murano, M. and Harwood, H.J. (1973) *J. Macromol. Sci. Chem.*, **7**, 1065.
91. Leon, N.H. and Broadbent, D. (1974) *J. Polym. Sci., C*, **12**, 693.
92. Kobayashi, N., Osawa, A., Shimizu, K., Hayashi, Y., Kimoto, M. and Fujisawa, T. (1977) *J. Polym. Sci., Polym. Lett. Ed.*, **15**, 137.
93. Yeager, H.L. and Steck, A. (1979) *Anal. Chem.*, **51**, 862.
94. Nyssen, G.A., Jones, M.M., Jeringan, J.D., Harbison, R.D. and MacDonald, J.S. (1977) *J. Inorg. Nucl. Chem.*, **39**, 1889.
95. Jones, M.M., Coble, H.D., Pratt, T.H.J. and Harbison, R.D. (1979) *J. Inorg. Nucl. Chem.*, **37**, 2409.
96. Huguet, J. and Vert, M. (1976) *J. Polym. Sci., A-1*, **14**, 1257.
97. Marvel, C.S. and Tarkoy, N. (1957) *J. Am. Chem. Soc.*, **79**, 6000; (1958) *J. Am. Chem. Soc.*, **80**, 832.
98. Bailar, J.C. and Goodwin, H.A. (1967) *J. Am. Chem. Soc.*, **83**, 2467.
99. Donaruma, L.G., Kitch, S., Walsworth, G., Edzwald, J.K., Depinto, J.V., Maslyn, M.J. and Niles, R.A. (1981) *Polym., Prepr., Am. Chem. Soc., Div. Polym. Chem.*, **22** (1), 149.
100. Lo, S.V.K. and Chow, A. (1981) *Talanta*, **28**, 157.
101. Koplow, S., Esch, T.E.H. and Smid, J. (1971) *Macromolecules*, **4**, 359; (1973) *Macromolecules*, **6**, 133.
102. Takaki, U. and Smid, J. (1974) *J. Am. Chem. Soc.*, **96**, 2588.
103. Smid, J., Shah, S., Wong, L. and Hurley, J. (1975) *J. Am. Chem. Soc.*, **97**, 5932.
104. Wong, L. and Smid, J. (1977) *J. Am. Chem. Soc.*, **99**, 5637.
105. Kopolow, S., Machacek, Z., Takaki, U. and Smid, J. (1973) *J. Macromol. Sci. Chem.*, **A-7**, 1015.
106. Kimura, K. and Smid, J. (1981) *Makromol. Chem. Rapid Commun.*, **2**, 235.
107. Jaycox, G.D. and Smid, J. (1981) *Makromol. Chem. Rapid. Commun.*, **2**, 299.
108. Tomoi, M., Abe, O., Ikeda, M., Kihara, K. and Kakiuchi, H. (1978) *Tetrahedron Lett.*, 3031.
109. Warshawsky, A. and Kahana, N. (1980) *Polym. Prepr., Am. Chem. Soc., Div. Polym. Chem.*, **21** (2), 114.
110. Blasius, E., Jangen, K.P., Adrion, W., Klautke, G., Lorscheider, R., Maurer, P.G., Nguyen, V.G., Nguyen, T., Scholten, G. and Stockerer, J. (1977) *Z. Anal. Chem.*, **284**, 337.
111. Kimura, K., Maeda, T. and Shono, T. (1981) *Makromol. Chem.*, **182**, 1579.
112. Varma, A.J., Majewicz, T. and Smid, J. (1979) *J. Polym. Sci., Polym. Chem. Ed.*, **17**, 1573.
113. Kimura, K., Maeda, T. and Shono, T. (1978) *Anal. Lett.*, **A-11**, 821; (1979) *Talanta*, **20**, 945; (1979) *Z. Anal. Chem.*, **298**, 363.
114. Shira, M., Orikata, T. and Tanaka, M. (1983) *Makromol. Chem. Rapid Commun.*, **4**, 65.
115. Kimura, K., Nakajima, M. and Shono, T. (1980) *Anal. Lett.*, **13**, (A9), 741.
116. Anzai, J.I., Ueno, A., Suzuki, Y. and Osa, T. (1982) *Makromol. Chem. Rapid. Commun.*, **3**, 55.
117. Manecke, G. and Kramer, A. (1981) *Makromol. Chem.*, **182**, 3017.
118. Molinari, H., Montanari, F. and Tundo, P. (1977) *J. Chem. Soc. Chem. Commun.*, 639.
119. Cinquini, M., Colonna, S., Molinari, H. and Montanari, F. (1976) *J. Chem. Soc. Chem. Commun.*, 394.

120. Djamali, M.G., Burba, P. and Lieser, K.H. (1980) *Angew. Makromol. Chem.*, **92**, 145.
121. Warshawsky, A., Kalir, R., Deshe, A., Berkovitz, H. and Patchornik, A. (1979) *J. Am. Chem. Soc.*, **101**, 4249.
122. Warshawsky, A. (1979) *Isr. J. Chem.*, **18**, 318.
123. Waddell, T.G. and Leyden, D.E. (1981) *J. Org. Chem.*, **46**, 2406.
124. Blasius, E., Janzen, K.P. and Neumann, W. (1977) *Mikrochim. Acta*, 279.
125. Blasius, E. and Maurer, P.G. (1977) *Makromol. Chem.*, **178**, 649.
126. Tabushi, I., Kuroda, H. and Kato, H. (1976) *Polym. Prepr. Jpn*, **25**, 26.
127. Maeda, T., Ouchi, M., Kimura, K. and Shono, T. (1981) *Chem. Lett.*, 1573.
128. Blasius, E., Adrian, W., Jangen, K.P. and Kautke, G. (1974) *J. Chromatogr.*, **96**, 89.
129. Gramain, P. and Frere, Y. (1979) *Macromolecules*, **12**, 1038.
130. Bormann, S., Brossas, J., Franto, E., Gramain, P., Kirch, M. and Lehin, J.M. (1975) *Tetrahedron*, **31**, 2791.
131. Mathias, L.J. and Al-Jumah, K. (1980) *J. Polym. Sci., Polym. Chem. Ed.*, **18**, 2911.
132. Cho, I. and Chang, S.K. (1981) *Makromol. Chem. Rapid Commun.*, **2**, 155.
133. Shinkai, S., Nakaji, T., Nishida, Y., Ogawa, T. and Manake, O. (1980) *J. Am. Chem. Soc.*, **102**, 5860.
134. Butler, G.B. and Lien, Q.S. (1981) *Polym. Prepr., Am. Chem. Soc., Div. Polym. Chem.*, **22** (1), 54.
135. Manecke, G. and Reuter, P. (1981) *Makromol. Chem.*, **182**, 1973.
136. Gokel, G.W. and Durst, H.D. (1976) *Synthesis*, 168.
137. Gramain, P. and Frère, Y. (1981) *Makromol. Chem. Rapid. Commun.*, **2**, 161.
138. Frost, L.W. (1976) US Patent 3 956 136.
139. Volkova, M.S., Kiselva, T.M. and Koton, M. (1977) *Vysokomol. Soedin., Ser. B*, **19**, 743; (1978) *Chem. Abstr.*, **88**, 38183-s.
140. Ohnishi, S., Nogami, T. and Mikawa, H. (1983) *Polym. J.*, **15**, 245.
141. Krespan, C.G. (1976) US Patent 3 952 105.
142. Sekor, R.M. (1963) *Chem. Rev.*, **63**, 297.
143. Boyle, P.H. (1971) *Q. Rev., Chem. Soc.*, **25**, 323.
144. Wilen, S.H. (1973) *Topics in Stereochemistry*, Vol. 6, Wiley Interscience, New York and London, p. 107.
145. Audebert, R. (1979) *J. Liq. Chromatogr.*, **2**, 1063.
146. Krull, I.S. (1978) *Adv. Chromatogr.*, **16**, 175.
147. Davankov, V.A. (1980) *Adv. Chromatogr.*, **18**, 139; (1982) *Pure Appl. Chem.*, **54**, 2159.
148. Blaschke, G. (1980) *Angew. Chem., Int. Ed. Engl.*, **19**, 13.
149. Manecke, G. and Lamer, W. (1965) *Naturwissenschaften*, **52**, 539.
150. Lott, J.A. and Rieman, W. (1966) *J. Org. Chem.*, **31**, 561.
151. Stevens, T.S. and Lottt, J.A. (1968) *J. Chromatogr.*, **34**, 480.
152. Kurganov, A.A., Zhuchkova, L.Y. and Davankov, V.A. (1979) *Makromol. Chem.*, **180**, 2101.
153. Blaschke, G. (1974) *Chem. Ber.*, **107**, 237.
154. Rammas, O. and Sammelson, O. (1974) *Acta Chem. Scand., Ser. B*, **28**, 955.
155. Mara, S. and Dobashi, A. (1979) *J. Liq. Chromatogr.*, **2**, 883.
156. Franck, H., Nicholson, G.J. and Bayer, E. (1977) *J. Chromatogr. Sci.*, **15**, 174; (1978) *Angew. Chem., Int. Ed. Engl.*, **17**, 363; (1978) *J. Chromatogr.*, **146**, 197.
157. Knudsen, G.A., Franzus, B. and Surridge, J.H. (1969) *Chem. Commun.*, 35.
158. Blaschke, G. and Donow, F. (1975) *Chem. Ber.*, **108**, 1188, 2792.
159. Schwanghart, A.D., Beckmann, W. and Blaschke, G. (1977) *Chem. Ber.*, **110**, 778.
160. Blaschke, G., Kraft, H.P. and Schwanghart, A.D. (1978) *Chem. Ber.*, **111**, 2732.
161. Blaschke, G. and Schwanghart, A.D. (1976) *Chem. Ber.*, **109**, 1967.
162. Ihara, Y., Nakano, H., Koga, J. and Kuroki, N. (1972) *J. Polym. Sci., Polym. Chem. Ed.*, **10**, 3569.

163. Lefebure, B., Audebert, R. and Quivoron, C. (1969) *Isr. J. Chem.*, **15**, 69.
164. Rogozhin, S.V. and Davankov, V.A. (1971) *J. Chem. Soc. Chem. Commun.*, 490.
165. Davankov, V.A. and Rogozhin, S.V. (1971) *J. Chromatogr.*, **60**, 280; (1974) *J. Chromatogr.*, **91**, 493.
166. Davankov, V.A., Rogozhin, S.V., Semechkin, A.V. and Sachkova, T.P. (1973) *J. Chromatogr.*, **82**, 359.
167. Davankov, V.A. and Zolotarev, Y.A. (1978) *J. Chromatogr.*, **155**, 285, 295, 303.
168. Zolotarev, Y.A., Mayasoedov, N.F., Penkina, V.I., Petrenik, O.R. and Davankov, V.A. (1981) *J. Chromatogr.*, **207**, 63.
169. Zolotarev, Y.A., Mayasoedov, N.F., Penkina, V.I., Dostovolov, I.N., Petrenik, O.V. and Davankov, V.A. (1981) *J. Chromatogr.*, **207**, 231.
170. Boue, J., Audebert, R. and Quivoron, C. (1981) *J. Chromatogr.*, **204**, 185.
171. Sugden, K., Hunter, C. and Jones, G.L. (1980) *J. Chromatogr.*, **192**, 228.
172. Petit, M.A. and Jozefonvicz, J. (1977) *J. Appl. Polym. Sci.*, **21**, 2585.
173. Vesa, V. (1973) *Zh. Obshch. Khim.*, **42** (12), 2780; (1973) *Chem. Abstr.*, **78**, 136637-f.
174. Manecke, G. and Lamer, W. (1972) *Angew. Makromol. Chem.*, **24**, 51.
175. Yamskov, I.A., Berezina, B.B. and Davankov, V.A. (1978) *Makromol. Chem.*, **179**, 2121.
176. Rogozhin, S.V. and Davankov, V.A. (1972) *Vysokomol. Soedin. Ser. B*, **14**, 472; (1972) *Chem. Abstr.*, **77**, 115180-r.
177. Rogozhin, S.V., Davankov, V.A. and Belov, Y.P. (1973) *Isv. Akad. Nauk SSSR, Ser. Khim.*, 955.
178. Belov, Y.P., Davankov, V.A. and Rogozhin, S.V. (1976) *Isv. Akad. Nauk SSSR, Ser. Khim.*, 1596; (1977) *Isv. Akad. Nauk SSSR, Ser. Khim.*, 1856.
179. Belov, Y.P., Rogozhin, S.V. and Davankov, V.A. (1973) *Isv. Akad. Nauk SSSR, Ser. Khim.*, 2320.
180. Gubitz, G., Jellenz, W. and Santi, W. (1981) *J. Chromatogr.*, **203**, 377.
181. Foucault, A., Caude, M. and Oliveros, L. (1979) *J. Chromatogr.*, **185**, 345.
182. Tsuchida, E., Nishikawa, H. and Terada, E. (1976) *Eur. Polym. J.*, **12**, 611.
183. Snyder, R.V., Angelici, R.J. and Meck, R.B. (1972) *J. Am. Chem. Soc.*, **94**, 2660.
184. Dotsevi, G., Sogah, Y. and Cram, D.J. (1975) *J. Am. Chem. Soc.*, **97**, 1259; (1976) *J. Am. Chem. Soc.*, **98**, 3038.
185. Wulff, G., Sarhan, A. and Zabrocki, K. (1973) *Tetrahedron Lett.*, 4329.
186. Wulff, G., Dederichs, W., Grotstollen, R. and Jupe, C. (1982) In *Affinity Chromatography and Related Techniques* (eds T.C.J. Gribnau, J. Visserr and R.J.F. Nivard), Elsevier, Amsterdam, p. 207.
187. Wulff, G. and Sarhan, A. (1982) In *Chemical Approaches to Understanding Enzyme Catalysis: Biomimetic Chemistry and Transition State Analogs* (eds B.S. Green, Y. Ashani and D. Chipman), Elsevier, Amsterdam, p. 106.
188. Wulff, G., Kemmerer, R., Vietmeier, J. and Poll, H.G. (1982) *Nouv. J. Chim.*, **6**, 681.
189. Wulff, G., Poll, H.G. and Minarik, M. (1986) *J. Liq. Chromatogr.*, **9**, 385.
190. Wulff, G. and Sarhan, A. (1972) *Angew. Chem.*, **84**, 364.
191. Wulff, G., Sarhan, A., Gimpel, J. and Lohmer, E. (1974) *Chem. Ber.*, **107**, 3364.
192. Sarhan, A. and Wulff, G. (1982) *Makromol. Chem.*, **183**, 85.
193. Wulff, G. and Gimpel, J. (1982) *Makromol. Chem.*, **183**, 2469.
194. Sarhan, A. and Wulff, G. (1982) *Makromol. Chem.*, **183**, 1603.
195. Wulff, G., Grobe-Einsler, R., Vesper, W. and Sarhan, A. (1977) *Makromol. Chem.*, **178**, 2817.
196. Wulff, G., Vesper, W., Grobe-Einsler, R. and Sarhan, A. (1977) *Makromol. Chem.*, **178**, 2799.
197. Wulff, G. and Hohn, J. (1982) *Macromolecules*, **15**, 1255.
198. Wulff, G. and Vesper, W. (1978) *J. Chromatogr.*, **167**, 171.
199. Wulff, G., Schulze, I. and Zabrocki, K. (1980) *Makromol. Chem.*, **181**, 531.

200. Wulff, G. and Akelah, A. (1979) *Makromol. Chem.*, **179**, 2647; (1980) Ger. Patents 2 835 225, 2 835 226, 2 848 967, 2 849 112.
201. Wulff, G., Lauer, M. and Disse, B. (1979) *Chem. Ber.*, **112**, 2845.
202. Pirkle, W.H., House, D.W. and Fin, J.M. (1980) *J. Chromatogr.*, **192**, 143.
203. Pirkle, W.H. and House, D.W. (1979) *J. Org. Chem.*, **44**, 1957.
204. Yuki, Y., Okamoto, Y. and Okamoto, I. (1980) *J. Am. Chem. Soc.*, **102**, 6356.
205. Newman, M.S. and Junjappa, H. (1971) *J. Org. Chem.*, **36**, 2606.
206. Lochmuller, C.H. and Ryall, R.R. (1978) *J. Chromatogr.*, **150**, 511.
207. Mikes, F., Boshart, G. and Gil-Av, E. (1976) *J. Chromatogr.*, **122**, 205; (1978) *J. Chromatogr.*, **149**, 455; (1978) *Chem. Soc. Chem. Commun.*, 173.
208. Lochmuller, C.H. and Scouter, R.W. (1975) *J. Chromatogr.*, **113**, 283.
209. Meditsch, J. de O. (1957) *Rev. Quim. Ind.*, **26**, 142.
210. Rubinstein, M., Amit, B. and Patchornik, A. (1975) *Tetrahedron Lett.*, 1445.
211. Rubinstein, M., Shechter, Y. and Patchornik, A. (1976) *Biochem. Biophys. Res. Commun.*, **70**, 1257.
212. Frechet, J.M.J., Farrall, M.J., Benezra, C. and Cheminat, A. (1980) *Polym. Prepr., Am. Chem. Soc., Div. Polym. Chem.*, **21** (2), 101.
213. Cheminat, A., Benezra, C.Z., Farrall, M.J. and Frechet, J.M.J. (1980) *Tetrahedron Lett.*, 617.
214. Krohn, E., Eberlein, K. and Gercken, G. (1978) *J. Chromatogr.*, **153**, 550.
215. Seymour, E. and Frechet, J.M.J. (1976) *Tetrahedron Lett.*, 3669.
216. Weith, H.L., Wiebers, J.L. and Gilham, P.T. (1970) *Biochemistry*, **9**, 4396.
217. Rosenberg, M., Wiebers, J.L. and Gilham, P.T. (1972) *Biochemistry*, **11**, 3623.
218. Fukuda, Y., Morikawa, Y. and Matsumoto, I. (1981) *Anal. Chem.*, **53**, 2000.
219. Peppard, T.L. and Halsey, S.A. (1980) *J. Chromatogr.*, **202**, 271.
220. Becker, N. and Unger, K. (1980) *Z. Anal. Chem.*, **304**, 374.
221. Okamoto, Y., Khojasteh, M., Hou, C.J. and Rice, J. (1980) *Polym. Prepr., Am. Chem. Soc., Div. Polym. Chem.*, **21** (2), 99.
222. Greber, G. and Merchant, S. (1969) *Kinetics and Mechanisms of Polyreactions, Int. Symp. Macromol. Chem. Prepr.*, **5**, 131; (1971) *Chem. Abstr.*, **75**, 64544-m.
223. Petronio, B.M., Lagaua, A. and Russo, M.V. (1981) *Talanta*, **28**, 215.
224. Hodge, P. and Waterhouse, J. (1983) *J. Chem. Soc., Perkin Trans. 1*, 2319.
225. Jost, W., Unger, K.K., Lipecky, R. and Gassen, H.G. (1979) *J. Chromatogr.*, **185**, 403.
226. Szyfter, K. and Langer, J. (1979) *J. Chromatogr.*, **175**, 189.
227. Kundu, S.K., Chakravarty, S.K., Roy, S.K. and Roy, A.K. (1979) *J. Chromatogr.*, **170**, 65.
228. Kawabata, N. and Taketani, Y. (1980) *Bull. Chem. Soc. Jpn*, **53**, 2986.
229. Mikes, O., Strop, P., Smrz, M. and Coupek, J. (1980) *J. Chromatogr.*, **192**, 159.
230. Bird, C.R. and Smith, T.A. (1981) *J. Chromatogr.*, **214**, 263.
231. Walton, H.F., Eiceman, G.A. and Otto, J.L. (1979) *J. Chromatogr.*, **180**, 145.
232. Hasegawa, E., Kanayama, T. and Tsuchida, E. (1977) *J. Polym. Sci., Polym. Chem. Ed.*, **15**, 3039.
233. Shigehara, K. and Tsuchida, E. (1979) *Polym. Prepr., Am. Chem. Soc., Div. Polym. Chem.*, **20** (1), 1057.
234. Ledon, H. and Brigandat, Y. (1979) *J. Organomet. Chem.*, **165**, C25–C27.
235. Ledon, H., Brigandat, Y., Primet, M., Negre, M. and Bartholin, M. (1979) *C.R. Acad. Sci. Paris, Ser. C*, **288**, 77.
236. Rollmann, L.D. (1975) *J. Am. Chem. Soc.*, **97**, 2132.
237. Rollmann, L.D. and Reed, C.A. (1973) *J. Am. Chem. Soc.*, **95**, 2048.
238. Rollmann, L.D., Gagne, R.R. and Kouba, J. (1974) *J. Am. Chem. Soc.*, **96**, 6800.
239. Shirakawa, H., Kagami, M. and Ikeda, S. (1976) *Polym. Prepr. Jpn*, **25** (1), 47.
240. Leal, O., Anderson, D.L., Bowman, R.G., Basolo, F. and Burwell, R.L. (1975) *J. Am. Chem. Soc.*, **97**, 5125.

241. Ledon, H. and Brigandat, Y. (1980) *J. Organomet. Chem.*, **190**, C87-C90.
242. Bohlen, H., Martens, B. and Wohrle, D. (1980) *Makromol. Chem. Rapid. Commun.*, **1**, 753.
243. Wohrle, D. (1980) *Polym. Bull.*, **3**, 227.
244. Aeissen, H. and Wohrle, D. (1981) *Makromol. Chem.*, **182**, 2961.
245. Sawodny, W., Grunes, R. and Reitzle, H. (1982) *Angew. Chem., Int. Ed. Engl.*, **21**, 775.
246. Kim, S.J. and Takizawa, T. (1975) *Makromol. Chem.*, **176**, 891, 1217.
247. Kraus, M. and Tomanova, D. (1974) *J. Polym. Sci., A-1*, **12**, 1781.
248. Koide, M., Suganuma, N., Tsuchida, E. and Kurimura, Y. (1978) *Polym. Prepr. Jpn*, **27**, 1028.
249. Koide, M., Tsuchida, E. and Kurimura, Y. (1979) *Polym. Prepr., Am. Chem. Soc., Div. Polym. Chem.*, **20** (1), 1059.
250. Kurimura, Y. and Tsuchida, E. (1979) *Kagaku No Ryoiki*, **33**, 587.
251. Mills, G.F. and Dickinson, B.N. (1949) *Ind. Eng. Chem.*, **41**, 2842.
252. Manecke, G. (1955) *Angew. Chem.*, **67**, 613.
253. Yamanoto, A. (1955) *Nippon Kagaku Kaishi*, **29**, 838.
254. Fulton, S.P. (1980) *Polym. Prepr., Am. Chem. Soc., Div. Polym. Chem.*, **21** (2), 98.
255. Gin, M.E. and Church, C.L. (1959) *Anal. Chem.*, **31**, 551.
256. Newburger, S.H. (1977) *Manual of Cosmetic Analysis*, 2nd edn, Association of Official Analytical Chemists, Washington, DC, pp. 18, 55.
257. Tsuchida, E., Kaneko, M. and Nishide, H. (1973) *Makromol. Chem.*, **164**, 203.
258. Kurimura, Y. Tsuchida, E. and Kaneko, M. (1971) *Bull. Chem. Soc. Jpn*, **44**, 1293.
259. Cook, M.M., Wagner, S.E., Mikulski, R.A., Demko, P.R. and Clements, J.G. (1978) *Polym. Prepr., Am. Chem. Soc., Div. Polym. Chem.*, **19** (2), 415.
260. Cuatrecasas, P. (1970) *J. Biol. Chem.*, **245**, 3059.
261. Cuatrecasas, P. and Anfinsen, C.B. (1971) *Ann. Rev. Biochem.*, **40**, 259.
262. Cuatrecasas, P. (1972) Affinity chromatography of macromolecules. In *Advances in Enzymology*, vol. 36 (ed. A. Meister), Wiley, New York.
263. Guilford, H. (1973) *Chem. Soc. Rev.*, **2**, 249.
264. Jakoby, W.B. and Wilcheck, M. (1974) *Methods in Enzymology*, vol. 34, Academic Press, New York.
265. Lowe, C.R. and Dean, P.D.G. (1974) *Affinity Chromatography*, Wiley Interscience, London.
266. Cuatrecasas, P. and Hollenberg, M.D. (1976) *Adv. Protein Chem.*, **30**, 251.
267. Epton, R. (ed.) (1978) *Chromatography of Synthetic and Biological Polymers*, vols 1 and 2, Ellis Horwood, Chichester.
268. Turkova, J. (1978) *Affinity Chromatography*, Elsevier, Amsterdam.
269. Sharma, S.K. and Mahendroo, P.P. (1980) *J. Chromatogr.*, **184**, 471.
270. Scouten, W.H. (1981) *Affinity Chromatography*, Wiley, New York.
271. Venter, J.C. (1982) *Pharmacol. Rev.*, **34**, 153.
272. Chaiken, I., Wilchek, M. and Parikh, I. (eds) (1984) *Affinity Chromatography and Biological Recognition*, Academic Press, New York.
273. Schott, H. (1985) *Affinity Chromatography: Template Chromatography of Nucleic Acids and Proteins*, Marcel Dekker, New York.

6
Organic synthesis on polymeric carriers

Reactive polymers may be used as carriers in the stepwise construction of large molecules. In this type of synthesis, the reactive polymer is used to block selectively a certain functional group in multifunctional compounds. The large excess of reagents usually used in each step to allow the complete conversion of the starting substrate to the product is easily isolated after each stage of the synthesis by filtration. In addition to the simplification and acceleration of the synthesis with such systems, large losses normally encountered during the isolation and purification of intermediates at all stages of the synthesis are avoided, leading to high yields of final products. In principle such processes usually involve coupling substrate to the polymer through the functionality followed by the desired chemical modification of the substrate and cleavage of the final product (which remains attached to the polymer after completion of the reaction) in a separate reaction step.

$$\text{(P)}-X + A\text{(P)}-XA \longrightarrow \text{(P)}-XAB \longrightarrow \text{(P)}-X + AB \tag{6.1}$$

There are a number of areas where reactive polymer molecules have been employed as convenient carriers. One example of this has already been described in section 3.6. Merrifields's solid-phase method for producing polypeptides [1] initiated this technique, which has subsequently been applied to both oligonucleotide [2] and oligosaccharide [3] synthesis. The same principle of using reactive polymers as carriers has also been used in a number of other reactions such as asymmetric synthesis, cyclization and related reactions, and also in the detection and isolation of reactive intermediates in chemical reactions.

6.1 ASYMMETRIC SYNTHESIS

The classical organic synthesis of optically active compounds always gives a racemic mixture in which the two enantiomers are in a 1:1 ratio. In contrast, asymmetric synthesis gives predominantly or even only one of the two enantiomers. Reactive polymers have recently been used for the transformation of prochiral substrates in many reaction types into the desired enantiomer product, to eliminate the problem of destroying the chirality of the product during the separation when a conventional soluble reagent or catalyst is used.

In the asymmetric synthesis technique, a homogeneous optically active compound is attached to a macromolecular support and then either a synthesis is carried out on the support or the optically active polymer is used as an asymmetric polymeric reagent or catalyst. The latter approach has been used in most applications but in principle, in order to achieve significant induction, at some stage in the reaction the substrate must become intimately associated with, or specifically bound to, the optically active centre. At this point the polymer acts effectively as an asymmetric species in which the polymer can provide chirality centres which bring about asymmetric reactions.

Stoichiometric asymmetric syntheses are reaction sequences in which prochiral substrates are bonded to chiral reagents to give intermediates and the chiral reagents are split off once again after incorporation of the desired chirality. In catalytic asymmetric synthesis chirality is induced catalytically and only a small amount of chiral catalyst is required to produce a large quantity of enantiomerically pure compounds.

One example where the synthesis occurs unambiguously on the support is in the preparation of atrolactic acid [4] using 1,2-O-cyclohexylidene-α-D-xylofuranose bound to a trityl chloride polymer 1: A polymer-supported

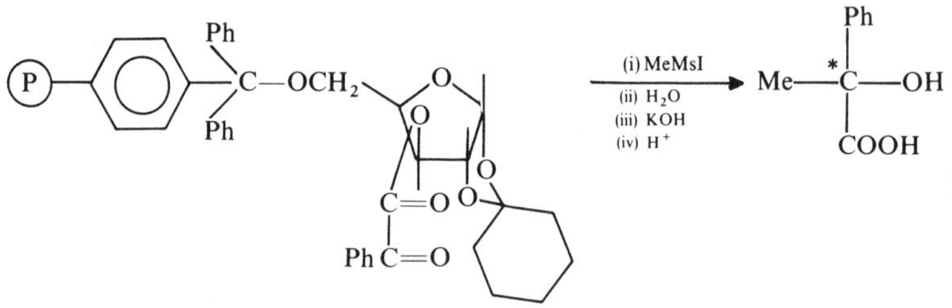

(6.2)

1

optically active primary alkoxyamine 2 has been used in a similar role in the synthesis of chiral 2-alkylcyclohexanones [5]:

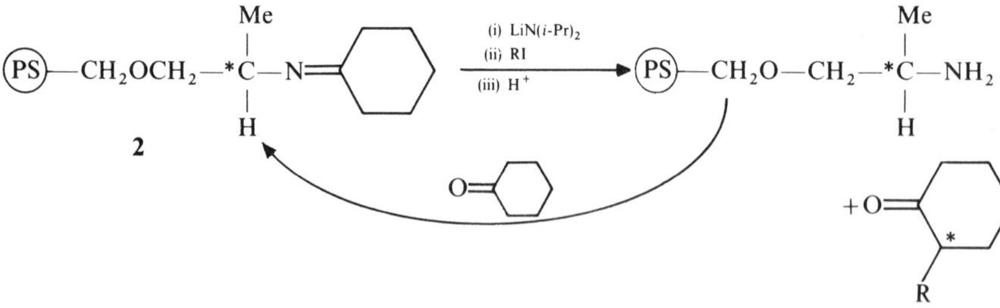

2

(6.3)

Table 6.1 Organic syntheses on polymeric carriers

Functional polymer	Application	Reference
Asymmetric syntheses		
(PS)—C(Ph)(Ph)—O—CH$_2$ ... C=O, R—C=O (cyclohexanone-fused cyclic carbonate structure) $\quad$ R ≡ Me, Ph	Asymmetric synthesis of α-hydroxy acids by reaction with **RMgX, RR'C(OH)COOH**	4
(PS)—CH$_2$OCH$_2$CH*—NH$_2$, with CH$_2$R $\quad$ R ≡ H, Me, Ph	Asymmetric synthesis of 2-alkylcyclohexanones	5, 9
(P)—COOCH$_2$ (pyrrolidine, N—R) $\quad$ R ≡ Me, PhCH$_2$	Asymmetric reduction of $\rangle$C=C=O with MeOH to —CH—COOMe	10–12
(P)—COOCHPh—CHMe—NMeR	Asymmetric acetylation of DL-1-phenylethanol	11–13
	Asymmetric reduction of ketene with MeOH	10
	Asymmetric acylation of DL-1-phenylethanol	11–13

Table 6.1 (*Contd.*)

Functional polymer	Application	Reference
(PS)—CH₂N⁺(Me₂)CHMe—C(OH)Ph Br⁻	Asymmetric Darzen reaction Asymmetric ethylation of phenylacetonitrile	14 15, 16
(P)—⟨C₆H₄⟩N⁺—CH₂—*CHMe—CH₂Me	Asymmetric ethylation of phenyl acetonitrile	17
(P)—R—Z—CH⟨quinoline, R'⟩ R ≡ R' ≡ H, MeO	Asymmetric reduction of ketene with MeOH	10, 13
	Asymmetric acylation of DL-1-phenylethanol	11, 12
Z = COO, —⟨C₆H₄⟩—CO(CH₂)₂COO	Asymmetric Michael reaction	18

$$\text{COOMe} + \diagup\!\!\!\diagdown\text{—COMe} \longrightarrow$$

$$\begin{array}{c}\text{COOMe}\\(\text{CH}_2)_2\text{COMe}\end{array}$$

Asymmetric epoxidation of chalcones   19

18

Asymmetric Michael reaction   20

$$\text{COOMe} + \diagup\!\!\diagdown \text{COMe} \longrightarrow$$

$$\text{CH}_2\text{CH}_2\text{COMe}$$

$$\text{COOMe}$$

$$R\text{—CH}_2\text{SH} + \text{PhCH}=\text{CHNO}_2 \longrightarrow$$

$$R\text{—CH}_2\text{S—CH—CH}_2\text{NO}_2$$
$$\qquad\qquad |$$
$$\qquad\qquad \text{Ph}$$

$$\text{(PS)—(CH}_2)_m \overset{+}{\text{—N}}\ X^-$$

$$\text{CHOH}$$

$$\text{OMe}$$

$$m = 1, 12; \ X^- \equiv \text{Cl}^-, \ \text{Br}^-$$

$$\text{HO}$$
$$\text{H}\cdots$$

$$\text{(PS)—CH}_2\text{—O}$$

$$\text{CHOH}$$

$$\text{P} \quad \text{CN}$$

$$R$$

$$R \equiv \text{H, MeO}$$

Table 6.1 (*Contd.*)

Functional polymer	Application	Reference
(PS)—CH₂—O—[ ]—NH₂, (PS)—CH₂—O—[ ]—NH₂	Asymmetric synthesis of 2-alkylcyclohexanones	21
(PS)—CH₂(OCH₂CH₂)₄—R R ≡ —OCO—[ ], —OCO—[ ] R ≡ Me, Et	Asymmetric reduction of ketones and epoxidation of chalcones	7, 8
	Asymmetric epoxidation of allylic alcohols	22
⁅CH₂—CH—N⁆ₙ R ≡ i-Pr; R¹ = H, Me R ≡ H; R¹ = Me	Asymmetric synthesis of esters from acid salts, chlorides and anhydrides	23

	Synthesis of optically active carboxylic acids and esters $R-CH-COO(H$ or $Et)$ $\quad\ \ \|$ $\quad\ \ Me$	24
$PS-CH_2OCO$ + $PS-CH_2O-CO-CHPhCl$	Synthesis of asymmetric cyclopropyl derivatives 	25
$[CH_2-CH-NH]_n$ $\qquad\ \|$ $\ \ \ CH_2CHMe_2 \quad \vec{B}H_3$	Asymmetric reduction of PhCOMe	26
	Asymmetric reduction of ketone	26
$PS-CH_2P-CH_2-\overset{*}{C}H-NMe_2$ $\qquad\ \|\qquad\qquad\ \|$ $\qquad\ Ph\qquad\qquad R$ $R \equiv i\text{-}Pr,\ PhCH_2$	Asymmetric cross-coupling of secondary alkyl Grignard reagent with vinyl bromide: $Ar(Me)CH-MgCl\ +$ $BrCH=CH_2 \longrightarrow Me\ {\underset{Ar}{\overset{H}{\diagdown}}}C-CH=CH_2$	27

Table 6.1 (*Contd.*)

Functional polymer	Application	Reference
(PS)—CH₂CH—(PS) with $\mid$ NH₂ → BH₃	Asymmetric reduction of ketone	26
(PS)—CH₂[NHCO(CH₂)₁₀]ₙ—N―□ H₂NCOCHNHCO—R ; $n = 0, 1$; R ≡ Me, PhCH₂	Asymmetric reduction of carbonyl compounds to alcohols	28
${-}$NH—CH—CO$-$ₙ $\mid$ R R ≡ Me, —(CH₂)ₘCOOCH₂Ph, (CH₂)₂COO(CH₂)₃Me; $m = 1, 2$	Asymmetric epoxidation of chalcone	29, 30
R ≡ valine or (S)-leucine	Asymmetric hydrogen of α-Me cinnamic acid to 2-methyl-3-phenyl propionic acid and α-acetamidocinnamic acid to phenylalanine	
	Asymmetric conversion of	31

PS—CH₂OCH₂—*CH—CH—O (crown ether), $n = 1, 2$ — 32

$X \equiv COO(CH_2)_2OH, COOCHMeCHMeOH^*$
$L = $ ligand

Asymmetric hydrogenation of olefins and hydrosilylation of ketones — 6, 33–38

Asymmetric hydroformylation of olefins — 39
Asymmetric hydroformylation of α-acetamidocinnamic acid — 40
Asymmetric reduction of acetophenone — 41

$X \equiv COO(CH_2)_2OH, COCHMeCHMeOH, CONMe_2$

Asymmetric hydrogenation of α, β-unsaturated — 40–43

Asymmetric hydrogenation of α-acetamidoacrylic acid and benzoyl formic acid — 44

Asymmetric hydrogenation of α-acetamidoacrylic acid — 44

Table 6.1 (*Contd.*)

Functional polymer	Application	Reference
$-[CH_2-CH_2NH]-$ $\quad\vert$ $\quad R$ $R \equiv Me, i-Bu$	Asymmetric addition of HCN to benzaldehyde	45
	Asymmetric addition of laurylmercaptan to $-C \equiv C - C \equiv O$	46
	Asymmetric hydrogenation of methylisobutyl ketone	47
$-[NH-CH-CO]_n-$ $\qquad\vert$ $\qquad R$		
$R \equiv -(CH_2)_4-NH_2 \cdot Cu$	Asymmetric hydrolysis of phenyl alanine ester	48
$R \equiv -(CH_2)_2-COOH \cdot Pd$	Asymmetric oxidation of 3,4-dihydroxyphenylalanine	49, 50
$R \equiv CH_2COOH \cdot Pd$	Asymmetric hydrogenation of α-acetamidocinnamic acid	51
$(C)-CH_2OP(Ph)_2 \cdot RhClL$	Asymmetric hydrogenation of α-phthalimidoacrylic acid derivative to alanine	52

† Chiral cavity

There are many more examples in which polymer-supported chiral catalysts, mainly based on transition metal complexes, are employed in asymmetric induction. For example, polymer-supported medium complexes bearing the optically active 4,5-bis-(diphenylphosphinomethyl)-1,3-dioxolane **3** as chiral ligand have been reported to catalyse the asymmetric hydrogenation of alkenes and the asymmetric hydrosilylation of ketones to give optically active hydrocarbons and alcohols [6]. However, the degree of asymmetric induction so far obtained by using such polymeric chiral catalysts is not high enough for practical purposes in the majority of cases.

3

Chiral polymeric phase transfer catalysts have been used recently in the transformation of prochiral substrates in many nucleophilic substitution reactions [7, 8]. Chiral reactive polymers and the corresponding asymmetric reactions in which they have been used to enhance the stereochemical selectivity of the synthesis are listed in Table 6.1.

6.2   CYCLIZATION REACTIONS

In classical cyclization reactions there are two competing reactions which lead either to intermolecular or intramolecular cyclization. Thus in conventional solution synthesis it is necessary to control the relative rates of the two reactions to favour the required cyclization. For achieving intramolecular cyclization reactions, high dilution is usually used in order to promote intramolecular reactions over intermolecular reactions, but this results in undesirable side reactions that decrease yields and make product separation difficult.

Recently, solid phase cyclization has been applied to replace the high dilution technique. In this method the molecules to be cyclized are attached to the backbone of the polymeric carrier at distances far from each other to prevent the competing intermolecular reactions. The favoured of the two competing reactions can be achieved through an appropriate choice of the polymeric carrier properties. In general, intramolecular reactions can be achieved by using a polymer with a low degree of functionalization, a high degree of crosslinking and ionic charges, and by suitable choice of the reaction conditions and the nature of the swelling solvent. When a rigid polymer support is used the concentration of molecules to be cyclized on the polymer can be made higher than is normal in solution.

Table 6.2 Cyclizations

Functional polymer	Application	Reference
(PS)—CH$_2$OCO—(CH$_2$)$_n$—COOR		
(a) $n = 1$	Dieckmann cyclization in Br(CH$_2$)$_5$Br to	59
(b) $n = 8$; R ≡ Me$_3$C—	Synthesis of CO—(CH$_2$)$_7$—CH$_2$ ... CH$_2$(CH$_2$)$_6$—CH—COOH	58
(c) $n = 14$	Dieckmann cyclization	59
(d) $n = 5$; R ≡ Et, t-Bu, Et$_3$C—	Dieckmann cyclization	58, 60
(PS)—CH$_2$OCOCOOEt$_3$ R ≡ H, Et	Dieckmann cyclization	57, 58
(P)—OR NO$_2$	Synthesis of cyclic peptides	61

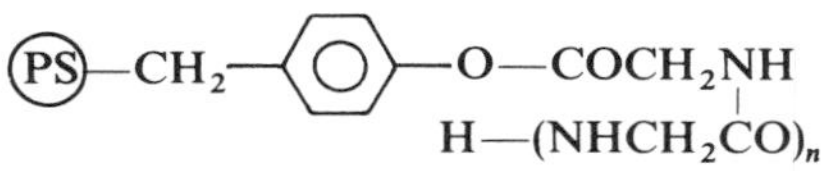

$PS-CH_2-$[benzene]$-O-COCH_2NH$
$H-(NHCH_2CO)_n$

Synthesis of cyclo(Gly)$_{n+1}$ 62

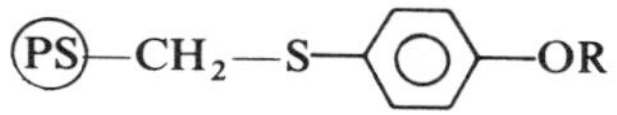

$PS-CH_2-S-$[benzene]$-OR$

Synthesis of cyclic peptides   53–55

$PS-Z-OR$

$Z \equiv CH_2-$, $-CH_2NHCOCH_2-$,

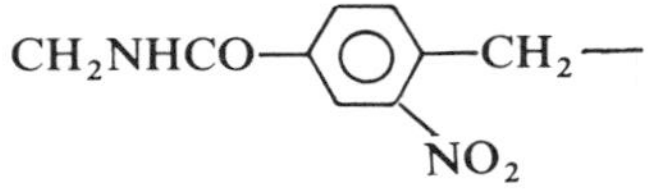

[benzene with CH_2NHCO and CH_2-NO_2]

Synthesis of cyclic dipeptides (diketopiperazines)   63, 64

$PS-CH_2SCO(CH_2)_{11}-OH$

Synthesis of $(CH_2)_2 C=O$ and

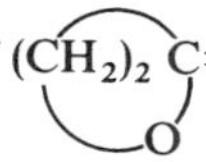

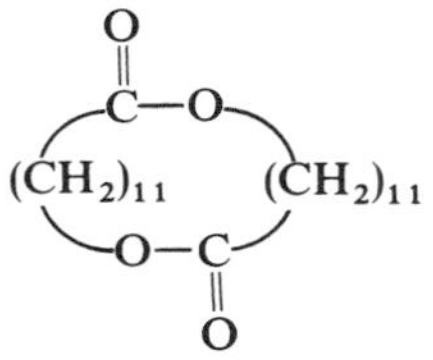

65

$PS-CH_2-Z-CO(CH_2)_n-OH$

(a) $Z \equiv S$; $n = 8$

(b) $Z \equiv O$

Synthesis of $CO-(CH_2)_7-CH-CN$
$NC-CH-(CH_2)_7-CO$ 58, 66

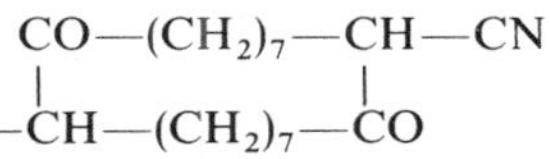

Synthesis of $C=O$ $CH-CN$ $(CH_2)_{n-1}$ 59

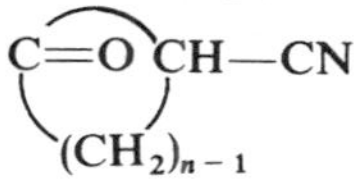

Table 6.2 (*Contd.*)

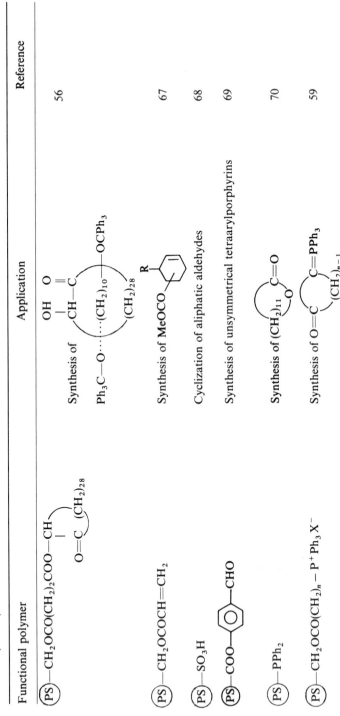

Functional polymer	Application	Reference
PS—CH₂OCO(CH₂)₂COO—CH, O=C (CH₂)₂₈	Synthesis of (structure) Ph₃C—O···(CH₂)₁₀—OCPh₃, (CH₂)₂₈	56
PS—CH₂OCOCH=CH₂	Synthesis of MeOCO (structure)	67
PS—SO₃H	Cyclization of aliphatic aldehydes	68
PS—COO—⟨ ⟩—CHO	Synthesis of unsymmetrical tetraarylporphyrins	69
PS—PPh₂	Synthesis of (CH₂)₁₁ C=O O	70
PS—CH₂OCO(CH₂)ₙ—P⁺Ph₃ X⁻	Synthesis of O=C C=PPh₃ (CH₂)ₙ₋₁	59

(PS)—CH₂P⁺(n-Bu)₃ ⁻OCO(CH₂)ₙOMe

$n = 7, 10, 11, 12, 14$

Synthesis of $(CH_2)_n$ $C{=}O$ 71

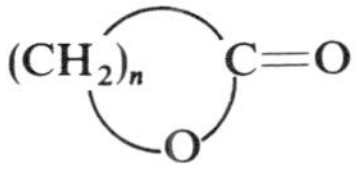

(PS)—CH₂OCH₂— [dioxane ring with Me, two (CH₂)ₙCN substituents]

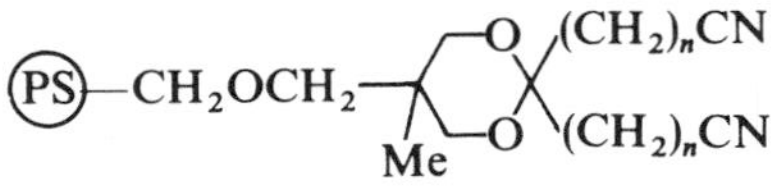

Synthesis of   59

Me ... O $(CH_2)_{n-1}$—C—CN

$HOCH_2$ ... O $(CH_2)_n$—C—NH$_2$

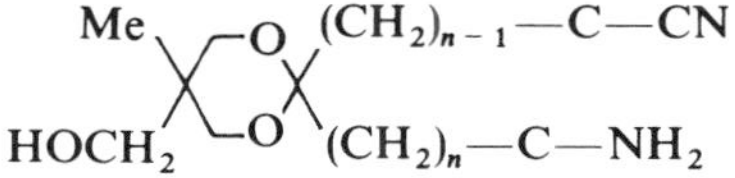

This solid phase cyclization technique has reduced the synthesis problems encountered in the preparation of macrocyclic compounds such as cyclic peptides [53–55] and threaded macrocycles [56]. Dieckmann [57, 58] and other cyclizations have also been improved by this approach, as shown in Table 6.2.

Various cyclic peptides have been prepared by stepwise rearrangement of the linear peptide on a polymeric carrier followed by intramolecular cyclization. For example, the preparation of diketopiperazine has recently been reported [63]:

$$(6.4)$$

The selective Dieckmann cyclization of mixed esters has not been reported using conventional techniques. However, application of a polymeric support does improve the situation. With such a technique, one product remains attached to the polymer whereas the other is obtained in solution, solving the problem encountered with separation of products in high dilution:

$$(6.5)$$

Insoluble polymer supports have also been used in the synthesis of catenanes and other threaded compounds. Undesirable non-threaded byproducts are simply washed from the threaded macrocycle attached to the polymer [56]. Subsequent cleavage yields the pure threaded compound:

$$
\text{(PS)}-CH_2OCOCH_2CH_2COO\underset{\displaystyle CH}{\overset{\displaystyle O=C}{\quad}} \quad (CH_2)_{28} \qquad (6.6)
$$

In addition to these intramolecular reactions, intermolecular cyclization of polymer-attached chains has also been reported [58]:

$$(6.7)$$

6.3 DETECTION AND ISOLATION OF REACTIVE INTERMEDIATES

Among the various advantages associated with functionalized insoluble polymers is the possibility of maintaining mutual separation of reactive functional groups bound to the same polymer support. This is particularly attractive and its exploitation has been described earlier. Thus, site isolation on a polymeric support, in which the reactive species attached to the rigid polymer chains are prevented from reacting with each other, favours either intramolecular reactions or reactions with reagents in solution and prevents the competing reactions such as intermolecular reactions or self-condensation; hence the formation of asymmetrical products or cross-products is promoted. However, site isolation can be achieved by decreasing the polymer chain mobility which is favoured by decreasing the swelling of the polymer.

Ester enolates in solution are known to undergo self-condensation and diacylation or dialkylation resulting in extremely complex mixtures and lower product yields. These side reactions may be avoided and pure products obtained by forming the enolates from polymer-bound esters. For example, the generation of enolates at room temperaure of 3-phenylpropanoic acid supported on 10% and 20% crosslinked polystyrenes has been described recently [87, 88]. The isolated enolates were acylated and alkylated to produce high yields of cross-product with the elimination of ester self-condensation:

$$
\text{(PS)}-CH_2O\overset{\overset{\displaystyle O}{\|}}{C}-CH_2CH_2Ph \xrightarrow{\phi_3CLi} \text{(PS)}-CH_2O-\overset{\overset{\displaystyle O}{\|}}{C}-\bar{C}H-CH_2Ph \quad Li^+
$$

$$
\xrightarrow[R'-Br]{} \quad \downarrow RCOCl
$$

$$
\underset{\underset{\displaystyle R'}{|}}{MeO\overset{\overset{\displaystyle O}{\|}}{C}-CH-CH_2Ph} \qquad R-\overset{\overset{\displaystyle O}{\|}}{C}-CH_2CH_2Ph \qquad (6.8)
$$

Derivatized polymeric beads are generally functionalized throughout their interiors and only a negligible fraction of reactive sites are at the surface of the bead. Hence, two highly reactive compounds may become inert towards each other when attached to two polymers and any reaction between them requires the existence of free reactive intermediates, as was demonstrated in the case of polymeric trityllithium 4 and a polymeric active ester 5. The two polymers are stable in the presence of each other, but on addition of an enolizable ketone reaction proceeded because the free carbanion was able to migrate from one resin to the other [89]:

$$
\text{(P)}-\underset{4}{\phi}-\underset{\underset{\displaystyle Ph}{|}}{\overset{\overset{\displaystyle Ph}{|}}{C^-}}\ Li^+ + CH_3CO-R \longrightarrow \bar{C}H_2CO-R \qquad Li^+
$$

$$
(6.9)
$$

$$
Li^+ \quad CH_2CO-R + \text{(P)}-\phi(NO_2)-OCOR' \longrightarrow R'-COCH_2CO-R
$$

5

This clear demonstration of the existence of a reactive intermediate in any chemical reaction by using a polymeric support has become known as the 'three-phase test'; other examples are listed in Table 6.3. In principle, the technique involves the generation of a reaction intermediate from an insoluble polymeric precursor by the addition of an appropriate reagent and the trapping of the liberated intermediate by a second solid phase reagent suspended in the same medium. The isolation and detection of the polymeric adduct provides positive evidence for the liberated free reactive intermediate. This technique has been applied successfully for the detection of reactive intermediates in nucleophilic catalysis, elimination reactions, phosphate transfer, Fries rearrangement and cyclobutadiene and singlet oxygen formation.

Table 6.3 Detection and isolation of reactive intermediates

Functional polymer	Application	Reference
$(PS)-SO_2NH \cdots Fe(CO)C_4H_4$ as precursor	Detection of cyclobutadiene	72
$(PS)-CH_2-N$ (succinimide) as trap		
$(PS)-CH_2-\phi-OCOR$, $(PS)-CH_2OCO-\phi$ with NO_2 and $HOOC$ substituents as precursor $(PS)-CH_2-NH_2$ as trap	Detection of intermediates in nucleophilic and elimination reactions	73, 74
$(PS)-CH_2-\phi-O-\overset{\overset{\displaystyle O}{\|}}{P}-(NHR)_2$, NO_2 $R \equiv C-C_6H_{11}$ as precursor	Detection of metaphosphates $R-NH-\overset{\overset{\displaystyle O}{\|}}{P}=M-R$	75, 76

Table 6.3 (*Contd.*)

Functional polymer	Application	Reference
(PS)—CH₂OCO—CH—NH₂ as trap		
(PS)—COO—$\overset{\text{O}}{\underset{\overset{\shortmid}{\text{O}^-}}{\overset{\shortparallel}{P}}}$—Ō as precursor	Detection of monomeric metaphosphate [PO₃⁻]	76, 77
(PS)—CH₂OCOCH₂NH₂ as trap		
(PS)—COO—$\overset{\text{O}}{\underset{\overset{\shortmid}{\text{O}^-}}{\overset{\shortparallel}{P}}}$—OPh as precursor	Detection of $[\text{Ph—O—}\overset{\text{O}}{\overset{\shortparallel}{\text{P}}}\text{=O}^-]$	75
(PS)—CH₂OCOCH₂—NH₂ as trap		
(PS)—CH—CH₂—$\overset{\text{O}}{\underset{\overset{\shortmid}{\text{OH}}}{\overset{\shortparallel}{P}}}$—OH as precursor (with —Cl)	Detection of monomeric metaphosphate [PO₃⁻]	78
(PS)—CH₂OCOCH₂—NH₂ as trap		
(PS)—COOCH₂ (benzotriazole)—NH₂ as precursor	Detection of benzyne	79

Structure	Application	Ref.
Ph, Ph, Ph, Ph cyclopentadienone (as trap)	Trapping of $^1O_2$	80
PS—Rose-bengal sensitizer		
PS—CH$_2$OCO—(CH$_2$)$_3$CH=CMe$_2$, CONH$_2$	Trapping of alloxan and ninhydrin radical anions	81
PS—CH$_2$—$^+$N(pyridinium) Cl$^-$	Trapping of alloxan and ninhydrin radical anions	81
P—N$^+$—CH$_2$Ph Br$^-$	Trapping of nitrilium ions	82
PS—SO$_3$H	Detection of oxocarbonium ion as intermediate in Fries rearrangement	83
PS—CH$_2$—(OR, NO$_2$)		
P— (R = H, Me)	As $^1O_2$ source	84

Table 6.3 (*Contd.*)

Functional polymer	Application	Reference
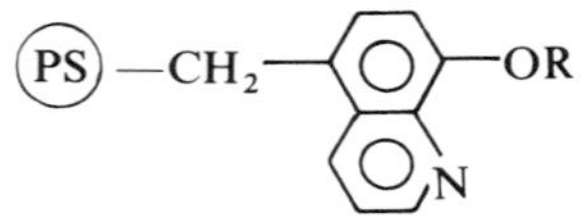 (PS)—CH$_2$— (aromatic ring) —OR $\quad$ R ≡ COMe, COPh, p-O$_2$N—C$_6$H$_4$—CO—, p-Cl—C$_6$H$_4$—CO—	Detection of oxocarbonium ion	83
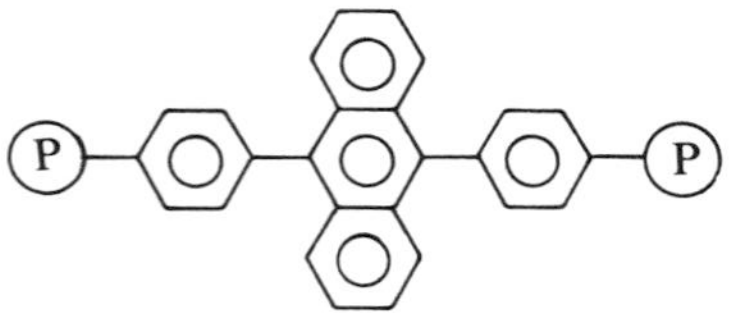 (anthracene bearing two (P)-phenyl substituents)	As carrier for ^{1}O$_2$	85
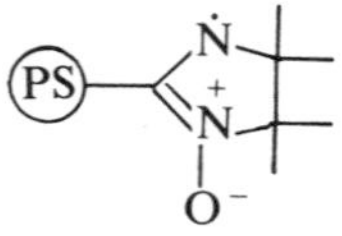 (PS) nitrone/aminoxyl structure	Stable radical	86

(P) ≡ polymer;
(PS) ≡ polystyrene;
(C) ≡ Cellulose;
(P) ≡ polystyrene, copolymer of styrene and CH$_2$ = CMeCOOCH$_2$CH$_2$OH, terpolymer of styrene, CH$_2$ = CMeCOOCH$_2$CH$_2$OH and CH$_2$ = CHCHMeOH.

Polymeric nicotinamide **6** and pyridinium salt derivatives **7** were reported recently to trap alloxan **8** and ninhydrin **9** radical anions [81]. The polymer-trapped intermediates were used for the reduction of several substrates.

While benzyne has been isolated as a stable entity in rigid matrices at very low temperature, its lifetime in solution and in the gas phase is limited by dimerization to form biphenylene. However, by generation of benzyne on a functionalized polymer [79], the ubiquitous dimerization reaction of this reactive molecule is suppressed and its lifetime is extended. This allows other products to be made more readily:

$$(6.10)$$

Polymeric naphthalene derivative **10** has been used to absorb $^1O_2$ to give endoperoxide polymers which on warming release $^1O_2$ again [84]. The storage of the electronically excited oxygen molecule within the polymeric support produced a useful $^1O_2$ source in oxygenation reactions of 2, 3-diphenyl-*p*-dioxene and citronellol.

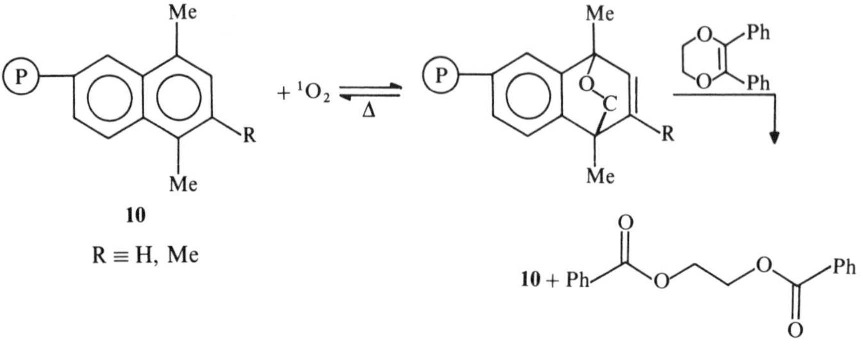

10

R ≡ H, Me

The results of studies on such three-phase tests for the detection of some other reactive intermediates in various reactions are shown in Table 6.3.

REFERENCES

1. Steward, J.M. (1980) In *Polymer-Supported Reactions in Organic Synthesis* (eds P. Hodge and D.C. Sherrington), Wiley, London, p. 343.
2. Mathur, N.K., Narang, C.K. and Williams, R.E. (eds) (1980) *Polymers as Aids in Organic Chemistry*, Academic Press, New York, p. 81.
3. Frechet, J.M.J. (1980) In *Polymer-Supported Reactions in Organic Synthesis* (eds P. Hodge and D.C. Sherrington), Wiley, London, p. 407.
4. Kawana, M. and Emoto, S. (1972) *Tetrahedron. Lett.*, 4855; (1974) *Bull. Chem. Soc. Jpn*, **47**, 160.
5. Worster, P.M., McArthur, C.R. and Leznoff, C.C. (1979) *Angew. Chem., Int. Ed. Engl.*, **18**, 221.
6. Dumont, W., Poulin, J.C., Dang, T.P. and Kagan, H.B. (1973) *J. Am. Chem. Soc.*, **95**, 8295.
7. Frechet, J.M.J., Kelly, J. and Sherrington, D.C. (1984) *Polymer*, **25**, 1591.
8. Kelly, J. and Sherrington D.C. (1984) *Polymer*, **25**, 1499.
9. McArthur, C.R., Worster, R.M., Jiang, J.L. and Leznoff, C.C. (1982) *Can. J. Chem.*, **60**, 1836.
10. Yamashita, T., Yasueda, H. and Nakamura, N. (1974) *Chem. Lett.*, 585.
11. Yamashita, T., Yasueda, H., Nakatani, N. and Nakamura, N. (1978) *Bull. Chem. Soc. Jpn*, **51**, 1183.
12. Yamashita, T., Yasueda, H., Miyauchi, Y. and Nakamura, N. (1977) *Bull. Chem. Soc. Jpn*, **50**, 1532.
13. Yamashita, T., Yasueda, H., Nakamura, N., Hermann, K. and Wynberg, H. (1978) *Bull. Chem. Soc. Jpn*, **51**, 1247.
14. Colonna, S., Fornasier, R. and Pfeiffer, U. (1978) *J. Chem. Soc., Perkin Trans. 1*, 8.
15. Chiellini, E. and Solaro, R. (1977) *J. Chem. Soc. Chem. Commun.*, 231.
16. Chiellini, E. and Solaro, R. (1978) *Chim. Ind. (Milan)*, **60**, 1006.

17. Chiellini, E., Solaro, R. and D'Antone, S. (1977) *Makromol. Chem.*, **178**, 3165.
18. Hermann, K. and Wynberg, H. (1977) *Helv. Chim. Acta*, **60**, 2208.
19. Kobayashi, N. and Iwai, K. (1981) *Makromol. Chem. Rapid Commun.*, **2**, 105.
20. Kobayashi, N. and Iwai, K. (1978) *J. Am. Chem. Soc.*, **100**, 7071; (1980) *J. Polym. Sci., Polym. Chem. Ed.*, **18**, 923; (1980) *Macromolecules*, **13**, 31; (1982) *J. Polym. Sci., Polym. Lett. Ed.*, **20**, 85.
21. Liu, J.H., Kondo, K. and Takemoto, K. (1984) *Makromol. Chem.*, **185**, 2125.
22. Farrall, M.J., Alexis, M. and Trecarten, M. (1983) *Neuv. J. Chim.*, **7**, 449.
23. Kawakami, Y., Sugiura, T., Mizutani, Y. and Yamashita, Y. (1980) *J. Polym. Sci., Polym. Chem. Ed.*, **18**, 3009.
24. Colwell, A.A., Duckwall, L.R., Brooks, R. and McManus, S.P. (1981) *J. Org. Chem.*, **46**, 3097.
25. Arkand, I. and Viout, P. (1981) *Tetrahedron Lett.*, 1009.
26. Yamashita, T., Mitsui, H., Watanabe, H. and Nakamura, N. (1980) *Makromol. Chem.*, **181**, 2563.
27. Hayashi, T., Nagashima, N. and Kumada, M. (1980) *Tetrahedron Lett.*, 4623.
28. Shinkai, S., Tsuji, H., Sone, T. and Manabe, O. (1981) *J. Polym. Sci., Polym. Lett. Ed.*, **19**, 17.
29. Julia, S., Masana, J. and Vega, J.C. (1980) *Angew. Chem., Int. Ed. Engl.*, **19**, 929.
30. Ueyanag, K. and Inoue, S. (1977) *Makromol. Chem.*, **178**, 235.
31. Shea, K.J. and Thompson, E.A. (1978) *J. Org. Chem.*, **43**, 4253.
32. Manecke, G. and Winter, H.J. (1981) *Makromol. Chem. Rapid Commun.*, **2**, 569.
33. Dang, T.P. and Kagan, H.B. (1973) Fr. Patent 2 199 756.
34. Poulin, J.C., Dumont, W. and Dang, T.D. (1973) *Compt. Rend., Ser. C*, **277**, 41.
35. Takaishi, N., Imai, H., Bertelo, C.A. and Stille, J.K. (1976) *J. Am. Chem. Soc.*, **98**, 5400.
36. Takaishi, N., Imai, H., Bertelo, C.A. and Stille, J.K. (1978) *J. Am. Chem. Soc.*, **100**, 264.
37. Masuda, T. and Stille, J.K. (1978) *J. Am. Chem. Soc.*, **100**, 268.
38. Grubbs, R.H., Kroll, L.C. and Sweet, E.M. (1973) *J. Macromol. Sci. Chem.*, **A-7**, 1047.
39. Bayer, E. and Schuring, V. (1976) *Chem. Tech.*, 212.
40. Baker, G.L., Fritschel, S.J. and Stille, J.K. (1981) *J. Org. Chem.*, **46**, 2960.
41. Kagan, H.B., Yamagishi, T., Motte, J.C. and Setton, R. (1978) *Isr. J. Chem.*, **17**, 274.
42. Achiwa, K. (1978) *Chem. Lett.*, 905.
43. Baker, G.L., Fritschel, S.J., Stille, J.R. and Stille, J.K. (1981) *J. Org. Chem.*, **46**, 2954.
44. Harada, K. and Yoshida, T. (1970) *Naturwissenschaften*, **57**, 131, 306.
45. Tsubeyama, S. (1962) *Bull. Chem. Soc. Jpn*, **35**, 1004.
46. Ohashi, S. and Inoue, S. (1971) *Makromol. Chem.*, **150**, 105; (1972) *Makromol. Chem.*, **160**, 69.
47. Hirai, H. and Furuta, T. (1971) *J. Polym. Sci., Polym. Lett. Ed.*, **9**, 459, 729.
48. Nozawa, T., Akimoto, Y. and Hatano, M. (1972) *Makromol. Chem.*, **158**, 21.
49. Hatano, M., Nozawa, T., Ikeda, S. and Yamamoto, T. (1969) *Kagyo Kagaku Zasshi*, **72**, 474; (1970) *Bull. Chem. Soc. Jpn*, **43**, 295; (1971) *Makromol. Chem.*, **141**, 1, 11.
50. Nozawa, T. and Hatano, M. (1971) *Makromol. Chem.*, **141**, 21, 31.
51. Beamer, R.L., Belding, R.H. and Fickling, C.S. (1969) *J. Pharm. Sci.*, **58**, 1142, 1419.
52. Paracejus, H. and Bursian, M. (1972) DDR Patent 92 031.
53. Marshall, G.R. and Liener, I.E. (1970) *J. Org. Chem.*, **35**, 867.
54. Wieland, T. and Birr, C. (1966) *Angew. Chem., Int. Ed. Engl.*, **5**, 310.
55. Flanigan, E. and Marshall, G.R. (1970) *Tetrahedron Lett.*, 2403.
56. Harrison, I.T. and Harrison, S. (1967) *J. Am. Chem. Soc.*, **89**, 5723.
57. Crowley, J.I. and Rapoport, H. (1970) *J. Am. Chem. Soc.*, **92**, 6363.
58. Crowley, J.I., Harvey, T.B. and Rapoport, H. (1973) *J. Macromol. Sci. Chem.*, **A-7**, 1117.
59. Patchornik, A. and Kraus, M.A. (1970) *J. Am. Chem. Soc.*, **92**, 5877.
60. Kraus, M. and Patchornik, A. (1978) *Isr. J. Chem.*, **17**, 298.

61. Fridkin, M., Patchornik, A. and Katchalski, E. (1965) *J. Am. Chem. Soc.*, **87**, 4646; (1966) *J. Am. Chem. Soc.*, **88**, 3164.
62. Rebek, J. and Trend, J.E. (1979) *J. Am. Chem. Soc.*, **101**, 737.
63. Giralt, E., Eritja, R. and Pedroso, E. (1981) *Tetrahedron Lett.*, 3779.
64. Isied, S.S., Kuehn, C.G. and Lyon, J.M. (1932) *J. Am. Chem. Soc.*, **104**, 2632.
65. Mohanraj, S. and Ford, W.T. (1985) *J. Org. Chem.*, **50**, 1616.
66. Crowley, J.I., Harvey, T.B. and Rapoport, H. (1972) *Polym. Prepr., Am. Chem. Soc., Div. Polym. Chem.*, **13**, 958.
67. Uedidia, V. and Leznoff, C.C. (1980) *Can. J. Chem.*, **58**, 1144.
68. Cossu, M. and Uccioni, E. (1973) *C.R. Acad. Sci., Ser. C*, **277**, 1145.
69. Leznoff, C.C. and Svirskaya, P.I. (1978) *Angew. Chem., Int. Ed. Engl.*, **17**, 947.
70. Amos, R.A., Emblidge, R.W. and Havens, N. (1983) *J. Org. Chem.*, **48**, 3598.
71. Regen, S.L. and Kimura, Y. (1982) *J. Am. Chem. Soc.*, **104**, 2064.
72. Rebek, J. and Gavina, F. (1974) *J. Am. Chem. Soc.*, **96**, 7112; (1975) *J. Am. Chem. Soc.*, **97**, 3433.
73. Rebek, J., Brown, D. and Zimmermann, S. (1975) *J. Am. Chem. Soc.*, **97**, 454.
74. Rebek, J., Horton, J.A. and Brown, D. (1978) *Isr. J. Chem.*, **17**, 316.
75. Rebek, J. and Gavina, F. (1975) *J. Am. Chem. Soc.*, **97**, 1591.
76. Rebek, J., Gavina, F. and Navarro, C. (1978) *J. Am. Chem. Soc.*, **100**, 8113.
77. Rebek, J. and Gavina, F. (1975) *J. Am. Chem. Soc.*, **97**, 3221.
78. Rebek, J. (1977) *Tetrahedron Lett.*, 3021.
79. Jayalekshmy, P. and Mazur, S. (1976) *J. Am. Chem. Soc.*, **98**, 6710.
80. Wolf, S., Foote, C.S. and Rebek, J. (1978) *J. Am. Chem. Soc.*, **100**, 7770.
81. Endo, T. and Okawara, M. (1980) *J. Polym. Sci., Polym. Chem. Ed.*, **18**, 2121.
82. Bewick, A., Mellor, J.M. and Pons, B.S. (1978) *J. Chem. Soc. Chem. Commun.*, 738.
83. Warshawsky, A., Kalir, R. and Patchornik, A. (1978) *J. Am. Chem. Soc.*, **100**, 4544.
84. Saito, I., Nagata, R. and Matsura, T. (1981) *Tetrahedron Lett.*, 4231.
85. Rosenthal, I. and Acher, A.J. (1974) *Isr. J. Chem.*, **12**, 897.
86. Miura, Y., Nakai, K. and Kinoshita, M. (1973) *Makromol. Chem.*, **172**, 233.
87. Chang, Y.H. and Ford, W.T. (1981) *J. Org. Chem.*, **46**, 3756.
88. Nambu, Y., Endo, T. and Okawara, M. (1980) *J. Polym. Sci., Polym. Chem. Ed.*, **18**, 2793.
89. Cohen, B.J., Kraus, M.A. and Patchornik, A. (1977) *J. Am. Chem. Soc.*, **99**, 4165.

Part Three   Biological Applications

Functionalized polymers containing biologically active groups have recently received considerable attention [1–12]. Chapter 7 focuses on controlled release formulations and their medically related applications, and describes applications of active polymers in agriculture. Chapter 8 is based on applications of functionalized polymers for food additives as well as sunscreens.

REFERENCES

1. Tanquary, A.C. and Lacey, R.E. (eds) (1974) *Controlled Release of Biologically Active Agents*, vol. 47, Plenum Press, New York.
2. Paul, D.R. and Harris, F.W. (eds) (1976) *Controlled Release Polymeric Formulations*, Am. Chem. Soc. Symp. Ser. 33, Washington, D.C.
3. Cardarelli, N.F. (1976) *Controlled Release Pesticides Formulations*, CRC Press, Boca Raton, FL.
4. Scher, H.B. (ed.) (1977) *Controlled Release Pesticides*, Am. Chem. Soc. Symp. Ser. 53, Washington, D.C.
5. Donaruma, L.G. and Vogl, O. (eds) (1978) *Polymeric Drugs*, Academic Press, New York.
6. Carraher, C.E., Sheats, J.E. and Pittman, C. (eds) (1978) *Organometallic Polymers*, Academic Press, New York.
7. Baker, R. (ed.) (1980) *Controlled Release of Bioactive Materials*, Academic Press, New York.
8. Kydonieus, A.F. (ed.) (1980) *Controlled Release Technologies: Methods, Theory, and Applications*, vols I and II, CRC Press, Boca Raton, FL.
9. Lewis, D.H. (ed.) (1981) *Controlled Release of Pesticides and Pharmaceuticals*, Plenum Press, New York.
10. Carraher, C.E. and Gebelein, C.G. (eds) (1982) *Biological Activities of Polymers*, Am. Chem. Soc. Symp. Ser. 186, Washington, D.C.
11. Akelah, A. (1984) *J. Chem. Technol. Biotechnol.*, **34-A**, 263.
12. Akelah, A. and Selim, A. (1987) *Chim. Ind. (Milan)*, **69**, 62.

7
Controlled release formulations

During the past few years, controlled release technology has emerged as an alternative approach with promises of solving the problems associated with the applications of some biologically active agents, whether pharmaceutical or agricultural chemicals. Conventionally, biologically active agents are normally administered to a target, either in medicinal drugs or in agrochemicals, systemically or topically at a site somewhat remote from the target, which results in non-specific and periodic applications. Both factors, in addition to increasing the cost of the treatment, produce undesired side effects either to the target or to the environment around the target. The aim of controlled release formulations is to protect the supply of the agent and to allow the continuous release of the agent to the target at controlled rates to maintain its concentration in the system within optimum limits over a specified period of time and hence to result in great specificity and persistence. In addition, it holds great promise for improving the undesirable side effects of pharmaceutical and agricultural chemicals.

Controlled release is based on the concept of combining biologically active substances with polymeric materials either by physical combination to act as a rate-controlling device or by chemical combination to act as carrier for the agent. The choice of the best system to release the active agent in sufficient quantity to exert the desired biological effect with minimum biological or ecological side effects is highly dependent on the biological and chemical properties of the active compound and on its physicochemical interactions in the system.

(i) Physical combinations. Two different approaches have been reported in physical combinations of biologically active agents with polymeric materials. The first approach is the encapsulation of the biologically active agent in a polymeric material in which the release of the active agent is controlled by Fickian diffusion through the micropores in the capsule walls. Fick's law of diffusion states that the rate of diffusion R_d depends on the dimensional factors A and h, which involve the geometry or dimensions of the device, and the

diffusional factors D, C_s, K and C_e, which involve active agent–polymer interactions:

$$R_d = \frac{\mathrm{d}M}{\mathrm{d}t} = \frac{A}{h}D(C_s - KC_e) \tag{7.1}$$

where A is the surface area of the membrane, h is the thickness through which diffusion occurs, D is the diffusion coefficient of the active agent in the polymer, C_s is the saturation solubility of the active agent in the polymer, K is the partition coefficient of the active agent between the polymer and the medium which surrounds the device and C_e is the concentration of released active agent in the environment.

The second approach involves heterogeneous dispersion or dissolution of the biologically active agent in a solid polymeric matrix, which can be biodegradable or non-biodegradable. The release of the active agent is generally controlled by diffusion phenomena through the matrix, by chemical or biological erosion, or by a combination of both diffusion and erosion. Release by erosion is a surface-area-dependent phenomenon, and the general expression which describes the rate of release R_r by an erosion mechanism is

$$R_r = \frac{\mathrm{d}M_t}{\mathrm{d}t} = k_E C_0 A \tag{7.2}$$

where k_E is the erosion rate constant, A is the surface area exposed to the environment and C_0 is the loading of the active agent in the erodible matrix.

The design of the physical combination is not necessarily influenced by the structure of the active agent, i.e. there is no need for a specific structural moiety within the biologically active agent molecule. Thus this technique has general applicability for the controlled release of a wide variety of active materials. It is not influenced either by the structure of the polymer matrix and a broad range of polymer matrices can be used. However, in some cases there are requirements on the polymer such as (a) compatibility with the active agent in which there are no undesirable reactions or physical interactions, (b) a low softening point to prevent thermal degradation of the active agent during mixing of an agent with a molten polymer, (c) low crystallinity since in a highly ordered matrix the release rate of dissolved material would be altered and (d) mechanical stability, ease of fabrication and low cost.

Such deposit systems have recently obtained some importance in their technical applications, as described in the literature [1–12]. Since it is not necessary for the polymer to contain any reactive functional group in these physical combinations, this subject is not involved in the present work which will concentrate only on the chemical combinations.

(ii) Chemical combinations. In the chemical combination type of controlled release technology, the active agent is chemically attached to a natural or synthetic polymeric material, an ionic or a covalent linkage. Obviously, only those biologically active agents that contain a structural moiety with at least one reactive functional group suitable for use as a link to the functionalized polymer can be used in this technique.

Polymers chemically bonding active agents can be prepared by two synthetic methods.

1. Chemical modification of a preformed polymer with the desired active agent via a chemical bond leads to a polymer with the active group linked to the main chain as a pendant:

$$\left(\!\!\left.R\right.\!\!\right)_n$$
$$\mid$$
$$Z$$

2. Polymerization of biologically active monomer(s) leads either to polymers with the active groups as repeat units in the main backbone, $\left(\!\!\left.R\!-\!Z\right.\!\!\right)_n$ through the polycondensation technique or to polymers that contain the active group as a pendant through the free radical and polycondensation techniques. The major advantage of the polymerization technique lies in the ability to control the molecular design of the polymer and the active agent–polymer ratio.

Ionic combination to prolong the effect of biologically active groups is based on the principle that positively or negatively charged bioactives combined with the appropriate ion exchange resins yield insoluble poly-salt resinates. The loading of bioactive groups into an ion exchange resin may be accomplished by two methods: (a) a highly concentrated bioactive solution is eluted through a bed or column of the resin (column process) until equilibrium is established, or (b) the resin particles are simply stirred with a large volume of a concentrated bioactive solution (batch process). In the batch process an equilibrium will occur, resulting in a reduced yield, while in the column process the liberated cation is driven downwards, thus avoiding competition.

The active material, which is attached to the polymeric substrate by a definite identifiable chemical bond, is released by slow degradation of the polymer itself or through cleavage of the active agent–polymer linkage by the environmental reactants. In this combination, the rate of release of the active group from the polymer matrix and the consequent efficacy and duration of the effective action is influenced by

1. the chemical characteristics of the active agent structure
2. the strength and type of the active agent–polymer bonds
3. the environmental conditions, i.e. the rate of chemical, biological or environmental (sunlight, moisture, micro-organisms) breakdown of the polymer–active group bonds

4. the chemical nature of the polymer backbone and the groups surrounding the active groups (a non-degrading active polymer could maintain its activity for a long period of time but could create new environmental problems; hydrophobic groups offer protection against rapid hydrolysis whereas hydrophilic groups assist hydrolysis and hence result in shortening the period of protection)
5. the dimensions and structures of the polymer molecule as governed by the degree of polymerization, comonomers, solubility, the degree of crosslinking and the stereochemistry (a stereoregular or crystalline polymer is less susceptible to hydrolytic attack than an amorphous or atactic polymer and an uncrosslinked polymer is much more susceptible to hydrolysis than a highly crosslinked one)
6. the temperature and pH of the surrounding medium
7. the length of the pendant side chain (an increase would enhance the hydrolysis of the bioactive–polymer bond since it would be removed from the hydrophobic backbone and less sterically hindered)

The persistence of activity of a particular formulation is determined by measuring the time at which the release of the active agent fails to make up the loss. The most common cleavage reaction employed is hydrolysis induced by water in the surrounding environment. The kinetic expressions which describe the rate of release of active agent from chemical combination systems depend on the type of polymeric backbone, i.e. on whether the cleavage reaction occurs on the surface of an insoluble particle or in solution. For a heterogeneous reaction on the surface of insoluble spherical particles, i.e. water-insoluble polymers, the rate of release of biocide is given by

$$\frac{\mathrm{d}M_t}{\mathrm{d}t} = nk_h 4\pi r^2 C_0 \tag{7.3}$$

where n is the number of spherical particles of average radius r at time t, k_h is the reaction rate constant for hydrolysis and C_0 is the concentration of active agent–polymer linkages, which is constant because, as one active agent molecule escapes from the surface, the water finds another combined active agent behind. For water-soluble delivery (homogeneous) systems, the rate of release of pendant active groups follows conventional first-order kinetics:

$$-\frac{\mathrm{d}C}{\mathrm{d}t} = k_2 C \tag{7.4}$$

$$-\frac{\mathrm{d}C}{C} = k_2 \mathrm{d}t$$

$$\ln\left(\frac{C_0}{C}\right) = k_2 t$$

where C is the concentration of active agent per unit weight at time t and k_2 is the degradation rate constant. When the active agent moiety is present as a comonomeric unit in the backbone, the release of active group occurs through depolymerization, and the chemical system may be of zero order if the mechanism of release comprises unzipping of the polymer chain. Despite the considerable potential of agrochemicals chemically bonded to polymeric materials, only few have been studied in these combinations.

The liberation of bioactive molecules from ion exchange resins occurs slowly by exchange with the ions present. The rate can be influenced by various factors. The release kinetics depend on (a) the resin characteristics such as the type and strength of ionogenic groups, i.e. acid or base strength, the degree of crosslinking, porosity and particle size, (b) the nature of the bioactive group and (c) the test conditions, e.g. the ionic strength of the dissolution medium, pH, competing ions.

Quantitative studies of ion exchange processes have been mainly concerned with equilibria rather than kinetics. With the exchange of small ions, the equilibrium is reached fairly rapidly. For large organic ions the equilibrium is reached only very slowly and kinetic considerations are important. In the exchange process one counter-ion must migrate from the solution into the interior of the ion exchanger, while another must migrate from the exchanger into the solution. The rate-controlling step has been shown to be diffusion either in the resin particle itself or in an adherent stagnant film. As particle and film diffusion are sequential steps, the slower of the two is rate controlling. In the case of particle diffusion, the concentration gradients in the resin particles will level out and on re-immersion the exchange rate will be higher than at the moment of interruption. With film diffusion, control of the rate depends on concentration differences across the film and these are not affected by the interruption. Hence there will be no effect on the rate.

If all resin particles are uniform spheres of radius r, and under conditions where particle diffusion is the rate-controlling step, the fraction F of bioactive released as a function of time is given by

$$F = \frac{Q_t}{Q_\infty} = 1 - \frac{6}{\pi^2} \sum_{n=1}^{\infty} \frac{\exp(-n^2 Bt)}{n^2} \tag{7.5}$$

where Q_t and Q_∞ are the amounts released at time t and at time ∞, $B = \pi^2 D/r^2$ and D is the effective diffusion coefficient of the exchanging ions in the resin particle. This equation holds only for conditions of infinite solution volume, obtained when a solution of constant composition is continuously passed through a thin layer of beads or in a batch experiment if the solution volume is very large. The rate of exchange will be inversely proportional to the square of the particle radius.

7.1 PHARMACOLOGICALLY ACTIVE POLYMERS

Polymers are already widely used to construct artificial organs [13]. However, the general interest in this field is directed towards the most serious problems encountered in the application of synthetic polymers to the human organism, such as the thrombogenic activity usually induced by the non-compatibility of the polymer surface with the blood and the tissue when they are in contact as well as the toxicity of low molecular weight substances [14].

Synthetic polymers also have a wide variety of uses in some other areas of medicine such as ophthalmology, e.g. contact lenses [15–17] and vitreous eye fluid [18]. In addition, polymeric materials are widely used in various dental applications such as denture bases, artificial teeth, cavity restorations, resilient linings, crown and bridge materials, dental cements, root canals and impression materials [19–22]. Acrylic resins are the principal plastics used and provide the best combination of physical, chemical and cosmetic properties for the large-volume applications of denture bases and teeth. The polymers which are used in dentistry not only must exhibit structural stability and provide the desired function, but also must perform over extended periods of time in the environment of the body, which is a very stringent requirement. Since the basic condition for artificial material used for these applications is its absolute inertness towards biological systems, i.e. the polymer must not contain any reactive functional group, this subject is not involved in this part but is discussed in detail in several reviews [13, 23–25], as is the physical combination of pharmacologically active substances with polymeric materials [26].

The therapeutic activity of polymeric drugs may be due either to the polymer–drug compound, mainly depending on its macromolecular nature and the bond is not attacked, or to the drug molecule in which the bond is gradually destroyed and the polymer is used only as carrier for a normal known pharmaceutical agent. The main idea behind using pharmacologically active polymers in chemotherapy is to try to achieve desirable effects: to increase the duration of the drug action and to enhance its specificity, i.e. the selective enhancement of the action responsible for the therapeutic effect [7, 27–40].

Because macromolecules diffuse slowly in solution and penetrate cell membranes very slowly, coupling of a biologically active molecule to polymeric material is one of several approaches used to enhance the specificity and duration of drug activity. This technique enhances these two factors by confining the drug's pharmacological action to the receptors, by restricting its penetration in body tissue and by increasing its retention in the body by reducing its absorption or excretion. For example, a derivative of penicillin bound to a copolymer **1** with vinylalcohol and vinylamine units shows an activity which is 30–40 times longer lasting than that of the free penicillin [41]:

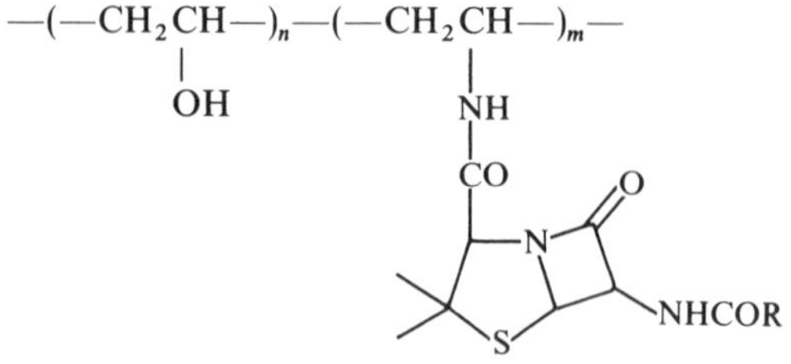

1

The design of polymeric drugs depends on the disease to be treated, the type of drug being considered, and the type of action being elicited. However, the nature of the polymer, the drug and the type of coupling chosen affect the pharmacodynamics of the drug through the effects on absorption, distribution in tissues, variations in metabolism or excretion, and possible cell-specific effects because of the variations in binding capacity to the cells and the receptor sites. For example, copolymers of various sulpha drugs with dimethylolurea (2) have been synthesized and show higher antibacterial activity than the corresponding monomeric sulpha drug [42]:

2

Thus in order to optimize the chemotherapeutic action of a polymeric drug by releasing pharmacologically active substances, with an improvement in both specificity and duration of action, structural variables that affect polymer properties and certain other factors must be taken into account.

(a)   Stereochemical configuration of the polymer

Coupling of an active moiety into a stereochemically restricted position on the polymer could result in maximization of the activity and hence lead to a more potent drug, and also result in protecting the active groups from loss of specificity.

(b)   Chemical nature of the polymer backbone

The choice of polymer type used in polymeric drugs as well as the molecular weight and dimensional structures of the polymer depend on whether its application is topical or systemic. A high polymer injected into a living organism is practically non-eliminable through the usual excretion routes. The molecular weight of the polymer is the most important factor for the excretion through the kidney. Equally, the same polymer will not be absorbed if taken orally or by other routes of administration. Consequently, if a polymeric drug is to be used systemically, its main backbone must be biodegradable so that its macro-molecular residues do not accumulate in the body and its degradation products must be non-toxic. The duration of action of biodegradable polymers can be increased by crosslinking them with biodegradable bonds to form a higher molecular weight polymer.

In the case of a polymeric drug to be applied topically, e.g. to the skin, eye or mouth, or to be taken orally for local treatment of the gastrointestinal tract, the non-degradability of the main backbone is obviously an advantage (since the degradability of the matrix is of no importance) if the drug is to be stable for long periods. In these applications, it is desirable to allow absorption of the released active substance but to prevent absorption of the polymeric drug (through skin or mucosal tissues) by selecting a suitably high molecular weight polymer.

(c)   Type of bonding

The type of bonding of the drug to the polymer matrix and the nature of the bond between them depends on the application of the adduct. In some cases it is necessary to prepare polymers in which the pharmacon is firmly attached to the polymer by means of a linkage which is completely stable under all normal body conditions, i.e. in body fluids and under enzymatic action. A stable linkage between drug and polymer is appropriate when the drug can exert its pharmacological action in drug–polymer complex form.

Alternatively, in some cases, it is required to use a linkage from which the drug can be released rapidly in the body either by hydrolysis or via an enzymatic process if the drug is active only in its free form; the rate of release of the drug depends on the nature of the bond. This is usually true for drugs that act intracellularly; the polymeric drug serves as a reservoir and the polymer can be considered as a carrier. For these purposes it is necessary to attach the drug at some distance from the polymer backbone via a permanent spacer group to separate the active drug from the backbone or coil and hence to prevent the backbone from interfering with the drug–receptor interaction and to achieve an enhanced rate of release.

(d)   Solubility and specificity

In most cases, non-active units are attached to the backbone to improve the physicochemical properties of polymeric drugs. In general, these units are added either for solubilizing purposes or to increase the affinity of the polymeric drug for a given target.

The homopolymers of many drugs are completely insoluble in water or in the aqueous environments of biological systems. Since the release of active substances from the polymer is expected to take place by hydrolysis of the polymer–drug bonds, hydrophobic properties would protect the degradable groups from cleavage. Consequently, the polymer must dissolve or at least swell in aqueous environments to enhance its hydrolysis. This can be achieved by introducing non-toxic hydrophilic units containing ionic groups or solubilizers to be compatible with living tissues. In other cases it is necessary to enhance the absorption of polymeric drugs in lipid phases at cell membranes. This can be achieved by introducing lipophilic groups, such as alkyl groups, to the polymer backbone.

Moreover, separate functional groups can be attached to the polymer to introduce specific adsorption of the polymeric drug to cell membranes of the target area. In general, the polymeric drug distributes itself over the available space where it is introduced, i.e. the drug is transported to the diseased site and to other tissues and organs which are not diseased, resulting in unwanted side effects. In order to promote adsorption and uptake of the drug in a specific area, e.g. in the case of cancer chemotherapy, it is necessary to attach to the polymer a transport system (called a homing device) that has a specific affinity for the membrane on the target cells. For example, poly(glutamic acid) **3**, modified with *p*-phenylenediamine as an antitumour agent and the protein immunoglobulin (Ig) as a homing device, has recently been prepared [43]:

$$-NHCHCO-NHCHCO-NHCHCO-$$

$$(CH_2)_2 \quad (CH_2)_2 \quad (CH_2)_2$$

COOH (solubilizer)   CO   CONH—Ig homing device

HN—⬡—N(CH_2)_2Cl

pharmacon

3

(e) Advantages of polymeric drugs

The potential benefits of pharmacologically active polymers are as follows.

1. They prolong activity by retarding absorption and excretion. Thus they facilitate the use of many drugs which are rapidly metabolized or extremely active in minute quantities if applied in the free form.
2. They alter or modify conventional drug activity. The biological activity of the pharmacon is expected to change as a result of introducing the polymer backbone due to alteration of solubility and the rates of diffusion or to the coupling of several types of drugs to the same polymer.
3. They increase non-absorbability which is desirable in cases where treatment is localized, e.g. infections on the skin.
4. They reduce unfavourable side effects which usually come from rapid metabolism or from hyperdosage.
5. They decrease toxicity. For each drug there is an upper plasma level above which undesirable toxic effects become apparent, and a lower level below which the drug becomes ineffective. A controlled release dosage achieved from the polymeric drug may improve the control of systemic blood levels, providing more consistent protection without toxic effects. Thus they decrease the toxicity of known drugs with toxic properties.

6. They localize drug action at a specific body organ. Normal drug therapy involves a gross overkill technique in which the entire body is subjected to drug when only a small local concentration is required. Because of the polymer's specific structural characteristics, such as high molecular weight, coil structure, copolymer composition, tacticity and variable polyelectrolyte charges, polymeric drugs are capable of inducing some cell-specific uptake by directing the drug's action to the target in the right amounts at the right time. Hence they reduce the patient's exposure to a massive excess of drug over that required at the desired site of action.

7.2 POLYMERIC AGROCHEMICALS AND RELATED BIOCIDES

Agriculture represents an important area with respect to international requirements for health, environmental pollution control, nutrition and economic developments. The rapidly growing demand for food is the main impetus behind the need for more efficient operation in both agricultural and industrial production to afford higher yields and better quality.

Synthetic polymers play an important role in agriculture as structural materials for creating a climate beneficial to plant growth, e.g. mulches, shelters or greenhouses, for fumigation and for irrigation in transporting and controlling water distribution. However, the principal requirement in the polymers used in these applications concerns their physical properties, e.g. transmission, stability, permeability or weatherability, as inert materials rather than as active molecules.

Agrochemicals are chemicals used to control either plant or animal life disadvantageous to humans and animals and to improve production of crops both in quality and quantity. Hence a major increase in the quantities of these costly and toxic chemicals will be necessary for achieving any substantial increase in farm production of foodstuffs. However, the potential hazards of agrochemicals to public health and wildlife result in stringent limitations on their use. Depending on the method of application and climatic conditions, as much as 90% of applied agrochemicals never reach their objective and result in non-specific and periodic applications. Both factors, in addition to increasing the cost of the treatment, produce undesired side effects on either the plant or the environment. Controlled release technology emerged recently as a means of reducing and minimizing problems associated with the use of several types of agrochemicals.

Although the use of polymers physically combined with the biocides has been investigated [44] for most classes of biocide chemicals such as herbicides, insecticides, antifoulants, fungicides, molluscicides, rodenticides, nematicides, algicides and repellants and a number of commercial products have already been introduced, these combinations have considerable disadvantages: there are drawbacks in their production, there are limitations on their period of effectiveness and a large amount of inert polymer must be employed as carrier, thus leaving residuals when the biocide has been exhausted. In spite of the considerable potential of using agrochemicals chemically bonded to polymeric materials, only a few have been studied in these combinations. In addition to improving the efficiency of some existing pesticides and eliminating the problems associated with the use of other conventional biocides, the method has several advantages:

1. prolongation of activity, which means that much lower amounts can be used than with conventional biocides as it releases the required amount of active agent over a long period
2. reduction of the number of applications through achieving a long period of duration of activity in a single application
3. reduction in cost because it eliminates the time and cost of repeated and excess application
4. reduction of mammalian toxicity
5. convenience, because it converts liquids to solids and hence results in easily handled and transported materials of reduced flammability
6. reduction of environmental pollution because it eliminates the need for widespread distribution of large amounts of biocides at one time
7. reduction of evaporation and degradation losses by environmental forces and of leaching by rain into the soil or waterways because of the macromolecular nature of the agrochemical–polymer
8. alteration or modification of the activity, i.e. it extends the duration of activity of less persistent or non-persistent biocides which are unstable in an aquatic environment by protecting them from environmental degradation and hence enhances the practical applicability of these materials
9. reduction of phytotoxicity by lowering the high mobility of the biocides in the soil and hence reducing their residues in the food chain
10. extension of herbicide selectivity to additional crops by providing a continuous amount of herbicide at a level sufficient to control weeds but without injuring the crop.

7.2.1 Polymeric herbicides

The cultivation of plants for economic or ornamental purposes requires an incessant struggle against losses from weeds, which reduce yields by competing for sunlight, water and soil nutrients. Weeds can be controlled by many techniques such as mowing or tilling the soil. However, weed seeds remain dormant in the soil and are unaffected by these techniques. Herbicides contribute significantly to the problem of weed control by selectively killing weeds without crop damage.

The main problem with the use of conventional herbicides to produce a desired biological response in plants at a precise time is the use over a long period of a greater amount of herbicide than that actually needed to control the pest because of the need to compensate for the herbicide wasted by environmental forces such as photodecomposition, leaching and washing away by rain and evaporation or biodegradation by micro-organisms which act to remove the active agent from the site of application before it can perform its function.

However, the application of large amounts of persistent herbicides is undesirable because of their frequent incorporation into the food chain. In addition, they result in a major contamination of the surrounding environment which may be hazardous for humans. For these reasons many of these herbicides have been phased out. However, less persistent herbicides that have greater specificity are ineffective in controlling herbs for a prolonged time because they are unstable in an aquatic environment. These herbicides have other disadvantages such as a high exposure of operators and farm workers and are very costly because of the expense of synthesis and the expense of the multiple applications necessary in view of their lower persistence. Hence their practical application is impossible. Furthermore, the effective lives of conventional herbicides are shortened by leaching out into subsoil and then into underground water sources and lakes, with subsequent damage to aquatic and wild life.

Thus the achievement of improved production of crops using smaller amounts

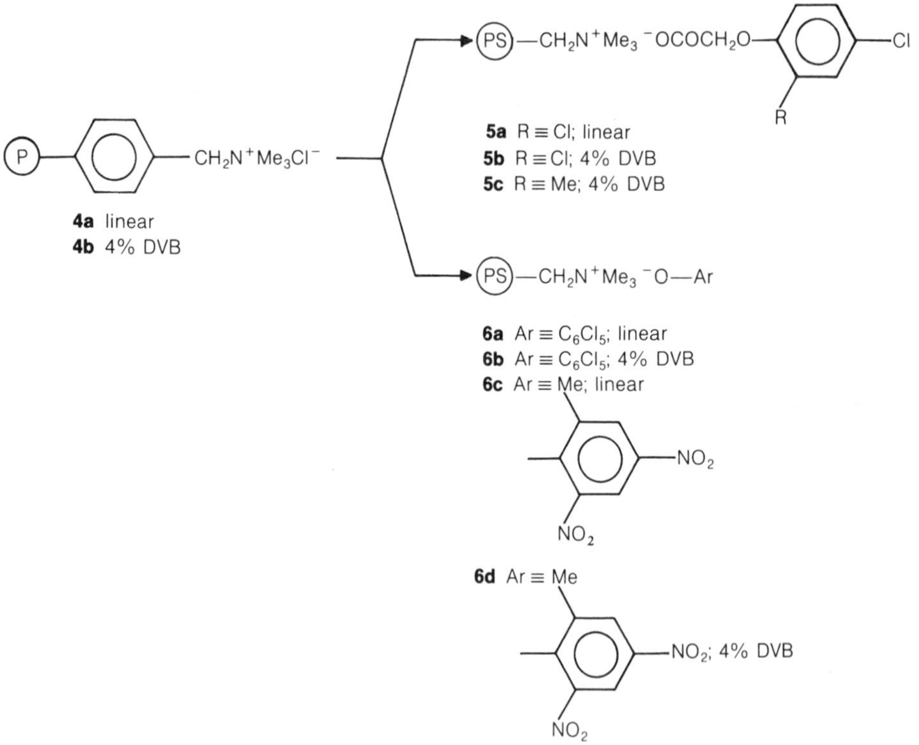

Fig. 7.1 Preparation of polymeric herbicides.

of herbicides with little or no detrimental effect on the surrounding environment but with high biological activity is necessary for agriculture. Recently, interest has grown in using controlled release technology that allows delivery of the herbicide to the plant at a controlled rate in the optimum quantitites required over a specified time [1–12, 45–50]. In most of the formulations the polymers containing the herbicide moities as pendant groups are prepared by chemical modification of preformed polymers. For example, polymeric herbicides containing active moieties ionically bonded to ammonium salt groups have been prepared by chemical modification of anion exchange resins, as shown in Fig. 7.1 [51].

The herbicide pendant groups have also been incorporated by free radical polymerization of the vinyl monomer type or by polycondensation of bifunctional monomers. For example, polymeric herbicides containing pentachlorophenol as the active pendant groups have been prepared by radical homopolymerization and copolymerization with hydrophilic and hydrophobic comonomers, as shown in Fig. 7.2 [52].

A few polymers that incorporate the herbicide moiety directly into the polymer backbone by step growth mechanisms have been reported. Release would involve either depolymerization or chain scission; the rates of release will be adequate when hydrophilic functionality or spacer groups are incorporated into the backbone to allow more rapid chemical or enzymatic hydrolysis by opening the structure. Various other polymers containing herbicide moieties are reported in Table 7.1. Recently, polymeric insecticides containing the trichlorfon **8** as the toxic agent for control of fire ants have also been prepared [88]:

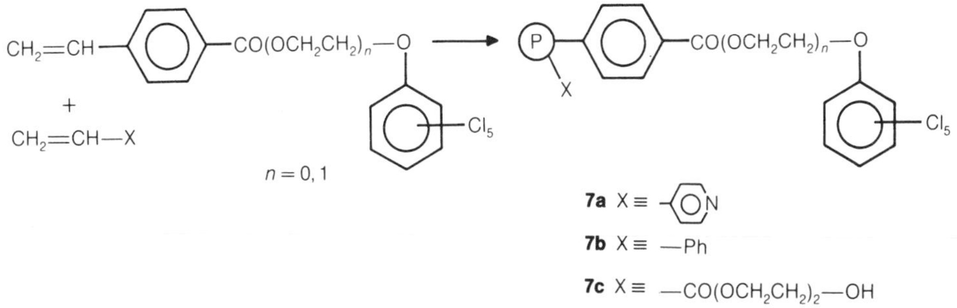
CH₂=CH
CO(OCH₂CH₂)ₙ—O
Cl₅
+
CH₂=CH—X
n = 0, 1
P
X
CO(OCH₂CH₂)ₙ—O
Cl₅
7a X ≡
N
7b X ≡ —Ph
7c X ≡ —CO(OCH₂CH₂)₂—OH

Fig. 7.2 Preparation of pentachlorophenol polymers.

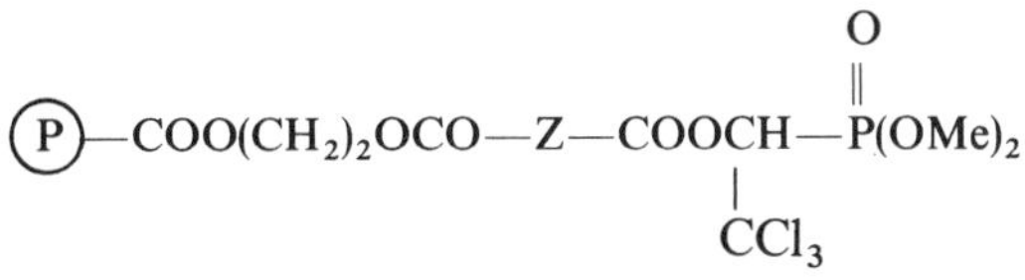
O
P
—COO(CH₂)₂OCO—Z—COOCH—P(OMe)₂
CCl₃

Z ≡ —(CF$_2$)$_2$—, —CH=CH—, —(CH$_2$)$_2$—

8

Table 7.1 Polymeric herbicides

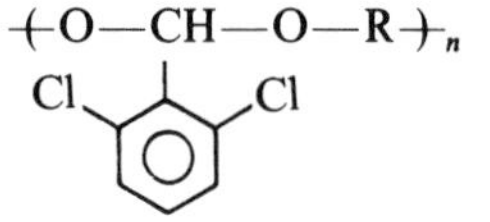

$$+O-CH-O-R+_n$$

with Cl and Cl on the dichlorophenyl ring attached to the CH.

Refs 53, 54

    (a) $R \equiv -(CH_2)_n-$; $n = 5, 6, 10$
    (b) $R \equiv \{-(CH_2)_2CO-(CH_2)_2\}_m$; $m = 1, 2$

$$+NH-(CH_2)_xNHCO-CH-CH-CO+_n$$

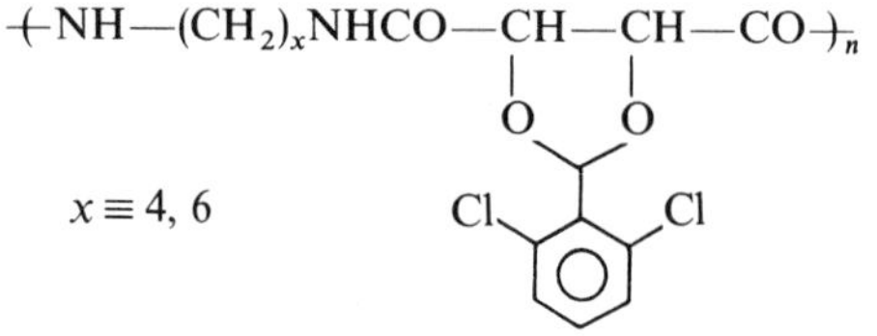

with the two CH groups bearing O and O linked to the dichlorophenyl ring.

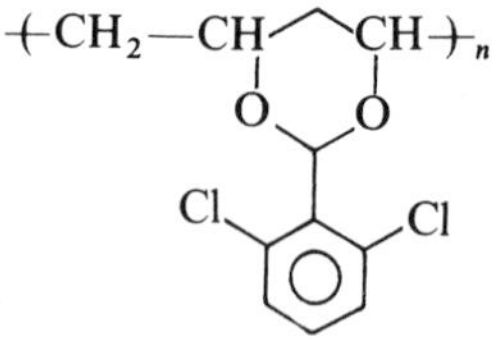

Ref. 55

    $x \equiv 4, 6$

$$+CH_2-CH \underset{O \quad O}{\overset{\frown}{}} CH+_n$$

with the dioxane ring bearing the dichlorophenyl substituent.

Ref. 54

$$+CH_2CH_2-N+_n$$
           |
          COR
    $R \equiv R^1, R^2, R^3, R^{15}, R^{16}$

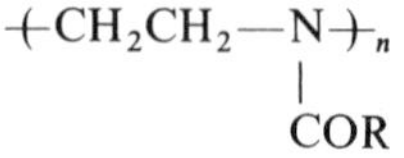

Refs 56, 57

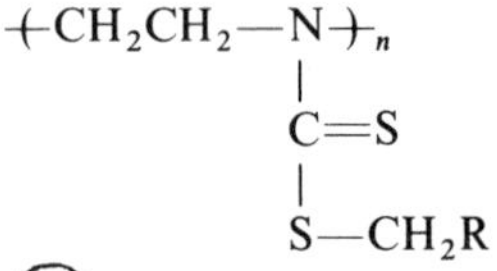

$$+CH_2CH_2-N+_n$$
          |
          $C=S$
          |
          $S-CH_2R$

Refs 58, 59

(PS)$-CH_2N^+Me_3\ ^-O-Ar$

Ref. 51

    $Ar \equiv -COR^2, -COR^4, R^{10}, R^{19}$

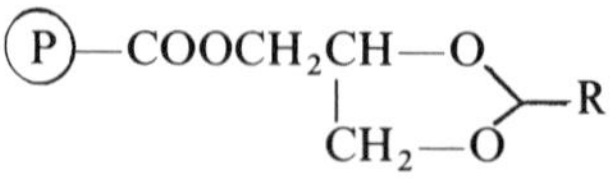

(P)$-CO(OCH_2CH_2)_n-OR^{10}$
  |
  X

Refs 52, 60

$n = 0, 1$; $X \equiv Ph, -\bigcirc N, -CO(OCH_2CH_2)_2OH$

(P)$-COOCH_2CH-O$
           | >-R
          CH_2-O

Ref. 61

    $R \equiv R^2, R^3, R^{13}, R^{14}$

(PS)$-Z-CO-R$

Refs 62, 63

    $Z \equiv -CH_2O-, -CH_2NH-,$
        $-CO(CH_2)_2COO-$;
    $R \equiv R^2, R^4, R^5, R^8$

Table 7.1 *(Contd.)*

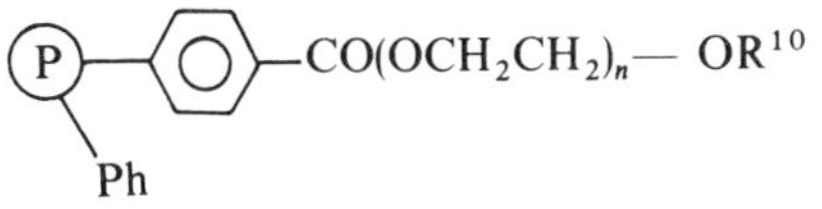 (P)—(O)—CO(OCH$_2$CH$_2$)$_n$— OR^{10} Ph	Refs 52, 60
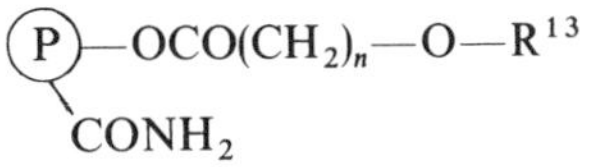 (P)—OCO(CH$_2$)$_n$—O—R^{13} CONH$_2$ $n \equiv 1, 3$	Ref. 64
(P)^a—OCO—R $R \equiv R^2, R^3, R^6, R^7, R^9, R^{11}, R^{17}, R^{18}$ 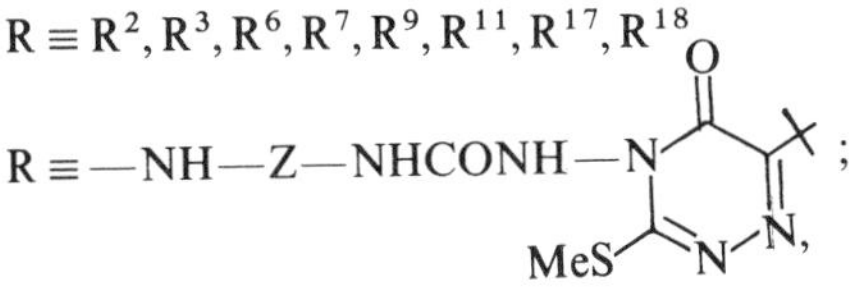	Refs 57, 59, 65–75

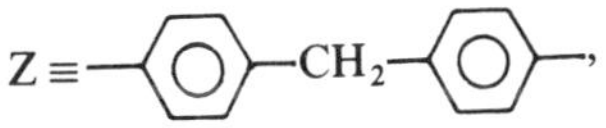

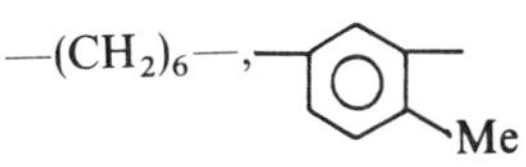

$$R \equiv -NH-Z-NHCONH-N\begin{array}{c} \end{array};$$

(with ring: MeS, N, N, O)

$$Z \equiv -\bigcirc-CH_2-\bigcirc-,$$ Refs 76, 77

$$-(CH_2)_6-, \quad \text{(ring)}-Me$$

(P)—COO(CH$_2$)$_n$—O—Z—R
X
    $X \equiv H, -COOH, -CON^-N^+Me_3,$
       $-CO(CH_2)_2OH$

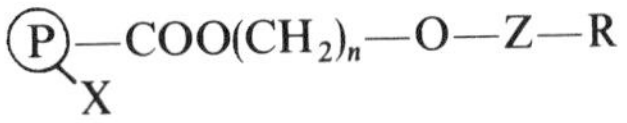

(a) $Z \equiv -CO-$, $n \equiv 2, 4$; $R \equiv R^2, R^3, R^8, R^9$	Refs 78–81
(b) $Z \equiv -CONHCO-$; $R \equiv R^3$; $n \equiv 2$	Ref. 82
(P)^{b,c}—COOR $R \equiv R^{10}, R^{12}$	Refs 83–86
(P)—S—CS—NR$_2$ $R \equiv C_nH_{2n+1}$; $n = 1-4$	Ref. 87

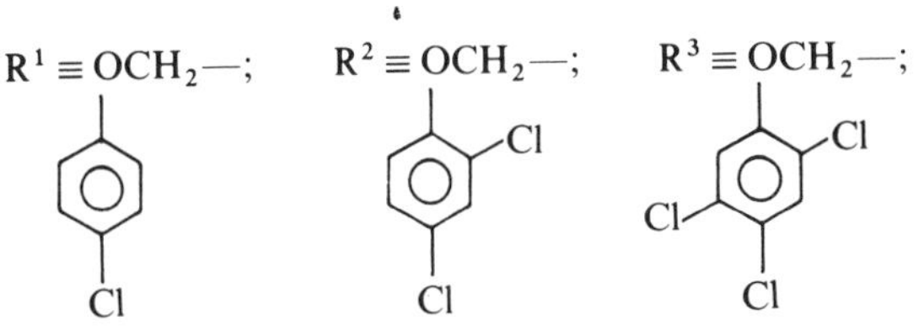

$R^1 \equiv OCH_2-$; $R^2 \equiv OCH_2-$; $R^3 \equiv OCH_2-$;

(ring with Cl para)   (ring with Cl, Cl)   (ring with Cl, Cl, Cl)

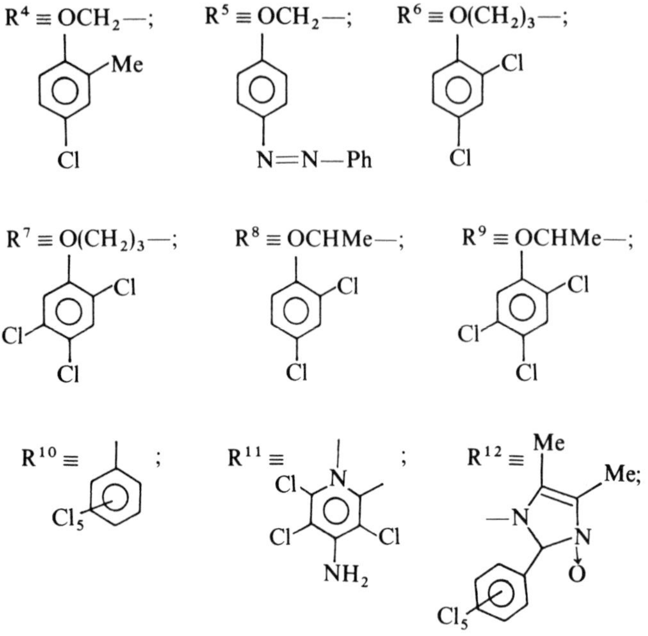

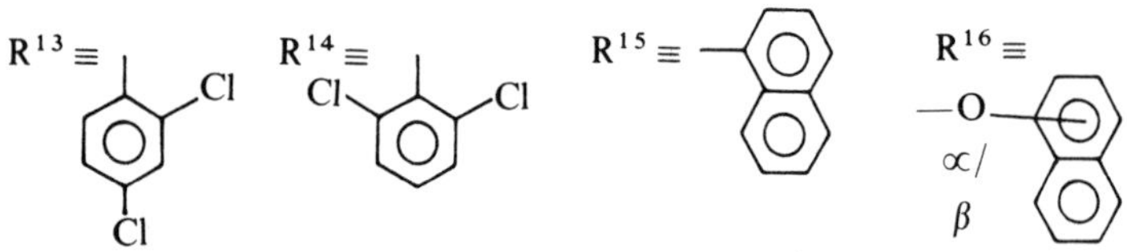

$R^4 \equiv OCH_2—$;

[structure: benzene ring with Me and Cl substituents]

$R^5 \equiv OCH_2—$;

[structure: benzene ring with $N=N—Ph$]

$R^6 \equiv O(CH_2)_3—$;

[structure: benzene ring with two Cl substituents]

$R^7 \equiv O(CH_2)_3—$;

[structure: benzene ring with three Cl substituents]

$R^8 \equiv OCHMe—$;

[structure: benzene ring with two Cl substituents]

$R^9 \equiv OCHMe—$;

[structure: benzene ring with three Cl substituents]

$R^{10} \equiv$ [structure: Cl_5 benzene ring] ;

$R^{11} \equiv$ [structure: pyridine ring with Cl substituents and NH_2] ;

$R^{12} \equiv$ [structure: imidazoline ring with Me, Me, $—N—$, N, O, Cl_5 phenyl];

$R^{13} \equiv$ [structure: benzene ring with two Cl substituents]

$R^{14} \equiv$ [structure: benzene ring with two Cl substituents]

$R^{15} \equiv$ [structure: naphthalene]

$R^{16} \equiv$ [structure: $—O—$ naphthalene α/β]

$R^{17} \equiv Cl_3C—$; $R^{18} \equiv MeCCl_2—$; $R^{19} \equiv Me$ [structure: benzene ring with two NO_2 substituents]

(PS)$— \equiv \text{+CH}_2\text{CH+}_n$ [with phenyl substituent]

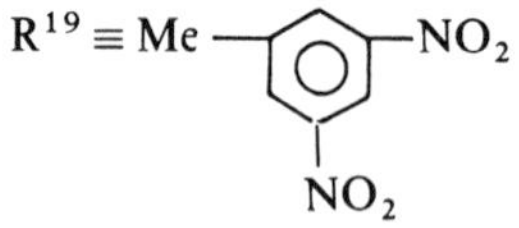

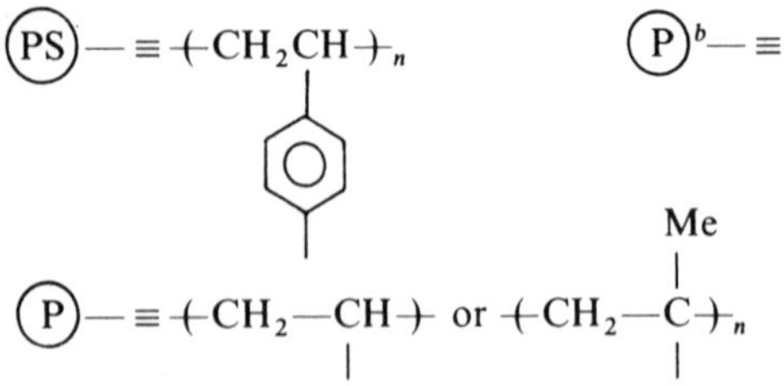

(P)$^b— \equiv$ homo-, Co—, or crosslinked polymer

(P)$— \equiv \text{+CH}_2—\text{CH+}$ or $\text{+CH}_2—\overset{Me}{\underset{|}{C}}\text{+}_n$

(P)$^c \equiv$ alkyd resin

(P)$^a— \equiv$ natural polymer such as bark, cellulose, kraft lignin, wood, sawdust, marine waste, Doylas fibark, sweitenia bark

The rates of release of the active moities from the polymers are affected by several factors such as the degree of hydrophilicity, the medium conditions, the temperature, the nature of the groups surrounding the active moities, the length of the spacer groups, the degree of crosslinking, the chemical nature of the linkage bond and the chemical characteristics of the active agent structure. However, the major drawback to the economical use of polymeric herbicides is the large amount of inert polymer that must be employed as a carrier for the herbicide and hence the large amount of residual material when the herbicide group has been exhausted, which is harmful for the soil and the plant. Some attempts to reduce this problem are mainly based on the concept of attaching the pesticides to biodegradable carriers of similar structures to agricultural residues, consisting of polysaccharides such as bark, sawdust, cellulose and other cellulosic wastes. However, the main disadvantages with the use of such naturally occurring polymers are the difficulties encountered with their chemical modification and the extremely low concentration of herbicide attached to them because of their insolubility in solvents suitable for modification. Hence the excessive amounts of such bioactive polymers that are usually necessary for herb control is inevitable. Furthermore, the hydrophilic and non-crosslinking nature of these polymers leads to a faster rate of hydrolytic cleavage of the pendant pesticide. Another factor is their rapid deterioration in soil by bio-degradation and the subsequent destruction of the polymeric matrix within a short period of time, which leads to a shorter period of effectiveness of the herbicide.

In order to eliminate or at least to reduce the disadvantage of using excessive amounts of inert polymers as carriers, in addition to the drawbacks of using soluble nitrogen fertilizers, we have recently used the principle of dual application of a controlled release herbicide–fertilizer combination. This principle is based on the use, as carriers, of appropriate polymers in which the residual products after degradation of the polymer become beneficial to plant growth and the soil by acting as fertilizer. For example, herbicide derivatives of bifunctional compounds have been prepared and polymerized under condensation polymerization conditions. A second attempt is based on the concept of attaching the herbicides to polymeric hydrogels in order to alter the basic character of sandy soil. In addition to the primary function of these polymers to control the rate of delivery of herbicides, they can play an important role in increasing the water retention by sandy soil through avoiding its rapid leaching. Hence the use of such dual combination of controlled release herbicide–water conserver can contribute positively to changing conventional agricultural irrigation, especially for sandy soil.

7.2.2 Polymeric molluscicides

Bilharzia is one of the most widespread endemic diseases in tropical countries. The opening up of large areas to perennial irrigation has increased the infection rate since new areas that are suitable habitats for the snails, which are the intermediate vectors of the parasite [89], have been created. Thus the fight against bilharzia is an important area of international need for health and economic development.

Since the chemotherapy of schistosomiasis has always met with toxicity problems, the application of molluscicides for combating the disease through eradication of the snails, has opened up the way to concepts in disease control by interrupting the cycle of transmission of snail-borne trematode parasites [90, 91].

Molluscs, including land slugs and snails, do not harm mammals directly but are alternative hosts for the schistosoma parasites, the causal agent for the debilitating human disease bilharzia. Since molluscicides kill various molluscs and offer rapid means for extermination of the causative organism, a great increase in the quantities of these costly and toxic chemicals will be necessary for any substantial increase in controlling and eradicating schistosoma snails. However, the main problem with the use of conventional molluscicides for producing the desired biological response is the relatively massive dosage. This overkill is essential in that a lethal quantity of the toxicant must reach each target snail prior to natural detoxification processes which reduce the active concentration. However, it is not practical in most situations to maintain a continuous toxicant concentration in the treated water. In addition, such chemicals result in a major contamination of the surrounding environment and make it toxic for aquatic plants, birds, fish and mammals, which is a serious handicap to their practical value. Furthermore, it is also difficult to achieve effective distribution of chemical molluscicides in the moving water and hence multiple applications are often used. Thus more effective elimination of snails with a smaller amount of molluscicides that have little or no detrimental effect on the surrounding environment but a high biological activity is necessary for combating bilharzia disease.

During the last few years, the combination of molluscicides with polymeric materials has emerged as a new approach for increasing the efficiency of molluscicides by allowing a continuous release of a lethal quantity of toxicant for controlling snail vectors of schistosomiasis. The technique, in addition to increasing the persistence of conventional molluscicide activity, eliminates the environmental and toxicological problems associated with their use. Furthermore, it is possible to incorporate an attractant and toxicant into the same polymeric matrix, so that snails are attracted by a genus-specific attractant and ingest the polymers containing the toxicant.

An attempt to increase the efficiency of copper(II) for the eradication of snails has been described, using ion exchange resin **9** as a substrate to hold copper(II) so that natural water-soluble salts would exchange away the copper(II) while the regenerable ion exchange was in a fixed accessible site [92].

As active molluscicide, niclosamide (5, 2-dichloro-4-nitrosalicylanilide), has been introduced by Bayer Co. under the trademarks Baylucide or Bayer 73 and

9

$R \equiv$ [benzene ring]—CH_2—[benzene ring]— , [benzene ring]—Y

$Y \equiv$ 4-Me, 4-t-Bu, 5-COOEt, 5-COOC$_{16}$H$_{33}$-n

used in Egypt for combating bilharzia. However, the use of great amounts of this compound has led to some economic and environmental toxicity problems. The chemical combination of molluscicides with functionalized polymers has been used in an attempt to facilitate the eradication of the snails and eliminate the side effects associated with the use of a relatively massive niclosamide dosage. Accordingly, molluscicide polymers containing niclosamide via covalent and ionic bonds have been prepared by chemical modification of polymers [93], as shown in Fig. 7.3. The hydrolytic release of niclosamide from the polymers indicate the following.

1. The hydrolysis of polymeric molluscicide containing niclosamide via ester bonds is slower than that of polymeric molluscicide containing the active moiety as a counter-ion associated with the ammonium salt group. This can be attributed to (a) the nature of the covalent ester groups which are more stable towards hydrolysis than the ionic ammonium salt groups and (b) the intramolecular interactions of the neighbouring hydrophilic ammonium salt groups which are not modified or generated during the hydrolysis.
2. The increase in the degree of crosslinking results in a decrease in the rate of exchange.

The main drawback with these polymers, however, is their low loading with active moieties. To eliminate this disadvantage, polymeric molluscicides have been prepared by polymerization of niclosamide monomers [94]. The niclosamide monomers and their salts were prepared according to Fig. 7.4, and were then homopolymerized and copolymerized with oligo(oxyethylene) monomers by a solution free radical technique.

7.2.3 Polymeric antifouling paints

In recent years, the development and applications of organometallic polymers with controlled release properties as antifouling paints have received

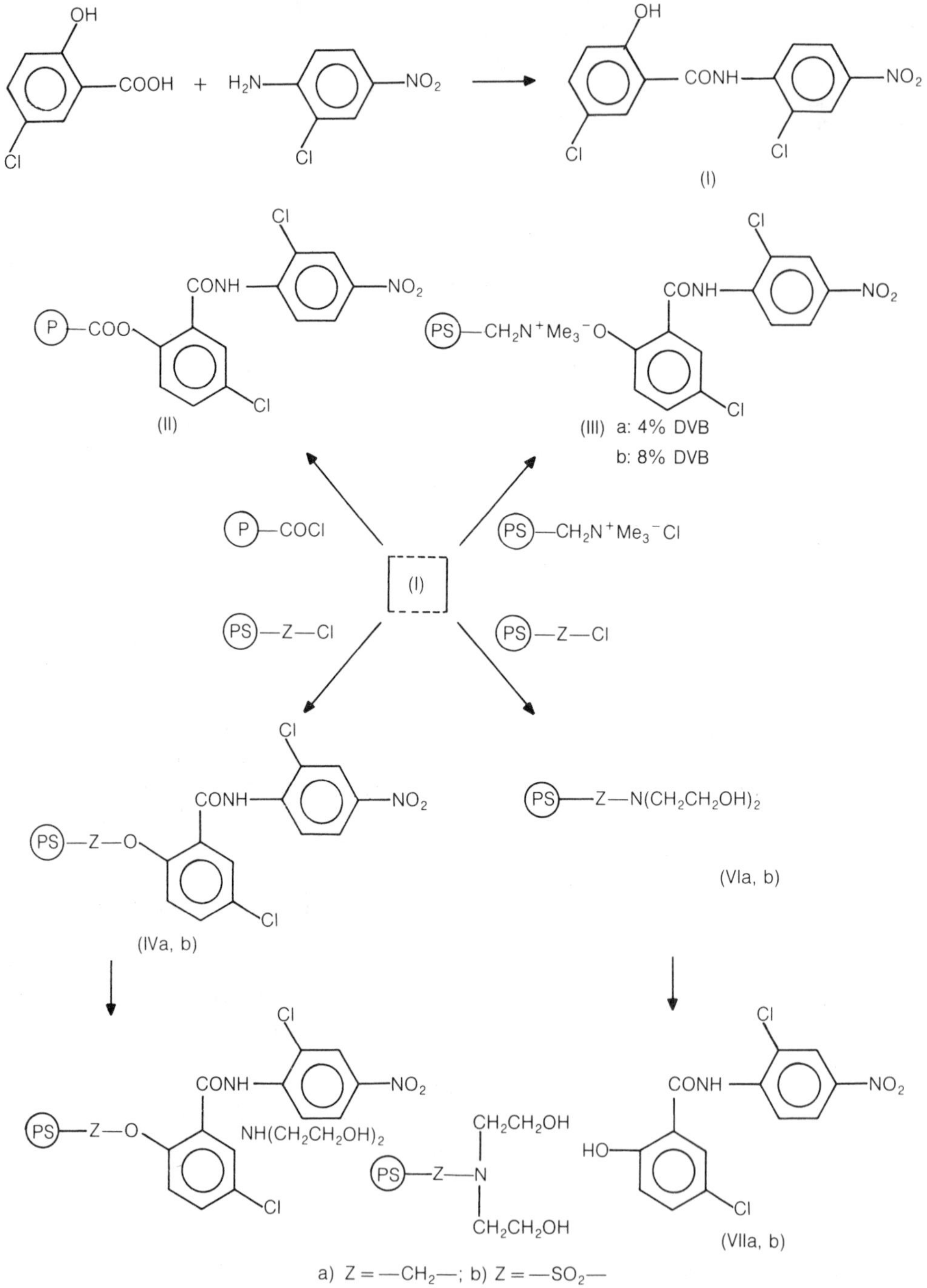

Fig. 7.3 Preparation of niclosamide-polymers by chemical modification.

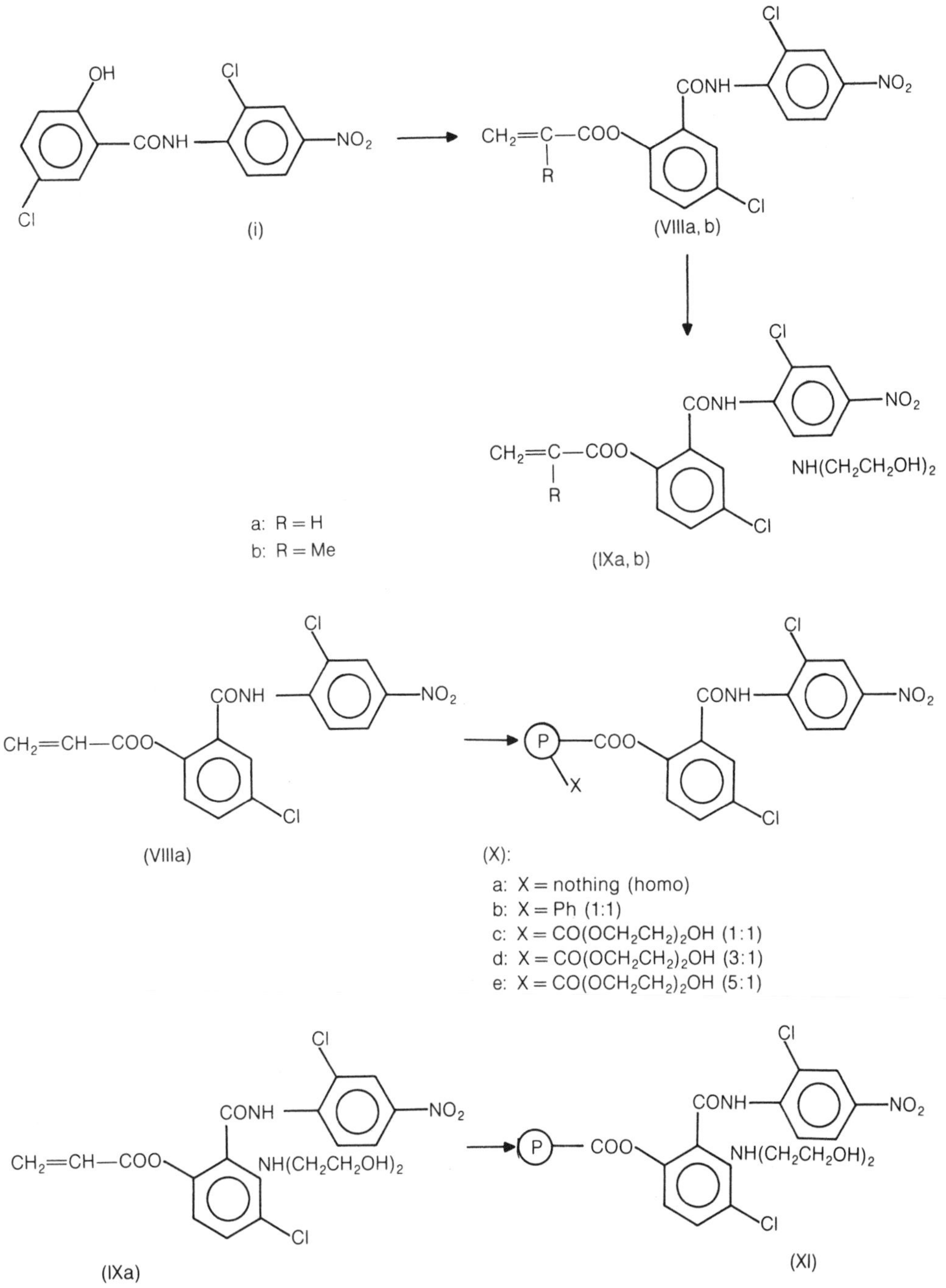

Fig. 7.4 Preparation of niclosamide monomers and their polymers.

considerable interest because fouling is one of the most serious problems in the marine environment. Fouling is the growth of marine organisms, vegetable or animal, which settle on submerged surfaces such as the bottoms of ships, submarines, buoys, sonic transmission equipment etc. These organisms destroy the smooth regularity of the hull's surface, thereby producing surface roughness which increases the frictional resistance to the boat's passage through the water, leading to reduced speed and increased fuel consumption. They also destroy the anticorrosion coating, thus leading to corrosion damage to the surface of marine equipment and causing an increase in the weight of submerged structures. In general, antifouling toxicants are applied to surfaces in continual contact with water as protective coatings. They are designed to prevent the attachment and growth of all fouling marine organisms by continuously releasing a biocide compound at the surface of the paint.

Various principles of formulation of antifouling paints have been described that give a continuous toxic release from the paint to kill the settled organisms. Since the biocides are simply dispersed into the paint which is a thin film and has a large surface area, the rate of water leaching, evaporation or migration is in excess of the amounts required to control fouling, and hence large amounts of biocide are wasted and the coating is left empty of toxin in a short period of time. Thus the biocide concentration drops below the critical level and the coating is free to interfere with the life processes of all organisms and hence is susceptible to fouling. Furthermore, whereas the antifouling action is needed mainly when the ship is in port, because the fouling organisms are at the sea shore, a very high percentage of biocide material is released when the ship is moving owing to the turbulent conditions around it.

A more recent important development in this field to decrease the rate of antifouling decay is the synthesis and use of polymeric materials containing organometallic toxicants chemically bound to the polymer backbone to provide a relatively low dose level of biocides and hence to extend the effective lifetime of antifouling protection [95–104].

Polymeric antifouling paints have considerable potential in applications as biocidal marine coatings for the following reasons.

1. Corrosion is absent;
2. There is no discoloration in water polluted by sulphide;
3. They degrade to non-toxic compounds in the environment;
4. They allow the use of agents that are highly water soluble or have high vapour pressures;
5. More than one biocide group can be incorporated in the polymer to achieve effective biocidal action against all marine organisms;
6. The period of effective action of antifouling agent is increased;
7. The amount of wasted biocide is decreased;
8. They prevent surface roughness which leads to reduced speed and hence decreases fuel consumption.

The duration of the effective action of polymeric antifouling agents is influenced by the factors which are discussed in the introduction, i.e. the structure and the properties of the polymer backbone and the bond linking the polymer to the active agent.

So far the most reported antifouling polymers are the organotin polymers or copolymers which contain the trialkyltin groups either as pendant substituents or as a part of the polymer backbone [105–110]. For example, a crosslinked polymer (10) with a variable density of the crosslinker has been prepared by the reaction of the carboxy groups of the partial tin esterified polymer with epoxy monomers [111]:

$$\text{(P)}\!-\!\text{COOH} \quad + \qquad \overset{}{\underset{O}{\triangle}}\!-\!R\!-\!\overset{}{\underset{O}{\triangle}}$$

$$\overset{|}{\underset{\text{COOSnBu}_3}{}}$$

$$\longrightarrow \quad \text{(P)}\!-\!\text{COOCH}_2\overset{\text{OH}}{\underset{|}{\text{CH}}}\!-\!R\!-\!\overset{\text{OH}}{\underset{|}{\text{CH}}}\text{CH}_2\text{OCO}\!-\!\text{(P)}$$

$$\overset{|}{\underset{\text{COOSnBu}_3}{}} \qquad\qquad \text{Bu}_3\text{SnOCO}$$

10

Poly(carboxystannyloxycarboalkylenes), **11**, have also been prepared by the interfacial polycondensation technique [112]:

$$\left(\!-\!\text{O}\!-\!\overset{R}{\underset{\underset{R}{|}}{\overset{|}{\text{Sn}}}}\!-\!\text{O}\!-\!\text{CO}\!-\!-\!-\!\text{(CH}_2)_4\text{CO}\!-\!\right)_n$$

11

R ≡ Bu, Ph

These polymers have an antifouling action due to slow hydrolysis of the organometallic carboxy groups. Other organometallic polymers of arsenic and mercury are very effective antifouling agents, but they are not used to any extent as toxic agents in marine antifouling coatings because of their effect on the environment (Table 7.2).

7.2.4 Wood preservation

Wood is one of the most important natural resources. It supplies material for many objects necessary to everyday living. Thus it is important to modify wood to improve its strength properties, appearance, resistance to penetration by water and chemicals and resistance to decay.

Controlled release formulations of antifouling moities have been used for the protection of wood against biodegradation [128–130]. They are designed to permeate the entire body, thereby protecting the interior as well as the exterior,

Table 7.2 Polymeric antifouling paints

$\widehat{P}^b$—COO—SnR$_3$ with X	Refs 113–124

(a) X ≡ H;   R ≡ C$_n$H$_{2n+1}$ (n ≡ 2–6), Ph   Refs 123, 113–118
(b) X ≡ Cl;   R ≡ Bu   Ref. 120
(c) X ≡ Ph;   R ≡ Bu   Ref. 121
(d) X ≡ COOMe;   R ≡ Bu   Ref. 122
(e) X ≡ H;   R ≡ Bu   Refs 117, 121, 123
(f) X ≡ CONHCH$_2$OH;   R ≡ Bu   Ref. 114

(g) X ≡ COCH$_2$ —[epoxide]— ; R ≡ Bu   Refs 114

(h) X ≡ OAc;   R ≡ Bu   Refs 115, 124

(i) X ≡ —N[pyrrolidinone ring]   R ≡ Bu   Ref. 115

(j) X ≡ H   R ≡ Pr   Ref. 98

$\widehat{PS}^c$—COOSnR$_3$ Ref. 123

$\widehat{P}$—OCO[benzene ring]COOSnBu$_3$ Ref. 125

$-(-O-Sn(R)(R)-O-R'-)_n-$ Ref. 112

R' ≡ —CO(CH$_2$)$_4$CO—, —(CH$_2$)$_4$—

$\widehat{P}$—COOPbR$_3$
R ≡ Bu, Ph   Ref. 126

$\widehat{P}$— CO—N[ring]As—Cl   Ref. 118

$\widehat{P}$—COO—[ring]—Cl$_5$ Ref. 127

Table 7.2 (*Contd.*)

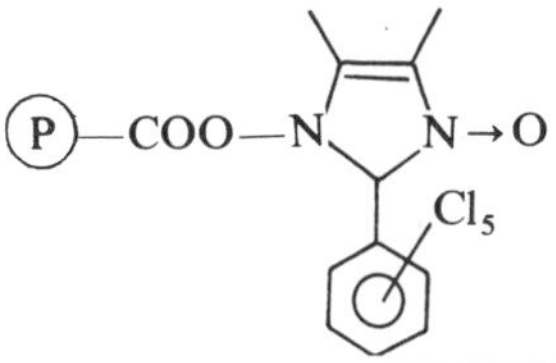

(P)—COO—N⟨⟩N→O   Ref. 127
      Cl$_5$

(PS) — ≡ +CH$_2$CH+$_n$ (P)^b— ≡ homo-, Co—, or crosslinked polymer

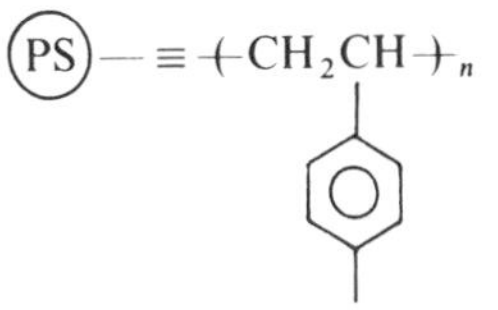

(P) — ≡ +CH$_2$—CH+ or +CH$_2$—C+$_n$ (P)^c ≡ alkyd resin

(P)^a— ≡ natural polymer such as bark, cellulose, kraft lignin, wood, sawdust, marine
       waste, Doylas fibark, sweitenia bark

and hence to minimize environmental hazards and improve other mechanical properties of wood at the same time. The long-term protection of wood against microbiological decay can be achieved by impregnating it with a solution of vinyl biocide monomer, comonomer and initiators and then heating it to initiate a copolymerization reaction within the wood, e.g. the *in situ* copolymerization of glycidylmethacrylate–tributyltin methacrylate impregnated in wood. As a result the accessible voids of the wood are impregnated with polymer and hence the amount of water that can be absorbed by the wood is decreased, thereby preventing the growth of marine organisms which cause rotting. It minimizes the alternate swelling and shrinking of wood and thus increases its mechanical and dimensional stability in water. The treatment, of wood with biocide chemically bound to the polymer chain also decreases the leach rate of the toxic moiety and hence increases its service life while ensuring minimal impact on the environment. In addition to *in situ* polymerization of wood, grafting of the polymeric biocide to wood by the reaction between the hydroxyl groups of wood and the functional groups in the polymer has also been described [129].

7.2.5 Polymeric fertilizers

Fertilizers are one of the most important products of the chemical industry. They are added to the soil to supply nutrients to sustain plants and promote their abundant and fruitful growth. In addition, they are important in adjusting the pH and tilth of the soil.

Whilst controlled release fertilizers are economically attractive for general farm use, particularly for long-term tropical crops, their use is still limited to non-farm markets because of their higher cost to season crops. However, research on the development of synthetic controlled release fertilizers for increasing the efficiency of fertilizers by controlling the appropriate dose of nutrients at the rate needed by growing plants has been conducted, mainly with nitrogen sources, for many years [131–140].

The three major fertilizers are based on nitrogen, phosphorus and potassium but the controlled release system is of no benefit for phosphorus fertilizer because soluble phosphorus nutrient is immobile and not subject to volatilization losses and repeat applications are unnecessary since most crops require a high concentration of available phosphorus early in their growth cycle and take it soon after the application. However, nitrogen fertilizers, unlike the other nutrients, are easily lost from soil, depending on the method of application, the soil, the climate, and the nature of the crop. Nitrogenous fertilizers usually contain nitrogen either as nitrate or in ammoniacal form which is normally converted in the soil to ammonium ions by hydrolysis. Nitrate ions are most readily absorbed by the crop through plant roots but they are not retained in the soil as well as the ammoniacal nitrogen.

Since soluble nitrogen fertilizers are readily absorbed and may often result in a high concentration of nitrogen in plant tissue to levels far greater than the actual crop requirement soon after fertilizer application, this large consumption of fertilizer results in less available nutrient for crop growth later, and also larger doses of fertilizer sometimes cause damage to the crop. Thus, split fertilizer application is used for achieving better utilization of fertilizer nutrient, but it results in increased costs of fertilizer material and extra cost for its application, besides causing water and air pollution. In addition, soluble nitrogen fertilizers have the following disadvantages.

1. They are highly mobile in sandy soil, especially in high rainfall conditions or under intensive irrigation, and hence their loss to drainage water by leaching without a growing crop on the land may be large.
2. Nitrogenous fertilizer, especially urea, may be lost from the dry soil by denitrification which transforms the ionic species into gaseous nitrogen and nitrogen compounds with the help of denitrifying bacteria in the soil.
3. Single heavy applications of nitrogen fertilizer may result in maximum losses of ammoniacal nitrogen to the atmosphere where release of NH_3 exceeds the capacity of crop or soil to absorb it.
4. Toxicity of soluble nitrogen fertilizers to many crops may be produced by high ionic concentrations resulting from rapid dissolution of soluble fertilizers or from evolution of NH_3 by hydrolysis of certain salts, particularly urea.

A variety of investigations have been reported on the controlled release of nitrogen fertilizers, especially urea, to regulate the desirable rate of nutrient to the plant. The two common routes developed to achieve this objective of slow release characteristics are as follows.

1. Chemically controlled release: compounds of low water solubility or polymeric organic compounds which yield available nitrogen at low rates upon biodegradation, dissolution or hydrolysis have been used, e.g. a variety of urea–formaldehyde condensates, isobutylidene diurea. Urea–formaldehyde condensate contains 38–42% nitrogen, is less hygroscopic than urea and does not have a caking tendency.
2. Physically controlled release: soluble fertilizer sources are encapsulated in a slightly porous insoluble coat to give a slow release rate of nutrient to the crop by diffusion through the pores or by erosion and degradation of the coatings [132, 133].

In general, controlled release fertilizers have several advantages:

1. a reduction in the number of applications to supply nutrient in accordance with normal crop requirement, particularly for long-term tropical crops such as sugar cane, tree crops and in forestry, and hence a reduction in the cost of application
2. an increase in nutrient uptake by crops and consequently increased crop yields as a result of the reduction in nutrient loss by leaching to drainage water under heavy rainfall conditions, chemical decomposition, denitrification, volatilization, large consumption or soil fixation
3. a reduction in environmental hazard from large applications or volatilization of soluble fertilizers
4. decreased toxicity

7.2.6 Miscellaneous

In addition to the uses of polymeric materials in packaging for fruits and vegetables [141], they are used in agriculture in significant and increasing amounts to grow more and better plants faster in less space and at lower cost [142]. An important application of hydrophilic polymers is in coatings for seeds to absorb water and thereby increase the rate of germination as well as the percentage of seeds that germinate [143]. They are also used for coating the root zone of seedlings before transplanting [144, 145]. Furthermore, hydrophilic polymers, e.g. polyacrylamide, are used as soil conditioners in the hydromulching technique [146–150].

Plant growth regulators are used to modify the crop by changing the rate of its response to the internal and external factors that govern all stages of crop development. They are applied directly to the plant to alter its life processes or structure in some beneficial way so as to enhance yield, improve quality or facilitate harvesting. Active polymers have been described as controlled release plant growth regulators [50].

Recently, functionalized polymers have been used to protect citrus trees from frost damage, e.g. a copolymer of methyl acrylate with N-vinylpyrrolidinone inhibits ice nucleation and suppresses the freezing temperature of aqueous solutions [151].

REFERENCES

1. Tanquary, A.C. and Lacey, R.E. (eds) (1974) *Controlled Release of Biologically Active Agents*, vol. 47, Plenum, New York.
2. Paul, D.R. and Harris, F.W. (eds) (1976) *Controlled Release Polymeric Formulations*, Am. Chem. Soc. Symp. Ser. 33, Washington, DC.
3. Cardarelli, N.F. (1976) *Controlled Release Pesticides Formulations*, CRC Press, Boca Raton, FL.
4. Scher, H.B. (ed.) (1977) *Controlled Release Pesticides*, Am. Chem. Soc. Ser. 53, Washington, DC.
5. Donaruma, L.G. and Vogl, O. (eds) (1978) *Polymeric Drugs*, Academic Press, New York.
6. Carraher, C.E., Sheats, J.E. and Pittman, C. (eds) (1978) *Organometallic Polymers*, Academic Press, New York.
7. Baker, R. (1980) (ed.) *Controlled Release of Bioactive Materials*, Academic press, New York.
8. Kydonieus, A.F. (ed.) (1980) *Controlled Release Technologies: Methods, Theory, and Applications*, vols I and II, CRC Press, Boca Raton, FL.
9. Lewis, D.H. (ed.) (1981) *Controlled Release of Pesticides and Pharmaceuticals*, Plenum, New York.
10. Carraher, C.E. and Gebelein, C.G. (eds) (1982) *Biological Activities of Polymers*, Am. Chem. Soc. Symp. Ser. 186, Washington, DC.
11. Akelah, A. (1984) *J. Chem. Technol. Biotechnol.*, **34-A**, 263.
12. Akelah, A. and Selim, A. (1987) *Chim. Ind. (Milan)*, **69**, 62.
13. Lyman, D.L. (1966) *Rev. Macromol. Chem.*, **1**, 355; (1974) *Angew. Chem.*, **82**, 367; (1978) *Pure Appl. Chem.*, **50**, 427.
14. Ferruti, P. (1981) *Makromol. Chem. Suppl.*, **5**, 1.
15. Tighe, B.J. (1978) *Br. Polym. J.*, **8**, 71.
16. Ng, C.O. and Tighe, B.J. (1978) *Br. Polym. J.*, **8**, 78, 118.
17. Ng, C.O., Pedley, D.G. and Tighe, B.J. (1978) *Br. Polym. J.*, **8**, 124.
18. Yamauchi, Y. (1981) *J. Synth. Org. Chem. Jpn*, **39**, 238.
19. Grant, A. (1978) *Br. Polym. J.*, **10**, 241.
20. Braden, M. (1978) *Br. Polym. J.*, **10**, 245.
21. Dichter, M. (1976) US Patent 3 956 480.
22. Gebelein, C.G. and Koblitz, F.F. (eds) (1981) *Biochemical and Dental Applications of Polymers, Polym. Sci. Technol.*, **14**, Plenum, New York.
23. Tighe, B.J. (1980) Biomedical applications of polymers. In *Macromolecular Chemistry*, vol. 1 (eds A.D. Jenkins and J.F. Kennedy), Royal Society of Chemistry, London, Ch. 10, p. 416.
24. Gregoriadis, G. (ed.) (1979) *Drug Carriers in Biology and Medicine*, Academic Press, London.
25. Goldberg, E.P. and Nakajima, A. (eds) (1980) *Biomedical Polymers: Polymeric Materials and Pharmaceuticals for Biomedical Use*, Academic Press, New York.
26. Graham, N.B. (1979) *Br. Polym. J.*, **10**, 260.
27. Khomyakov, K.P., Virnik, A.D. and Rogovin, Z.A. (1964) *Russ. Chem. Rev.*, **33**, 462.
28. Donaruma, L.G. (1975) *Prog. Polym. Sci.*, **4**, 1.
29. Ringsdorf, H. (1975) *J. Polym. Sci. Symp.*, **51**, 135.
30. Batz, H.G. (1977) *Adv. Polym. Sci.*, **23**, 25.
31. Samour, C.M. (1978) *Chem. Tech.*, 494.
32. Gebelein, C.G. (1978) *Polym. News*, **4**, 163.

33. Ferruti, P., Tanzi, M.C., Maggi, F., Marchisio, M.A., Vaccaroni, F., Martsuscelli, E., Riva, F. and Provenzale, L. (1980) *Chim. Ind. (Milan)*, **62**, 109.
34. Gros, L., Ringsdorf, H. and Schupp, H. (1981) *Angew. Chem., Int. Ed. Engl.*, **20**, 305.
35. Bearn, A. (ed.) (1981) *Better Therapy with Existing Drugs: New Uses and Delivery Systems*, Merck, Biomedical Information, New York.
36. Bruck, S.D. (ed.) (1983) *Controlled Drug Delivery: Basic Concepts*, CRC Press, Boca Raton, FL.
37. Chien, Y.W. (1982) *Novel Drug Delivery Systems*, Marcel Dekker, New York.
38. Langer, R. and Wise, D. (eds) (1984) *Medical Applications of Controlled Release*, CRC Press, Boca Raton, FL.
39. McClosky, J. (ed.) (1983) *Drug Delivery Systems*, Aster, Springfield, OR.
40. Roseman, T.J. and Mandorf, S.Z. (eds) (1983) *Controlled Release Delivery Systems*, Marcel Dekker, New York.
41. Ushakov, S.N. and Panarin, E.F. (1962) *Dokl. Akad. Nauk SSSR*, **147**, 1102; (1963) *Chem. Abstr.*, **58**, 11168; (1963) *Dokl. Akad. Nauk SSSR*, **149**, 334; (1963) *Chem. Abstr.*, **59**, 6201.
42. Domhroski, J.R., Donaruma, L.G. and Razzano, J. (1971) *J. Med. Chem.*, **14**, 993.
43. Rowland, C.F., O'Neil, G.J. and Davies, D.A.L. (1975); *Nature*, **255**, 487; (1976) *Chemotherapy*, **8**, 11.
44. Zweig, G. (1977) In *Controlled Release Pesticides* (ed. H.B. Scher), Am. Chem. Soc. Symp. Ser. 53, Washington, DC, p. 37.
45. Allan, G.G., Chopra, C.S., Neogi, A.N. and Wilkins, R.M. (1971) *Nature*, **234**, 349.
46. Allan, G.G., Chopra, C.S., Friedhoff, J.F., Gara, R.I., Maggi, M.W., Neogi, A.N., Roberts, S.C. and Wilkins, R.M. (1973) *Chem. Tech.*, 171.
47. Allan, G.G., Cousin, M.J., McConnell, W.J., Powell, J.C. and Yahlaoui, A. (1977) *Polym. Prepr., Am. Chem. Soc., Div. Polym. Chem.*, **18**(1), 566.
48. Allan, G.G., Friedhoff, J.F., McConnell, W.J. and Powell, J.C. (1976) *J. Macromol. Sci. Chem.*, **A-10**, 223.
49. Friedhoff, J.F., Allan, G.G., Powell, J.C. and Roberts, S.C. (1974) *Polym. Prepr., Am. Chem. Soc., Div. Polym. Chem.*, **15**(1), 377.
50. Allan, G.G., Cousin, M.J. and Mikels, R.A. (1979) *Polym. Prepr., Am. Chem. Soc., Div. Polym. Chem.*, **20**(1), 341.
51. Akelah, A. and Rehab, A. (1985) *J. Polym. Mater.*, **2**, 149.
52. Akelah, A., Hassanein, M., Selim, A. and Rehab, A. (1987) *J. Chem. Technol. Biotechnol.*, **37-A**, 169.
53. Schacht, E.H., Desmarets, G. and Pierre, T.S. (1978) *Makromol. Chem.*, **179**, 543.
54. Schacht, E.H., Pierre, T.S. and Desmarets, G.E. (1977) *Polym. Prepr. Am. Chem. Soc., Div. Polym. Chem.*, **18**(1), 590.
55. Schacht, E.H., Desmarets, G. and Bogaert, Y. (1978) *Makromol. Chem.*, **179**, 837.
56. Bartulin, J. and Rivas, B.L. (1981) *Makromol. Chem. Rapid Commun.*, **2**, 375.
57. Wilkins, R.M. (1976) *Proc. Int. Controlled Release Pesticides Symp.*, 7-1; (1979) *Chem. Abstr.*, **91**, 85069-f.
58. Naruse, H. and Meakawa, K. (1977) *J. Fac. Agric. Kyushu Univ.*, **21**, 107, 153, 167.
59. McCormick, C.L., Anderson, K.W., Pelezo, J.A. and Lichatowich, D.K. (1981) In *Controlled Release of Pesticides and Pharmaceuticals* (ed. D.H. Lewis), Plenum, New York, p. 147.
60. Akelah, A., Selim, A. and Rehab, A. (1986) *J. Polym. Mater.*, **3**, 37.
61. Kamogawa, H., Haramoto, Y., Nakazawa, T., Sugiura, H. and Nanasawa, M. (1981) *Bull. Chem. Soc. Jpn.*, **54**, 1577.
62. Shambhu, M.B., Digenis, G.A., Gulati, D.K., Bowman, K. and Sabharwal, P.S. (1976) *J. Agr. Food Chem.*, **24**, 666.
63. Jakubka, H.D. and Busch, E. (1973) *Z. Chem.*, **13**(3), 105.
64. Georgieva, M. and Georgieva, E. (1978) *Angew. Makromol. Chem.*, **66**, 1.

65. Allan, G.G., Beer, J.W. and Cousin, M.J. (1977) In *Controlled Release Pesticides* (ed. H.B. Scher), Am. Chem. Soc. Symp. Ser. 53, Washington, DC, p. 94.
66. Allan, G.G. and Neogi, A.N. (1977) US Patent 4 062 855; (1978) *Chem. Abstr.*, **88**, 132022-k.
67. Allan, G.G. (1969) Fr. Patent 1 544 406; (1969) *Chem. Abstr.*, **71**, 100731-y; (1971) Can. Patent 855 181; (1974) US Patent 3 813 236.
68. Allan, G.G., Chopra, C.S., Neogi, A.N. and Wilkins, R.M. (1971) *Tappi*, **54** (8), 1293.
69. Allan, G.G., Beer, J.W., Cousin, M.J. and Powell, J.C. (1978) *Tappi*, **61** (1), 33.
70. McCormick, C.L. and Lichatowich, D.K. (1979) *J. Polym. Sci., Polym. Lett. Ed.*, **17**, 479.
71. McCormick, (1981) US Patent 4 278 790.
72. Allan, G.G., Chopra, C.S., Neogi, A.N. and Wilkins, R.M. (1972) *Int. Pest. Control*, **14** (2), 15; (1973) *Int. Pest. Control*, **15** (3), 8; (1975) *Int. Pest. Control*, **17** (2), 4.
73. Mehltretter, C.L., Roth, W.B., Weakley, F.B., McGuire, T.A. and Russell, C.R. (1974) *Weed Sci.*, **22**, 50, 415.
74. Thayumanuavan, B., Jagtap, H.S., Das, A.B. and Tilak, B.D. (1978) Indian Patent 144 674; (1980) *Chem. Abstr.*, **92**, 17189-r.
75. Kuo, P.C. (1980) *Chem. Abstr.*, **92**, 1533-n.
76. McCormick, C.L. and Fooladi, M. (1977) In *Controlled Release Pesticides* (ed. H.B. Scher), Am. Chem. Soc. Symp. Ser., 53, p. 112.
77. McCormick, C.L. (1981) US Patents 4 267 280, 4 267 281 (1981) *Chem. Abstr.*, **95**, 11646-r, 92387-u.
78. Harris, F.W. and Post, L.K. (1975) *J. Polym. Sci., Polym. Lett. Ed.*, **13**, 225; (1975) *Polym. Prepr., Am. Chem. Soc., Div. Polym. Chem.*, **16** (1), 622.
79. Harris, F.W., Aulabaugh, A.E., Case, R.D., Dykes, M.K. and Feld, W.A. (1976) In *Controlled Release Polymeric Formulations* (eds D.R. Paul and F.W. Harris), Am. Chem. Soc. Symp. Ser. 33, p. 222.
80. Harris, F.W., Dykes, M.R., Baker, J.A. and Aulabaugh, A.E. (1977) In *Controlled Release Pesticides* (ed. H.B. Scher), Am. Chem. Soc. Symp. Ser. 53.
81. Harris, F.W. and Arah, C.O. (1980) *Polym. Prepr., Am. Chem. Soc., Div. Polym. Chem.*, **21** (1), 107.
82. Feld, W.A. and Friar, L.L. (1978) *Polym. Prepr., Am. Chem. Soc., Div. Polym. Chem.*, **19** (2), 667.
83. Akagane, K. and Matsura, K. (1972) *Shikizai Kyokaishi*, **45** (2), 69; (1972) *Chem. Abstr.*, **77**, 128180-r.
84. Allan, G.G. and Halabisky, D.D. (1970) *J. Appl. Chem. Biotechnol.*, **21**, 190.
85. Faerber, S. (1960) Br. Patent 826 831.
86. Wang, C.S. and Sheetz, D.P. (1972) US Patent 3 660 353; (1972) *Chem. Abstr.*, **77**, 75732-p.
87. Naruse, H. and Maekawa, K. (1977) *J. Fac. Agr. Kyushu Univ.*, **21** (4), 173; (1977) *Chem. Abstr.*, **87**, 146932-y.
88. Meyers, W.E., Lewis, D.H., Van der Meer, R.K. and Lofgren, C.S. (1981) In *Controlled Release of Pesticides and Pharmaceuticals*, (ed. D.H. Lewis), Plenum, New York, p. 171.
89. Sterling, C. (1972) In *Our Chemical Environment* (eds J.C. Giddings and M.B. Monroe), Canfield Press, San Francisco, CA, pp. 84–90.
90. Mostofi, F.K. (ed.) (1967) *Bilharzia*, Springer, New York.
91. Chang, T.C. (ed.) (1974) *Molluscicides in Schistosomiasis Control*, Academic Press, New York.
92. Donaruma, L.G., Kitch, S., Depinto, J.V., Edzwald, J.K. and Muslyn, M.J. (1982) In *Biological Activities of Polymers* (eds C.E. Carraher and G.G. Gebelein), Am. Chem. Soc. Symp. Ser. 186, p. 55.

93. Akelah, A., Selim, A. and Rehab, A. (1987) *J. Polym. Mater.*, **4**, 117.
94. Akelah, A. and Rehab, A. (1986) *J. Polym. Mater.*, **3**, 83.
95. Dyckman, E.J. and Montemarano, J.A. (1973) *Am. Paint. J.*, **58**(5), 66.
96. DeLacourt, F.H. and DeVries, H.J. (1973) *Prog. Org. Coat.*, **1**, 375.
97. Phillip, A.T. (1973–4) *Prog. Org. Coat.*, **2**, 159.
98. Montemarano, J.A. and Dyckman, E.J. (1975) *J. Paint Technol.*, **47**, 59.
99. Castelli, V.J. and Yeager, W.L. (1976) In *Controlled Release Polymeric Formulations* (eds D.R. Paul and F.W. Harris), Am. Chem. Soc. Symp. Ser. **33**, Washington, DC, p. 239.
100. Subramanian, R.V., Garg, B.K. and Corredor, J. (1978) In *Organometallic Polymers* (eds C.E. Carraher, J.E. Sheats and C.U. Pittman), Academic Press, New York, p. 181.
101. Rzaev, Z.M.O. (1979) *Chem. Tech.*, 58.
102. Kronstein, M. (1980) *Polym. Prepr., Am. Chem. Soc., Div. Polym. Chem.*, **21**(1), 115.
103. Thayer, J.S. (1981) *J. Chem. Ed.*, **58**, 764.
104. Subramanian, R.V. and Somasekharan, K.N. (1981) *J. Macromol. Sci. Chem.*, **A-16**, 73.
105. Carraher, C.E. and Dammeier, R.L. (1970) *Makromol. Chem.*, **135**, 107; (1972) *J. Polym. Sci., A-1*, **10**, 413.
106. Carraher, C.E. and Winter, D.O. (1971) *Makromol. Chem.*, **141**, 237; (1972) *Makromol. Chem.*, **152**, 55.
107. Carraher, C.E. and Scherubel, G. (1971) *J. Polym. Sci., A-1*, **9**, 983; (1972) *Makromol. Chem.*, **152**, 61; **160**, 259.
108. Carraher, C.E. (1972) *Inorg. Makromol. Rev.*, **1**, 271.
109. Carraher, C.E., Jorgensen, S. and Lessek, P.J. (1976) *J. Appl. Polym. Sci.*, **20**, 2255.
110. Carraher, C.E., Giron, D.J., Woelk, W.K., Schroeder, J.A. and Fedderson, M.F. (1979) *J. Appl. Polym. Sci.*, **23**, 1501.
111. Subramanian, R.V. and Anand, M. (1977) In *Chemistry and Properties of Crosslinked Polymers* (ed. S.S. Labana), Academic Press, New York, p. 1.
112. Carraher, C.E. and Dammeier, R.L. (1970) *J. Polym. Sci., A-1*, **8**, 3367.
113. Deeks, A.S., Hudson, R.W., James D.M. and Sparrow, B.W. (1968) *2nd Int. Congr. on Marine Corrosion and Fouling*, Technical Chamber of Greece, Athens, p. 549.
114. Garg, B.K., Corredor, J. and Subramanian, R.V. (1977) *J. Macromol. Sci. Chem.*, **A-11**, 1567.
115. Messiha, N.N. (1981) *Polymer*, **22**, 807.
116. Dyckman, E.J., Montemarano, J. and Fischer, E.C. (1973) *Nav. Eng. J.*, **85**, 33; (1974) *Chem. Abstr.*, **81**, 34454-p.
117. Montemroso, J.C., Andrews, T.M. and Marinelli, L.P. (1958) *J. Polym. Sci.*, **32**, 523.
118. Akagane, K. and Allan, G.G. (1973) *Shikizai Kyokaishi*, **46**, 437; (1974) *Chem. Abstr.*, **80**, 61104-k.
119. Ghanem, N.A., Messiha, N.N., Ikaldious, N.E. and Shaaban, A.F. (1979) *Eur. Polym. J.*, **15**, 823.
120. M. & T. Chemicals Inc., (1969) Ger. Patent 1 300 700; (1969) *Chem. Abstr.*, **71**, 82157.
121. Kochkin, D.A. (1969) Fr. Patent 1 561 245; (1969) *Chem. Abstr.*, **71**, 126 099.
122. James, D.M. (1968) Br. Patent 11 242 897.
123. Lubrick, J.R. (1965) US Patent 3 167 473.
124. Kochkin, D.A., Rzaev, Z.M., Suchareva, L.A. and Zubov, P.I. (1968) USSR Patent 255 564.
125. Kanasai Paint Co. (1971) Fr. Patent 2 026 091.
126. Kochkin, D.A., El'Khaner, G.E., Smutkina, Z.S. and Zubov, P.I. (1970) *Zh. Fiz. Khim.*, **44**, 2984.
127. Akagane, K. (1972) *Shikiza Kyakaish*, **45**, 69; (1972) *Chem. Abstr.*, **77**, 128180-r.

128. Smith, P. and Smith, L. (1975) *Chem. Br.*, **11**, 208.
129. Subramanian, R.V., Mendoza, J.M. and Garg, B.K. (1978) in *Proc. 5th Int. Symp. on Controlled Release of Bioactive Materials*, Gaithersburg, MD, 14–16 August, p. 6.8.
130. Meyer, J.A. (1981) *Wood Sci.*, **14**(2), 49.
131. Lunt, O.R. (1971) *J. Agr. Food Chem.*, **19**, 797.
132. Blouin, G.M., Rindt, D.W. and Moore, O.E. (1971) *J. Agr. Food Chem.*, **19**, 801.
133. Allen, S.E. and Mays, D.A. (1971) *J. Agr. Food Chem.*, **19**, 809.
134. James, B.L. (1971) *J. Agr. Food Chem.*, **19**, 813.
135. Smith, W.H., Underwood, H.G. and Hays, J.T. (1971) *J. Agr. Food Chem.*, **19**, 816.
136. Nobell, A. (1973) US Patent 3 759 687.
137. Cropp, J.A.D., D'Ouville, E.L. and Messman, H.C. (1974) Br. Patent 1 378 938.
138. Goertz, H.M. (1977) US Patent 402 539.
139. Jackson, L.P. (1977) US Patent 4 055 974.
140. Trivodi, R.N. and Pachaiyappan, V. (1979); *Fert. News*, **24** (10), 19; (1980) *Chem. Abstr.*, **92**, 162636-p.
141. Seymour, R.B. (1980) *Polym. News*, **6**, 101.
142. Dubois, P. (1978) *Plastics in Agriculture*, Applied Science, London.
143. Porter, F.E. (1978) *Chem. Technol.*, (5), 285.
144. Bryan, H. (1979) *Am. Veg. Grow.*, (5), 30.
145. Taylor, A.G., Motes, J.E. and Price, H.C. (1978) *Hort. Sci.*, **13**, 481.
146. Shrader, W.D. and Mostejeran, A. (1977) *Am. Chem. Soc., Div. Coat. Plast., Prepr.*, **37**, 683.
147. Miller, D.E. (1979) *Soil Sci. Soc. Am. J.*, **43**, 628.
148. Hemyari, P. and Nofziger, D.L. (1981) *Soil Sci. Soc. Am. J.*, **45**, 799.
149. Stefanson, R.C. (1975) *Soil Sci.*, **119**, 426.
150. Azzam, R.A. (1980) *Commun. Soil Sci. Plant Anal.*, **11**, 767.
151. Allegretto, B. (1984) *Chem. Technol.*, (3), 152.

8

Polymeric supports for active groups

Functionalized polymers attaching the desired reactive groups by chemical bonding possess a combination of the physicomechanical properties of a high polymer and the chemical properties of the attached group and hence lead to a polymer that combines the advantages of a conventional reactive moiety and of a polymer. There are a number of areas where functionalized polymers have been employed as a convenient support upon which to attach active groups such as food additives and sunscreens.

8.1 POLYMERIC FOOD ADDITIVES

Increasing demands for food additive safety have led to the application of functionalized polymers in the food industry to rid foods of problems with artificial additives while maintaining product appearance, texture, flavour and cost [1, 2]. In addition to the typical applications of immobilized enzymes on polymeric supports in the food industry [3,4], e.g. cheese making [5,6], stabilization of milk [7,8] (dairy industry) and clarifying fruit juices and wines [9], another important and successful application is the development of safe food additives. Such additives are added for a specific purpose other than nutritional purposes.

This concept is based on the attachment of the functional group of food additives to appropriate functionalized polymer molecules to produce additives of large molecular size which cannot be adsorbed through the intestinal wall, i.e. they cannot migrate across the membranes and therefore cannot pass into the blood stream or be metabolized. Consequently, the large molecule of the additive would not contact the usual organs such as the kidney and liver but would be excreted in the faeces without any metabolization, thus eliminating the possibilities of side effects resulting from the absorption of soluble additive materials. However, the charged or modified activity and toxicity of polymeric food additives are influenced by the nature of the functional groups and the polymeric matrix. Hence the following factors are important and must be taken into consideration during the design of polymeric food additives so that they are non-absorbable.

(i) Stability. The chemical linkage attaching the food additive to the polymer and also the polymeric backbone must be resistant to breakdown by chemical or biological environments, under food processing, shipping or storage conditions including light and heat exposure and under the enzymatic and microbiological conditions of the gastrointestinal tract. This stability is required in order to eliminate any possibility of the formation of low molecular weight species, by depolymerization, degradation, digestion or hydrolysis, which will give rise to absorbable fragments. In addition, chemical stability is also important to preserve functionality. A simple hydrocarbon backbone is especially stable under product processing conditions or under the conditions of metabolism and does not interfere with the additive properties.

(ii) Solubility. The choice of chemical nature of the polymer backbone often depends on the degree of water or oil solubility of the final polymeric food product. In some cases, e.g. in polymeric food dyes, it is desirable to incorporate water-solubilizing groups either into the backbone or into the food functional groups to obtain the clear solutions necessary for food processing. Generally, water solubility is achieved by incorporating hydrophilic polar groups to 10% or more. Conversely in some cases it is desirable to increase the oil and fat solubility of the polymeric food additives, as in the case of antioxidants used for the stabilization of oils and fats. This property can be achieved by the incorporation of non-polar oleophilic groups such as hydrocarbon chains into the polymeric food additive.

(iii) Molecular weight. To achieve the desired non-absorption through the intestinal wall and hence to eliminate any risk of systemic toxicity, the chemical backbone of the polymeric food additive must have a sufficiently large molecular weight and size. This is generally achieved when the polymer has a molecular weight of at least 10 000.

(iv) Type of bonding. In addition to the resistance of the chemical linkage to rupture, it must be tasteless and odourless, give no colour, interact only mildly with food components and not interfere with the properties of the food activity.

(v) Compatibility and blendability. The polymeric additives must be compatible and blendable with the other food components.

   Three types of non-conventional polymeric food additives have been developed including polymeric food colorants, antioxidants and sweeteners.

8.1.1 Polymeric food colorants

Coloured materials are widely used in the food industry to enhance and improve the appearance and appeal of processed food products, but their presence can be hazardous to health [10]. One of the solutions to this problem is the use of a polymeric backbone chemically attached to conventional food chromophores. The molecular weight of the polymer complex is sufficiently large to permit the colour to pass through the walls of the gastrointestinal tract without absorption into the body [1, 11–14]. Under these conditions, conventional chromophores, which are not suitable for food colouring purposes because of their water insolubility or toxicity, can be used in a polymeric form since they will achieve improved solubility and non-absorbability and hence will be non-toxic. In addition, polymeric food dyes must have good water solubility with colour purity in aqueous media, which can be achieved by introducing anionic groups. For example, the water-soluble polymers **1** with water-insoluble anthraquinone chromophores were prepared by treating poly(vinylamine) with 0·5 equivalent of bromoanthraquinone and converting the unreacted amines into water-solubilizing sulphamate groups by treatment with Me_3NSO_3 [15]:

1

Polymeric food dyes have been prepared either by polymerization of monomeric chromophores or by chemical modification of preformed polymers through a suitable functional group. For example, the methacrylamide derivative **2** was polymerized to give the polymeric food colorant **3** [16]:

Table 8.1 Polymeric food colorants

Functionalized polymers	Colour	Reference
$R \equiv H, Me;$ $R' \equiv H, Ph, COMe, COOEt$	Red	11, 15, 17–20
		15, 19
$R \equiv -NH-\bigcirc-Me$	Yellow, orange	15, 18, 19
$R' \equiv H, Me; R'' \equiv Me, OC_nH_{2n+1};$ $n = 1-4$ $R''' \equiv H, COOEt$		
$R \equiv NO_2, SO_3H$	Yellow	11
(a) $Z \equiv -NH-; R \equiv H, SO_3Na;$ $R' \equiv H, Me$	Blue	11, 15, 19

Table 8.1 (*Contd.*)

Functionalized polymers	Colour	Reference
(b) $Z \equiv -NH(CH_2)_2SO_2-\bigcirc-NH;$		21
$R \equiv -SO_3Na; \ R' \equiv H$		
ⓟ$-N-$ [anthraquinone structure] with X^1, Me	Purple	19
ⓟ$-Z-CH+(\bigcirc)-NR_2)_2$		22
$R \equiv Me, Et$		
(a) $Z \equiv -\bigcirc-$ with NH_2/NMe_2		
(b) $Z \equiv -\bigcirc-NHCH_2-\bigcirc-$	Blue, violet, green	
(c) $Z \equiv -\bigcirc-$	Blue	
ⓟ$-Z-\bigcirc-N=N-R$ with X		
(a) $Z \equiv -NHSO_2-; \ X \equiv X^2$		
$\quad R \equiv R^1, R^2$	Orange	19, 16, 23–26
$\quad R \equiv R^3$	Yellow	11, 23, 24
$\quad R \equiv R^4$	Burgundy	23–25
$\quad R \equiv R^5$	Red	11
$\quad R \equiv R^6$	Black	11
(b) $Z \equiv -NHSO_2-; \ X \equiv X^4$		
$\quad R \equiv R^7$	Amaranth	23–25
(c) $Z \equiv -SO_2-; \ X \equiv X^3$		
$\quad R \equiv R^1, R^2, R^5$	Burgundy	11, 23–25
ⓟ$-NHCH_2-\bigcirc(OH)-N=N-\bigcirc-SO_3Na$		19
	Orange	

Table 8.1 (*Contd.*)

Functionalized polymers	Colour	Reference
	Red, orange	16

$Z \equiv O, NH; R \equiv R' \equiv H, Me, SO_3Na$

2   3

A variety range of chromophore classes including azo groups, which do not meet the requirements of biological stability because they are cleaved in the gut to yield absorbable aromatic amines, have been reported to be incorporated into polymers. Other chromophore groups such as anthraquinones, anthrapyridones, anthrapyridines, benzanthrones, nitroanilines and triphenylmethanes have also been incorporated into polymers, as shown in Table 8.1.

8.1.2   Polymeric food antioxidants

The use of antioxidants has recently received considerable interest in the food industry for a wide range of foodstuffs that are especially exposed to deterioration, e.g. products containing oils and fats. Antioxidants increase the stability of foodstuffs in storage as well as increasing the retention of nutritional and flavour values by delaying rancidity. Although several phenolic compounds have been used as antioxidants in the food industry, they are not satisfactory because of the toxicity of many phenolic derivatives and because they lose their inhibitory action by evaporation during product processing [27].

Polymeric antioxidants have recently been employed to overcome the problems of conventional phenolic compounds. For example, polymeric antioxidants **4** were prepared by polymerizing the vinyl group of α-(2-hydroxy-3, 5-dialkylphenyl)ethylvinylbenzene [28–32]:

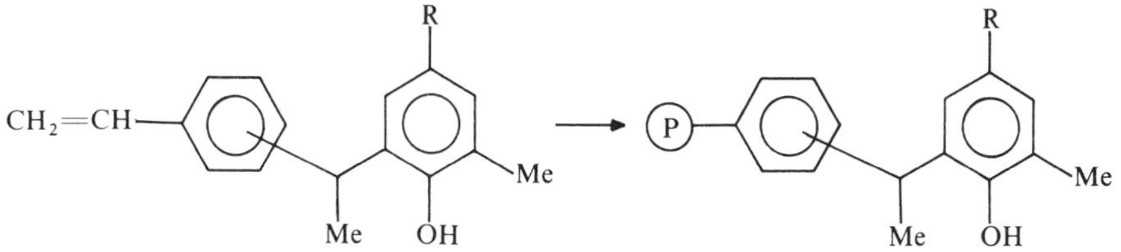

R ≡ H, Me, Et

4

Because of the high molecular weight of the polymer they are non-volatile and hence keep their inhibitory action in the finished food products to the desired degree without any addition of more antioxidants. In addition, they prevent the absorption of the antioxidant group through the intestinal wall, thereby eliminating any risk of toxicity. Since antioxidants are used in oil and fat food products and also in high temperature operations, oil solubility properties and thermal stability are particularly important. Based on the good thermal stability of condensation polymers, non-absorbable antioxidants have been prepared recently by condensation polymerization of active monomers with the desired functionality. For example, divinylbenzene was reacted with various phenols and hydroquinones such as hydroxyanisole, tertiary butyl phenol, p-cresol, bisphenol A and tertiary butyl hydroquinone in the presence of an aluminium catalyst to give polymeric antioxidants 5 with the desired properties [33–39]:

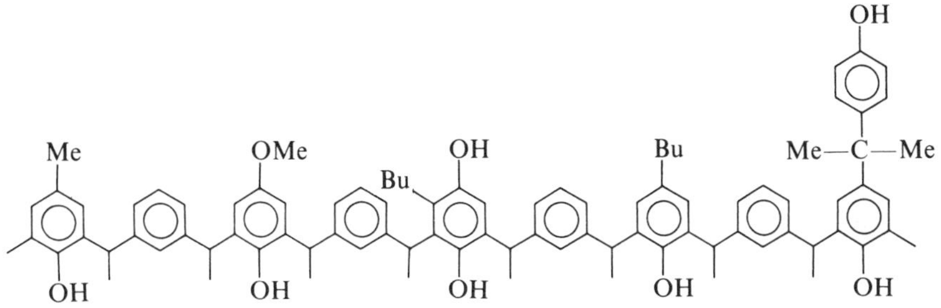

5

8.1.3 Polymeric non-nutritive sweeteners

Despite the advantages of using synthetic sweeteners for producing a sensory response of sweetness with no food value for diet control, in an attempt to remove the unwanted physiological disadvantages of naturally occurring carbohydrates such as obesity and tooth decay, they present another problem, toxicity, and in some cases cause chromosome damage and bladder trouble.

Polymeric sweeteners [40], prepared by chemically bonding a naturally or synthetic active sweetening group to a polymer backbone, have recently been used to produce the sweetening effect with non-nutritive value and without any appreciable absorption, hence eliminating all the problems associated with naturally occurring and artificial sweeteners. For example, a non-nutritive sweetener **6** comprising saccharin covalently bonded to agarose derivative has been prepared as follows:

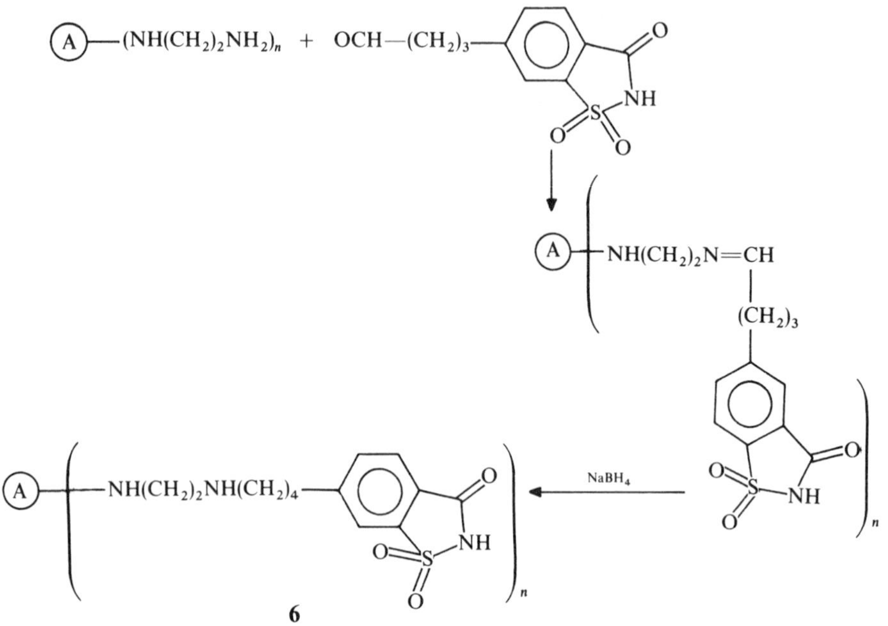

6

8.1.4   Polymeric feed additives for animals

Because of the importance of meat as a food product for human consumption it is desirable to increase the nutritional efficiency of feed supplied to domesticated animals such as poultry, cattle and sheep generally raised as sources of meat [41, 42]. An improved animal feed is effective when the rate of growth of the animal and the amount of growth per unit weight of feed devoured by the animal are improved. For example, polyvinylpyrrolidone incorporated in the feed of domesticated animals at relatively low concentrations, 0.01%–0.05% by weight, has produced a desirable stimulation in growth and improvement of feed efficiency [41]. In addition it promoted the rate of growth and was also capable of counteracting some of the undesirable effects of toxic agents, such as 3-nitro-4-hydroxyphenyl arsenic acid, incorporated into feeds for various medicinal purposes.

8.2   POLYMERIC SUNSCREEN

The beneficial effects of sunlight on the human organism are well known. Ultraviolet radiation produced by the sun or by sun lamps in contact with the

Table 8.2 Polymeric sunscreen

(P)—COOCH$_2$CHCH$_2$O—⟨○⟩—COPh (with OH at top of ring, and OH below CHCH$_2$)   Ref. 43

(P)—O(CH$_2$)$_2$OCOCH=CH—Ph
COO(CH$_2$)$_2$N$^+$Me$_3$MeSO$_4^-$ Ref. 44

(P)—CONHCH$_2$— (coumarin structure with OH)   Ref. 45

(P)—⟨○⟩—OR
COOR′   Ref. 46

(P)—CONHCH$_2$—⟨○⟩—COPh (with HO)
COOMe   Ref. 47

—(—CH$_2$CH$_2$—N——)$_n$—
CO—⟨○⟩—NR$_2$
R ≡ C$_m$H$_{2m+1}$; m = 1—3   Ref. 48

—(—CO—(CH$_2$)$_m$—N———)$_n$—
CONHR
m = 5, 11; R ≡ (naphthalene) (biphenyl)   Ref. 49

(P with CH$_2$ ... CH$_2$ ring)—N—⟨○⟩—COOR′
R
X
R′ ≡ H, Et; R ≡ H, 2-Me; n = 2, 3;
X ≡ COOH and COO(CH$_2$)$_n$OH   Refs 50, 51

(P)—CONR$_2$ Ref. 52

R ≡ C$_n$H$_{2n+1}$(n = 1–4), Ph, —(CH$_2$)$_m$(OR′ or CN)

Table 8.2 (*Contd.*)

$-(-CO-R-CONH(CH_2)_mNH-)_n-$ Ref. 52
$m = 2, 3$

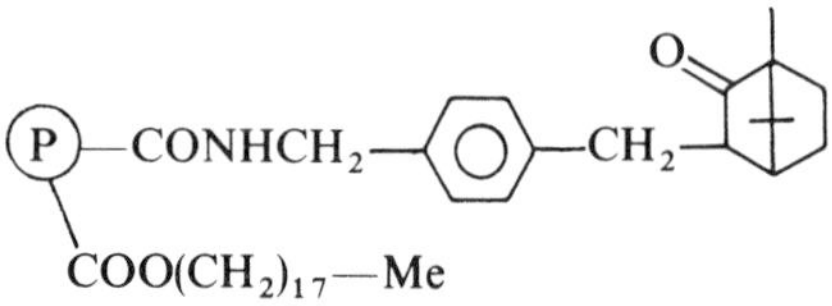

Ref. 53

$\text{(P)} \equiv -(CH_2-CH)_n- \text{ or } -(CH_2CMe)_n-;$

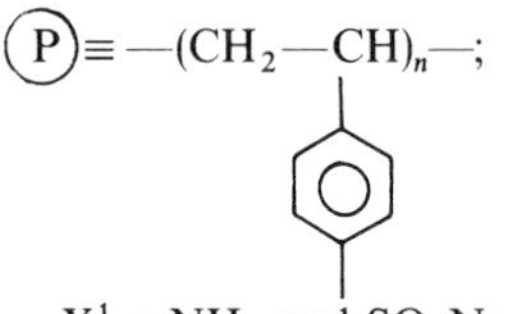

$\text{(P)} \equiv -(CH_2-CH)_n-;$

$X^1 \equiv NH_2 \text{ and } SO_3Na;$

$X^2 \equiv COOH \text{ and } NH_2 \text{ and } SO_3-\!\!\!\bigcirc\!\!\!-NH_2;$

$X^3 \equiv SO_2-\!\!\!\bigcirc\!\!\!-NH_2;$

$X^4 \equiv COOH \text{ and } NH_2 \text{ and } NHSO-\!\!\!\bigcirc\!\!\!-NH_2;$

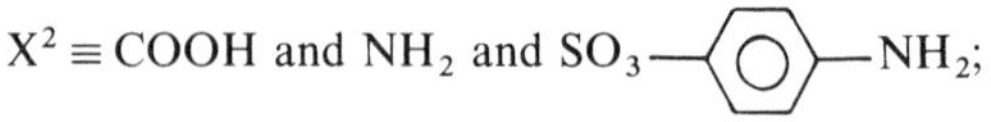

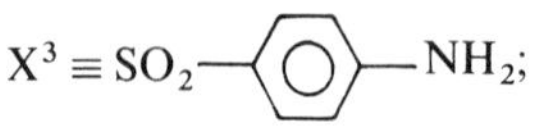

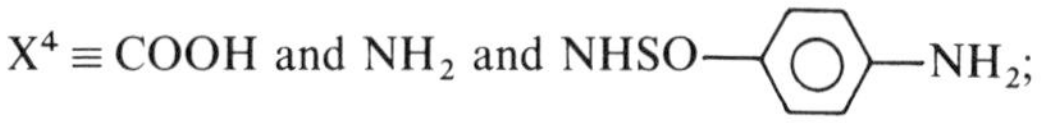

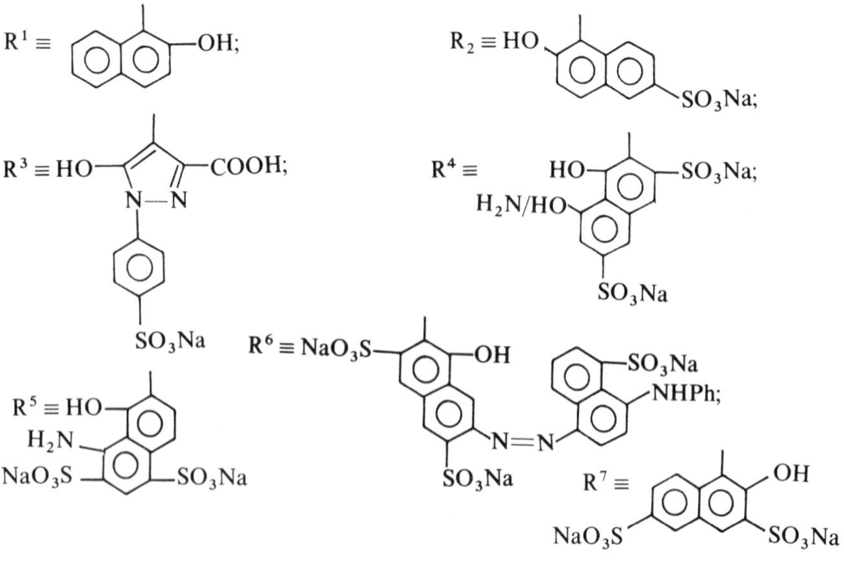

skin yields vitamin D which assists the body in warding off disease. However, rays with wavelengths of 260–320 nm are responsible for sunburn which is suffered on overexposure, while ultraviolet radiation up to 420 nm causes the desirable tanning coloration of the skin due to darkening of preformed pigment. The prevention of sunburn is only possible if a large percentage of these rays are not allowed to come into contact with the skin; this is impossible under the conditions most commonly found at the sea shore.

Sunscreening compositions containing compounds which can absorb the majority of the undesirable harmful radiation and re-emit it in a non-harmful form have been made and used to diminish or eliminate the discomforts and dangers of sunburn while promoting the development of a tanned appearance. Various factors other than the duration of exposure must be fulfilled by the ideal sunscreening agent.

1. *Stability*: since sunscreening compositions are generally used in hot weather at beach or water facilities and the heat causes the user to perspire, the active compound must be resistant to breakdown resulting from storage, light, heat or air oxidation and not easily removed from the skin by water and perspiration.
2. *Non-toxicity:* since sunscreen compositions are topically applied, they must not be harmful to the skin and must not penetrate it.
3. *Compatibility:* it must be compatible with the skin surface and with the other substances present in the sunscreen composition.

However, conventional compositions useful in filtering out the radiation are not ideal and lack one or more of these desirable characteristics. In an attempt to overcome the disadvantages of using conventional active agents, functionalized polymers chemically bonding the protective agents have been synthesized and used in sunscreen formulations. Such polymeric products, when topically applied to the skin, form a film which resists removal from the skin under the action of water or perspiration. Hence it is not necessary to reapply it after each contact with water owing to body perspiration. In addition, the use of a polymeric sunscreen reduces the possibility of undesirable skin penetration because of its high molecular weight and hence alleviates the problem of toxicity. Some functionalized polymers applied as protective agents in sunscreen formulations are shown in Table 8.2.

REFERENCES

1. Moore, K.K. (1977) *Food Prod. Dev.*, **11**(4), 63, 80; (1977) *Chem. Abstr.*, **87**, 37399-j.
2. Leonard, W.J. (1978) Macromolecular control of food additives. In *Polymeric Delivery Systems* (ed. R.J. Kostelnik), Gordon and Breach, New York, pp. 269–90.
3. Komaki, T. (1977) *New Food Ind.*, **19**(11), 2; (1978) *Chem. Abstr.*, **88**, 87662-f.
4. Kilara, A. and Shahani K.M. (1979) *CRC Crit. Rev. Food Sci. Nutr.*, **12**(2), 161.

260   Polymeric supports for active groups

5. Richardson, T. and Olson, N.F. (1974) In *Immobilized Enzymes in Food and Microbial Processes* (eds A.C. Olson and C.L. Cooney), Plenum, New York, p. 19.
6. Hicks, C.L., Ferrier, L.K., Olson, N.F. and Richardson, T. (1974) *J. Dairy Sci.*, **58**, 19.
7. Shipe, W.F., Senyk, G.F. and Weetall, H.H. (1972) *J. Dairy Sci.*, **55**, 647.
8. Le, E.C., Senyk, G.F. and Shipe, W.F. (1974) *J. Dairy Sci.*, **58**, 473.
9. Vieth, W.R. and Venkatasubramanian, K. (1973) *Chem. Tech.*, 677.
10. Noonan, J. (1975) Color additives in food. In *CRC Handbook of Food Additives*, 2nd edn (ed. T. Furia), CRC Press, Boca Raton, FL.
11. Dawson, D.J. (1981) *Aldrichimica Acta*, **14** (2), 23.
12. Dawson, D., Gless, R. and Wingard, R.E. (1976) *Chem. Tech.*, 724.
13. Bellanca, N. and Leonard, W.J. (1977) In *Current Aspects of Food Colorants* (ed. T.E. Furia), CRC Press, Cleveland, OH, p. 49; (1978) *Chem. Abstr.*, 145100-e.
14. Furia, T.E. (1977) *Food Tech.*, **31** (5), 34.
15. Dawson, D.J., Otteson, K.M., Wang, P.C. and Wingard, R.E. (1978) *Macromolecules.*, **11**, 320.
16. Ida, T., Takahashi, S. and Utsumi, S. (1969) *Yakugaku Zasshi*, **89** (4), 517; (1969) *Chem. Abstr.*, 51214-w.
17. Otteson, K.M. and Dawson, D.J. (1977) Ger. Patent 2 655 438; (1977) *Chem. Abstr.*, 137300-p.
18. Wang, P.C., Wingard, R.E. and Bunes, L.A. (1981) US Patent 4258189; (1981) *Chem. Abstr.*, **94**, 2102994-r.
19. Wang, P.C. and Wingard, R.E. (1977) US Patent 4 051 138; (1978) *Chem. Abstr.*, **88**, 24247-d.
20. Bunes, L.A. (1980) US Patent 4 182 885; (1980) *Chem. Abstr.*, **92**, 182565-t.
21. Wingard, R.E. and Dawson, D.J. (1978) Ger. Patent 275 162; (1978) *Chem. Abstr.*, **89**, 112385-y.
22. Ida, T., Takahashi, S. and Hashimoto, T. (1967) *Chem. Abstr.*, **66**, 19843-m; (1973) Jpn. Patent 144 33.
23. Dawson, D.J. and Rudinger, J. (1976) US Patent 4 000 118; (1977) *Chem. Abstr.*, **87**, 40725-t.
24. Dawson, D.J. and Rudinger, J. (1975) Ger. Patent 2 456 356; (1975) *Chem. Abstr.*, **83**, 149087-e.
25. Dawson, D.J., Gless, R.D. and Wingard, R.E. (1976) *J. Am. Chem. Soc.*, **98**, 5996.
26. Bellanca, N. and Furia, T.E. (1979) US Patent 4 167 422; (1980) *Chem. Abstr.*, **92**, 43267-j.
27. Stuckey, B.M. (1975) Antioxidants as food stabilizers. In *CRC Handbook of Food Additives*, 2nd edn (ed. T. Furia) CRC Press, Boca Raton, FL.
28. Dale, J.A. and Leonard, W.L. (1975) US Patent 3 930 047.
29. Zaffaroni, A. (1976) US Patent 3 994 828; (1978) *Chem. Abstr.*, **89**, 60475-f.
30. Zaffaroni, A. (1978) US Patent 4 104 196.
31. Zaffaroni, A. (1975) Ger. Patent 2 427 627; (1975) *Chem. Abstr.*, **82**, 169100-m.
32. Dale, J.A. and Leonard, W.L. (1977) US Patent 4 028 342; (1977) *Chem. Abstr.*, **87**, 85747-t.
33. Weinshenker, N.M., Bunes, L.A. and Davis, R. (1976) US Patent 3 996 199.
34. Wang, P.C. and Dale, J.D. (1976) US Patent 3 996 198.
35. Weinshenker, N.M. (1977) *Polym. Prepr., Am. Chem. Soc., Div. Polym. Chem.*, **18** (1), 531.
36. Weinshenker, N.M. (1979) *Polym. Prepr., Am. Chem. Soc., Div. Polym. Chem.*, **20** (1), 344.
37. Dale, J.A. and Ng, S.Y.W. (1978) US Patent 4 078 091.
38. Weinshenker, N.M. and Dale, J.A. (1977) US Patent 4 054 676.
39. Kolka, A.J., Napolitano, J.P., Filbey, A.H. and Ecke, G.G. (1957) *J. Org. Chem.*, **22**, 642.

40. Zaffaroni, A. (1975) US Patent 3 876 816.
41. Dawe, V. (1962) US Patent 3 015 564.
42. Wu, S.H., Dannelly, C.C. and Kormarek, R.J. (1981) In *Controlled Release of Pesticides and Pharmaceuticals* (ed. D.H. Lewis), Plenum, New York, p. 319.
43. Furendal, A.R.B. (1969) Fr. Patent 1 580 281; (1970) *Chem. Abstr.*, **93**, 123443-m.
44. Jacquet, B., Papantoniou, C., Dufaura, P. and Mahieux, C. (1974) Ger. Patent 2 333 306; (1974) *Chem. Abstr.*, **80**, 134227-q.
45. Jacquet, B., Papantoniou, C., Dufaura, P. and Mahieux, C. (1975) Fr. Patent 2 237 912.
46. Bailey, D., Tirrell, D. and Vogl, O. (1976) *J. Polym. Sci., Polym. Chem. Ed.*, **14**, 2725.
47. Jacquet, B., Papantoniou, C., Dufaure, P. and Mahieux, C. (1975) Ger. Patent 2 333 305; (1975) *Chem. Abstr.*, **83**, 84713-j.
48. Ciaudelli, J.P. (1975) US Patent 3 864 473.
49. Naruse, N., Yasumano, R. and Inoue, S. (1973) Jpn. Patent 7 325 423; (1974) *Chem. Abstr.*, **80**, 121859-x.
50. Skoultchi, M. and Meier, E.A. (1974) US Patent 3 795 733.
51. Skoultchi, M. and Meier, E.A. (1974) US Patent 3 836 571.
52. Karg, G. (1975) US Patent 3 895 104.
53. Oreal, S.A. (1974) Neth. Appl. 7 309 147; (1975) *Chem. Abstr.*, **83**, 48095-a.

Part Four   Other Technological Applications

9
Technological applications

Following the successful applications of polymers in the laboratory and in industrial processes, the introduction of active functional groups, other than those previously discussed, into polymers will be discussed in the present chapter. The uses of functionalized polymers in technological applications are reviewed in an attempt to explore the possible contributions of polymer science to the development of a number of new polymeric materials for a broad area of utilization. Accordingly, the application of reactive polymers is subdivided into various main areas.

9.1 CONDUCTIVE POLYMERS

Metals are defined as elements which possess certain characteristic chemical properties, such as ease in forming positive ions by chemical processes, and certain characteristic solid state physical properties, such as high electrical conductivity, good reflectivity of light, good thermal conductivity and ductility. Despite these collective properties, metals have some drawbacks such as high cost of fabrication, poor mechanical properties (stiffness, cut and abrasion resistance) and physical properties (high density and weight).

A significant property of many organic polymers is their ability to withstand high electric fields with negligible conduction. This property makes polymers the material of choice in a wide range of applications. The absence of conductivity is due to large energy differences between localized valence electron states and the conduction band. However, there is continuing interest in the synthesis of polymers possessing conductive properties. The following discussion deals with various phenomena concerned with the conductivity of polymers.

9.1.1 Photoconductive polymers

The photoconductivity of polymers is one of the commercially significant photoresponses of polymeric systems and plays a central role in the development of a generation of organic photoconductors which have the advantages of ease of fabrication and toughness. Since the first report on photoconductive polymers [1], considerable interest has been focused on these materials and the subject of polymer photoconduction has been extensively reviewed in recent years [2–8].

Photoconductivity is defined as a significant increase in conductivity caused by illumination. This increase is attributed to an increase in the number of charge carriers (electrons or holes) as a direct result of electronic excitation. The phenomenon of photoconductivity involves two distinctly different steps: charge generation and charge transport.

For a polymer to be photoconductive the absorption of a photon must lead to the generation of a mobile charge carrier, followed by energy transfer processes. The photons of the radiation interact with the semiconductor to promote an electron directly from the valence band to the conduction band to give an electron–hole pair, thereby enhancing the generation of charge carriers. For this to be possible the photon energy must overcome the band gap. Generation of carriers by photon absorption, however, often proceeds in a more indirect manner. The first step may be the production of excitons, which are localized but mobile excited electronic states which cannot by themselves transport charge. Two excitons may subsequently collide to produce an electron–hole pair or an exciton may migrate to the surface and react with a surface state to inject a carrier of one sign into the bulk (photoinjection of carriers). When the light is switched off, the photoconduction will decay as the carrier population gradually returns to equilibrium.

Hole transport is defined as a process in which cationic species are involved. By a series of electron transfers the positive charge migrates to the cathode where it is neutralized. Electron transport involves anions. A series of electron transfers moves a negative charge from the cathode to the anode where it is neutralized. The charge of charge carriers can be controlled by consideration of ionization potentials (low values being desirable for holes) and electron affinities (high values being desirable for electron transport). The rate of transport is related to the concentration of the active transport species and the behaviour to electron transfer. The accepted mechanism for photogeneration of charge carriers in polymers involves localization of migrating excitation energy at a trapping site followed by electron transfer to a neighbouring group. The resulting charged geminate pair may then separate and either a positive charge, an electron or both may migrate in a polarizing field. Thus, photogeneration is considered to be the formation of a radical anion–radical cation pair upon excitation in the presence of an electric field, i.e.

$$D + A \xrightarrow{h\nu} [D^+ A^-]^* \xrightarrow[\text{field}]{\text{electric}} D^{\cdot +} + A^{\cdot -}] \qquad (9.1)$$

Improved conductivity can be achieved by (a) the addition of a small molecular dopant or (b) chemical modification of the polymer. The dopant is a charge transfer agent such that an electron transfer reaction can occur between it and the polymer matrix. For example, dyes such as rose bengal, methyl violet, methylene blue and pinacyanol are commonly employed to extend the spectral response of photoconducting polymers (sensitization). The mechanisms of the dye sensitization process involve both energy transfer from the dye to the polymer and/or electron transfer between the excited state of the dye and the polymer transport matrix leading to the formation of a charge carrier. The direction of electron transfer depends upon the relative energy levels of the dye and the polymer. Electron transfer from the polymer to the excited dye is the situation usually encountered in practice. The improvement in the photogeneration of charge carriers by the donor–acceptor interaction between the hole-transporting polymer and the dopant (acceptor) leads to a charge transfer band in both the absorption and the action spectra. The technique of chemical modification appears to be of general applicability for improving both the spectral response and the increase in charge carrier photogeneration efficiency of hole-transporting systems. In addition, it has the following advantages.

1. Photoconductivity and its component processes, charge generation and transport, can be controlled.
2. The spectral range associated with the charge photogeneration process can be controlled by surface or bulk attachment of dyes or charge acceptors onto the polymer.
3. The charge transport process can be influenced in terms of the sign of the majority carriers and the rate of transport. The introduction of functionalities with a low ionization potential, such as carbazole, triarylamine or pyrene, by chemical reaction of functional polymers, can control the sign of the charge carriers and is successful in generating hole (cationic) transporting systems.
4. It is also useful in improving the mechanical properties, particularly the solubility, film formation flexibility, and impact resistance, of photoconductive polymer systems.
5. It eliminates deleterious reactions with water or dopants.

Carbazole-containing polymers are the subject of many investigations with respect to their photoconductive properties [9,10]. However, by changing the chemical structure while keeping the activity group, carbazole, some advantages can be achieved [11]. One idea is the copolymerization of a carbazole containing monomer with another monomer having the sensitizer (an acceptor group) linked to the double bond through a spacer.

$$PVK \xrightarrow{hv} [PVK^+ + e] \xrightarrow{trap} PVK^+ + trap^- \tag{9.2}$$

exciton

$$[PVK^+ + e] + TNF \xrightleftharpoons{sensitization} PVK^+ + TNF^+ \tag{9.3}$$

The advantages of such a structure (a) the possibility of designing donor to acceptor group ratios and sequences on the side chain, (b) the improvement of film properties, (c) the availability of a relatively large choice of acceptor groups with different electronic affinities and therefore exhibiting various charge transfer bands in the visible domain and (d) the influencing of electric absorption, fluoresence, emission and nuclear magnetic resonance spectra by the relative orientations and interactions of carbazole units in polymers. Some photoconductive polymers based on vinyl derivatives of polynuclear aromatic compounds are listed in Table 9.1.

(a) Xerography

Recently there has been an increasing interest in photoconductive polymers because of their potential utility in the electroreprographic industry as well as a number of related processes [51]. The process by which two bodies in contact with each other become equally and oppositely charged plays a central role in the xerographic copying process [52–54]. Photoconductive polymers have become important in the change in technology associated with copiers that have greater speed and better resolution. Organic photoconductive polymers are employed for the generation of useful developer materials for replacement of systems based on amorphous selenium (As_2Se_3) which are difficult to manipulate as the film is applied by vacuum sublimation and is brittle as well as expensive.

In xerography four basic steps are involved.

1. The surface of a metal drum coated with a photoconductive material is first negatively charged in the dark by spraying ions formed from molecules in the air under a high electric field, i.e. by a corona discharge (corotron). In the absence of light this leads to a uniform charge distribution on the surface of the photoconductor with the appropriate number of countercharges on the back (grounded) surface.
2. The coated drum is then exposed to a bright image of the item to be copied. The image to be copied is projected onto the photoconductor, thereby discharging the light areas. Thus exposure of the photoconductor to actinic radiation in an imagewise pattern (e.g. by reflection and projection from an original document) results in a loss of charge to the earthed metal drum underneath in areas where the light falls. Charge is retained in areas not illuminated, so that a pattern of charge is obtained that corresponds to the exposure pattern.
3. After the imaging step the electrostatic image is developed. The remaining charged areas, originally dark, are able to attract the black positively charged toner and it is held and subsequently transferred to negatively charged paper. The developer usually consists of two components: carrier and toner [55]. The carrier usually comprises metal beads, while the toner

Table 9.1 Photoconductive polymers

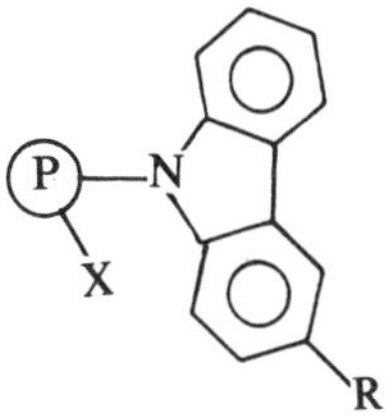

(a) $X \equiv H$; $R \equiv H$	Refs 9–12
$X \equiv H$; $R \equiv$ —CHMeEt	Ref. 13
$X \equiv H$; $R \equiv$ —C(CN)=C(CN)$_2$	Ref. 14
$X \equiv H$; $R \equiv$ Br, I	Refs 15–19
$X \equiv H$; $R \equiv$ NO$_2$	Refs 20, 21
$X \equiv H$; $R \equiv$ —SO$_3^-$ (methylene blue)$^+$	Ref. 22

$$X \equiv H;\ R \equiv\ ^+C\ (\ \underset{\underset{R}{|}{N}}{\text{[indole]}}\ )_2\ X^-$$

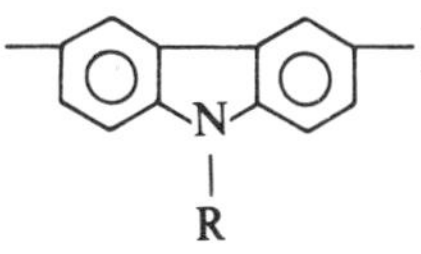

Ref. 23

(b) $R \equiv H$; $X \equiv$ —O-menthyl(–)	Ref. 24
$R \equiv H$; $X \equiv$ —COO-menthyl(–)	Ref. 25
$R \equiv H$; $X \equiv$ —NH—[ring]—NO$_2$ 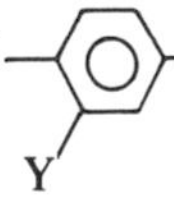	Ref. 26

Y

$R \equiv H$; $X \equiv$ —O—R^1 (R$^1 \equiv$ H, COMe)	Ref. 27
$R \equiv H$; $X \equiv$ —COR^2, —CO(CH$_2$)$_2$—R^2,	Ref. 27
—COOR^2, —O(CH$_2$)$_2$OCO-R^2,	
—R^3, —R^4	

(a) $X \equiv H$; $Z \equiv$  —[ring]—CH$_2$—	Ref. 28
$X \equiv H$; $Z \equiv$ —(CH$_2$)$_n$—; $n \equiv 3, 4$	Ref. 29
$X \equiv H$; $Z \equiv$ —O(CH$_2$)$_2$—, —CO—	Ref. 30
$X \equiv H$; $Z \equiv$ —CONH—(CH$_2$)$_3$—	Ref. 31
(b) $X \equiv H$; 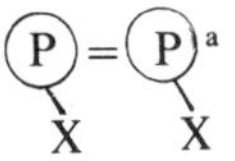 $\overset{|}{\underset{X}{P}}=\overset{|}{\underset{X}{P}}{}^a$;	
$Z \equiv$ —(CH$_2$)$_n$COO(CH$_2$)$_2$—; $n = 1, 2$	Refs 32, 33
$Z \equiv$ —CH$_2$—S—(CH$_2$)$_2$—	Refs 32, 34
(c) $X \equiv$ —CH$_2$CHMeEt, —O—menthyl,	Ref. 35
—COO-menthyl;	

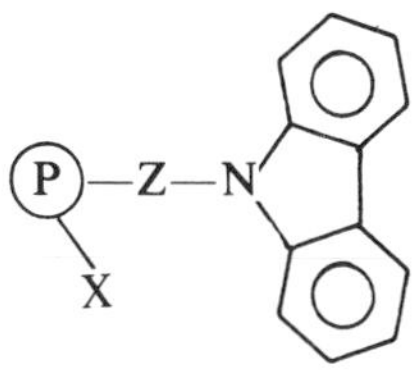

Table 9.1 (*Contd.*)

$$Z \equiv -\!\!\!\bigcirc\!\!\!-CH_2-; \quad \bigcirc\!\!\!-CH_2-,$$

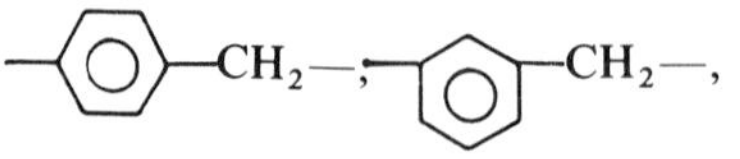

$$-COO(CH_2)_2-,$$
$$-COO(CH_2)_2-OCOCH_2-$$

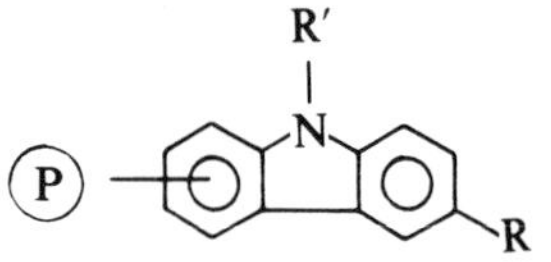	
(a) $R \equiv H$; $R' \equiv -CH_2CHMeEt$	Ref. 13
(b) $R \equiv H$, $R' \equiv Et$	Ref. 23
(c) $R \equiv -C(CN)\!=\!C(CN)_2$; $R' \equiv H$;	Ref. 14

$$\textcircled{P} \equiv \textcircled{P} \text{ or } \textcircled{P}^b$$

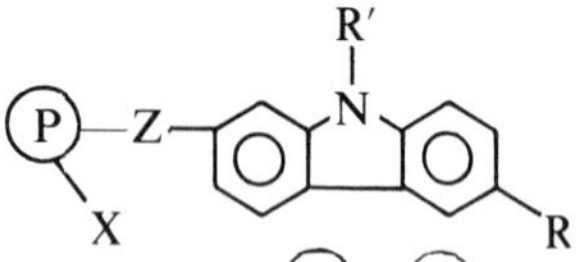

(a) $X \equiv H$; $\textcircled{P} \equiv \textcircled{P}^a$; $R \equiv -C(CN)\!=\!C(CN)_2$; $R' \equiv H$;

    $Z \equiv -CONH-$ Ref. 14

    $R \equiv H$; $R' \equiv Et$; $Z \equiv -\!\!\bigcirc\!\!-CH_2O-\!\!\bigcirc\!\!-CH\!=\!N-$ Ref. 36

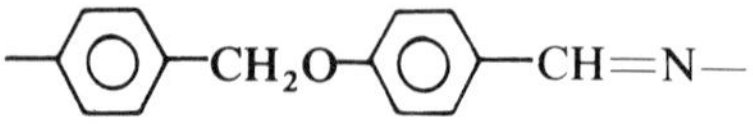

    $R \equiv H$; $R' \equiv Et$; $Z \equiv -(CH_2)_2COO-$ Ref. 37

(b) $X \equiv -COO(CH_2)_mOCOR^2$; $R \equiv H$;   Ref. 38

    $Z \equiv -COOCH_2-$; $R' \equiv Me$;

    $m = 2, 3$

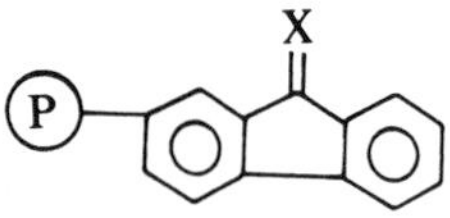

$X \equiv 0$, $C(CN)_2$ Ref. 39

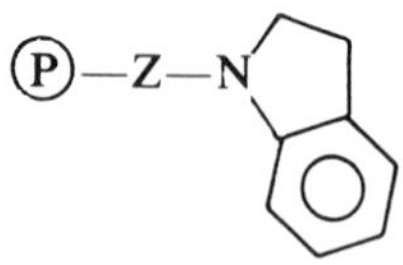

$Z \equiv$ nothing   Refs 40, 41

$Z \equiv -CO-$ Ref. 42

Table 9.1 (*Contd.*)

Z ≡ S	Ref. 40
Z ≡ O	Ref. 40

(P)—CO—N   Ref. 43

(PS)—CH$_2$—N   Ref. 28

R′ ≡ H, MeCO; R ≡ —⟨O⟩—N—Ph$_2$ Refs 14, 44

(P)—   Ref. 23

(PS)—CH$_2$—   Ref. 45

Table 9.1 (*Contd.*)

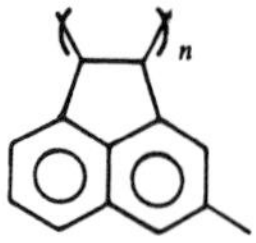 $X \equiv NO_2$	Refs 21, 46
$X \equiv I, Br$	Refs 18, 19

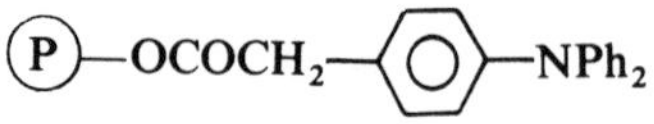

(P)—OCOCH$_2$—⟨O⟩—NPh$_2$ Ref. 44

(PS)—(CH$_2$)$_m$—N Ph$_2$ Refs 47, 48
$m = 0, 2$

(P)—NPh$_2$ Refs 41, 17, 40, 49

(P)—Z—CH—⟨naphthoxazole⟩
        |
       R

(a) $R \equiv H$; $Z \equiv$ —⟨O⟩—   Ref. 8, 28

(b) $R \equiv OCOPh$; $Z \equiv$ —⟨O⟩—,   Ref. 28

—⟨O⟩—CH$_2$O—⟨O⟩—

(PS)—CH=CH—⟨O⟩—CH=CH—⟨O⟩—N=NR^5 Ref. 50
    R^5—N=N

$R^2 \equiv$ —⟨benzene with NO$_2$, NO$_2$⟩ ; $R^3 =$ ⟨tetrachloro-benzoquinone with Cl, Cl, Cl⟩ ;

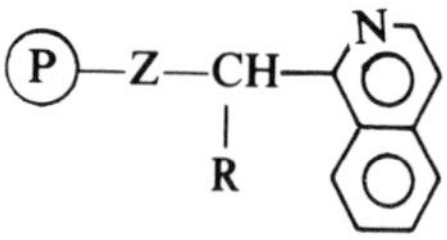

$R^4 =$ —N⟨phthalimide⟩ ; $R^5 =$ ⟨naphthalene with HO, CONH—Ph⟩

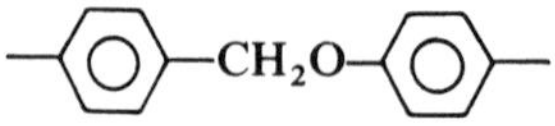

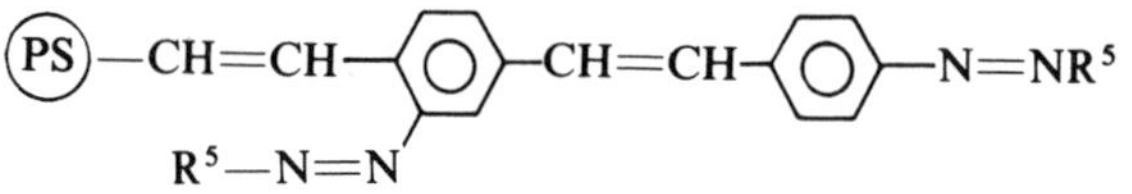

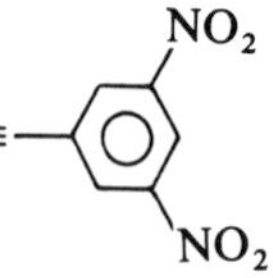

(P)— $\equiv$ ⟨CH$_2$CH—CH$_2$—CH⟩$_n$
  | | |
  X   X

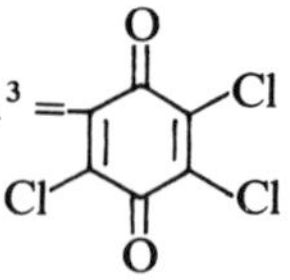

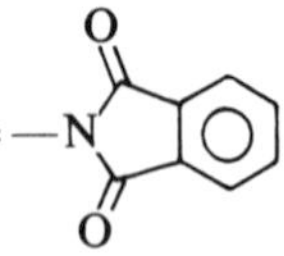

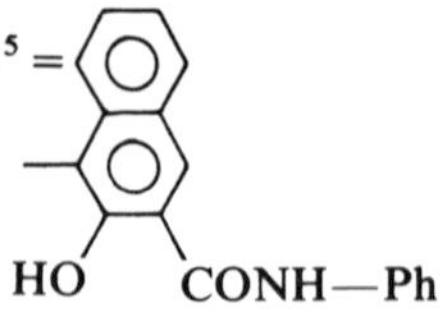

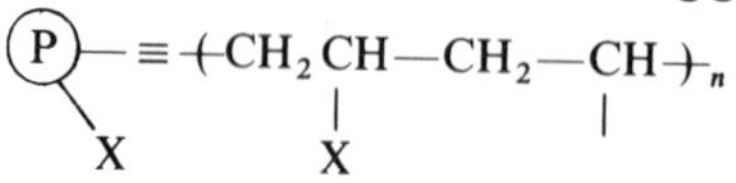

is a polymer containing a colorant (dye or pigment) to provide contrast. Agitation of the carrier and toner results in charging that leads to toner particles of sign opposite to the imagewise pattern on the photoconductor. When the developer is exposed to the image, the toner particles are electro-statically attracted to the charged areas, forming a visual positive image.

4. The image is then transferred and fixed to paper by heating to sinter the resin.

Poly(N-vinylcarbazole) (PVK) which is a good dark insulator, has been used to replace selenium in xerography [56, 57]. It absorbs ultraviolet light (360 nm) forming an exciton state which ionizes in an electric field, and the electron is stabilized in trapping levels of low potential leaving only the radical cation PVK^+ to act as charge carrier. PVK remains an insulator in visible light but can be sensitized with an electron acceptor, e.g. 2, 4, 7-trinitrofinorenone (TNF) which by forming charge transfer states shifts the adsorption into the visible and renders the material photoconductive. For example, the polymeric dye **1** has been used as a toner in xerography [58–60]:

$$\text{(P)}-CONH-(CH_2)_6-O-dye$$

$$Ph \qquad \mathbf{1}$$

$$\text{(P)}^a-\equiv +CH_2-\underset{\underset{X}{|}}{CH}-CH-\underset{\underset{|}{|}}{\overset{\overset{Me}{|}}{C}}+_n$$

$$\text{(PS)}-\equiv +CH_2-CH+_n$$

$$\text{(P)}-\equiv +CH_2-CH+_n$$

$$\text{(P)}^a-\equiv +CH_2-\underset{\underset{|}{|}}{\overset{\overset{Me}{|}}{C}}+_n$$

$$\text{(P)}^b\equiv +OCH_2CH+_n$$

$$\underset{CH_2}{|}$$

9.1.2   Electrically conductive polymers

Interest in electrically conductive polymers stems from the potential for combining in one material the advantageous properties of polymers and the electrical properties of semiconductors or metals [61–77]. The advantageous properties of polymers are their mechanical properties (such as flexibility, stretchability and impact resistance) and their ease and low cost of preparation and fabrication. Low density and weight saving is an important advantage of polymeric materials, especially for batteries. The most promising synthetic semi-conductors and metals for practical electronic applications are those which can combine solution or melt processibility to thin films, which are environmentally, electrically and mechanically stable and possess high conductivity.

In general, conduction in polymers can be either ionic, e.g. in salts of organic polymers, or electronic. Electronic conductivity of polymers may be achieved by one of three means.

1. By introducing delocalization: a variety of elimination reactions can be employed to generate conjugated polyene structures, i.e. polymers with a high degree of conjugation in the main chain, from a range of functionalized polymers, e.g. polyacetylene, polyphenylene. Electrical conductivity appears to be sensitive to the degree of conjugation along the chain backbone which, in turn, increases with increasing chain planarity.
2. By introducing electrically active pendant groups, e.g. substituted aromatic amines or large polynuclear aromatic groups with large π-electron systems.
3. By electrochemical doping: doping the polymers, which are normally excellent electrical insulators, with electrically active species of low molecular weight leads to highly electrically conducting materials by charge transfer interactions.

Electronic mobility in polymers is greatly enhanced along a polymer molecule with conjugated bonds, i.e. with delocalized π-orbitals. Instead of the typical insulator value of less than $10^{-14} \, \Omega^{-1} \, cm^{-1}$, the conductivity may be in the range 10^{-5}–$10^{-9} \, \Omega^{-1} \, cm^{-1}$. However, when such a polymer is doped with electron donors (e.g. halogens, AsF_5) or acceptors (e.g. alkaline metals) it becomes semiconducting-to-metallic in conduction $(10^{-6}$–$10^4 \, \Omega^{-1} \, cm^{-1})$. Doping techniques include exposure of films to the vapours of dopants such as I_2, H_2SO_4, AsF_5 or $SbCl_5$, or to solutions of dopants such as NO_2SbF_6, I_2 or sodium naphthalide. The enhanced conductivity can be attributed to the presence of cations and anions formed via electron transfer from the donor species to the acceptor. Transport of the charge may then occur via either cationic (hole) or anionic (electron) states. The electrical conductivity of a doped polymer can also depend on (a) the chemical structure of the polymer [78–81], (b) degree of doping since the conductivity increases with increasing dopant concentration and (c) the nature of the dopant ion. The p- and n-type conductivities depend on the dopant nature, i.e. doping of a polymer with an electron acceptor or a donor gives respectively a p- or n-type semiconductor [82–83].

Acceptor-doped polymers are characterized by a broad valence band and low ionization energy for formation of p-type conductors, which show positively charged carriers (hole conductor). Donor-doped polymers are characterized by a broad conduction band and high electron affinity for the formation of n-type conductors, which show negatively charged carriers (electron conductor).

Polymeric solid batteries

Batteries are electrochemical energy sources in which chemical energy results from chemical reactions that give products of low energy content. Batteries consist of negative and positive electrodes, an ionically conducting electrolyte and a separator which prevents direct contact between the negative and positive electrodes and retains the electrolyte within its structure. During the discharge of an electrochemical cell, i.e. the conversion of chemical energy to electrical energy, an oxidation reaction (production of electrons) occurs at the negative electrode (anode) and a reduction reaction (consumption of electrons) occurs at the positive electrode (cathode).

Batteries, as electrical energy storage devices, are either rechargeable (secondary) or non-rechargeable (primary). A rechargeable battery incorporates a highly reversible chemical–electrochemical reaction to generate electrical energy, in which the higher energy compounds are formed by putting electrical energy (produced by a dynamo or electricity generating device) into the battery and electrical energy is later withdrawn with formation of lower energy compounds.

The electrochemical generation of electricity in fuel cells (primary cells in which the oxidation of the fuel generates electricity) offers the possibility of achieving higher efficiencies than are obtainable with electrochemical generators driven by heat engines. Fuel cells are just the opposite of electrochemical processes and utilize electricity to generate chemicals. In electrolysis, for example, H_2O is decomposed by electricity into hydrogen and oxygen. In a fuel cell, hydrogen reacts with oxygen to form H_2O, heat and a direct current.

Fuel cells can be classified as direct or indirect, and reversible or irreversible. *Direct* fuel cells are those utilizing carbon or hydrocarbon as the fuel together with oxygen as the oxidizer. *Indirect* fuel cells utilize indirect fuel such as steam or H_2. *Reversible* fuel cells are those in which fuel and oxygen can be produced by reversing the supply of electrons to the cell so that electrons are removed from the oxidizing electrode and fed into the fuel electrode. The H_2-O_2 cell is an example of an *indirect reversible* cell where

$$2H_2 + O_2 \longrightarrow 2H_2O \qquad E_r = 1.23\ V \qquad (9.4)$$

at 20°C and 1 atm or, when reversed,

$$2H_2O \longrightarrow 2H_2 + O_2 \qquad (9.5)$$

The $C-O_2$ cell is direct and irreversible because CO_2 is not readily converted back into carbon and O_2. The fuel electrode produces electrons from its supply of fuel while the oxidizing electrode consumes electrons.

The rigidity of the inorganic materials is their main disadvantage in their use in solid state batteries. Because of dimensional changes in the electrodes during charging and discharging, it is difficult to maintain intimate contact between the electrodes and the solid electrolyte. Functionalized polymers can significantly change battery fabrication technology and have gained considerable commercial interest in promoting their use as (a) polymer-modified electrodes, (b) polymeric solid electrolytes, (c) separators in cells containing several electrolyte compartments, (d) reaction-product-removing depolarizers and (e) a source of reacting ions for the cell reactions.

(a) Polymer-modified electrodes A conductor surface is usually modified in order to change the chemical or physical properties of the electrode material, e.g. to catalyse a particular electrochemical reaction or to slow the photocorrosion of the semiconductor anode in a photocell. Electrodes can be modified by introducing a modifying species at the conductor–solution interface.

Chemical modification of electrodes is a major area of current research interest in electroanalytical and photoelectrochemical applications [84–87]. Polymer films containing electroactive species are used for modifying electrodes to act as mediators in the electron transfer process between the electrode and solution species.

Rechargeable batteries using polymers for both electrodes in different oxidation states are of considerable commercial interest because of their relatively light weight compared with the familiar lead battery [88, 89]. Polymers, e.g. polyacetylene, can act either as an electron source or an electron sink depending upon how they are doped, i.e. whether they are reduced or oxidized. In a double-polymer battery, the redox reactions at the two modified electrodes are as follows (Fig. 9.1).

$$\text{Anode:} \quad P^-D^+ \underset{+\,e,\,\text{charge}}{\overset{-\,e,\,\text{discharge}}{\rightleftharpoons}} P^0 + D_s^+ \tag{9.6}$$

$$\text{Cathode:} \quad P^+A^- \underset{-\,e,\,\text{charge}}{\overset{+\,e,\,\text{discharge}}{\rightleftharpoons}} P^0 + A_s^- \tag{9.7}$$

Here P^-, P^+ and P^0 are the negatively charged, positively charged and neutral polymer, A^- and A_s^- are anionic dopant counter-ion and its solvated counterpart in the electrolyte, D^+ and D_s^+ are the cationic dopant counter-ion and its solvated counterpart, P^-D^+ is the donor-doped polymer (anode) and P^+A^- is the acceptor-doped polymer (cathode). In the discharge process, the doped polymers revert back to the neutral state and the dopant counter-ions diffuse in and out of the polymer without change of oxidation state.

In a battery cell employing one conducting polymer electrode and one

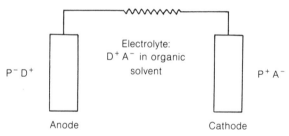

Fig. 9.1 Polymer battery.

conventional battery electrode, the redox reactions are as follows.

$$\text{Anode:}\quad Li^0 \underset{+e,\,charge}{\overset{-e,\,discharge}{\rightleftharpoons}} Li_s^+ \tag{9.8}$$

$$\text{Cathode:}\quad P^+A^- \underset{-e,\,charge}{\overset{+e,\,discharge}{\rightleftharpoons}} P^0 + A_s^- \tag{9.9}$$

For example, a solid state battery comprising a lithium anode and an I_2-polyvinylpyridine complex as the cathode and forming LiI electrolyte during cell discharge has been reported [90]. Such a polymer is used in a lithium cell on a commercial basis to overcome the disadvantages in using iodine, e.g. its high electrical resistance and the difficulties in moulding it. Other examples, iodine–nylon-6, poly(N-vinylpyrrolidone) and poly(vinyl alcohol), have also been used as positive electrodes in galvanic cells [91]. Two techniques are widely employed as a means of modification of electrodes with polymers.

(*i*) *Casting of polymer films on electrodes.*   In addition to the use of redox active polymers for coating electrodes, electrochemically inert polymers can be coated onto the electrode surface and an electroactive molecule or ion can then be introduced into the polymer film. Modes available for binding electroactive species to polymer films include metal–ligand coordination, non-metal covalent bond formation and simple electrostatic interactions. The electrode modified by coating can be used in solvent–electrolyte systems in which the polymer is insoluble.

(*ii*) *Covalent attachment of polymer films on electrodes.*   The polymer is formed from a suitable monomer directly on the electrode surface. Thus polymer solubilities do not restrict the selection of the solvent–electrolyte system. Only a few examples of electrodes modified by redox polymers have been reported [92,93] and they were prepared by oxidative or reductive electropolymerization of vinyl monomers.

(*b*) *Polymeric electrolytes* Polymeric solid electrolytes [94] are polymer–salt complexes that have significant ionic conductivity. These materials consist of alkali metal salts dissolved in a solid polar polymer, e.g. poly(ethylene oxide). The absence of water, solvent or reactive functional groups in the polymer electrolytes makes them excellent for use in batteries of high energy density.

Several advantages make functionalized polymers extremely promising materials for use as electrolytes in solid state batteries.

1. Unlike hard inorganic electrolytes, polymeric electrolytes can flow and deform, thus maintaining interfacial contact with the electrodes even if the electrodes grow or shrink.
2. The casting of polymeric electrolytes in thin film form minimizes the size and the electrical resistance of the electrolyte. Thus they are able to store much more energy per unit weight and volume than a conventional lead–acid battery.
3. Polymer-based electrolytes may be cost effective in rechargeable power units, and especially as micropower sources in electronic device applications.
4. A variety of polymer electrolytes can be prepared not only by varying the anion and cation of the salt but also by varying the polar polymers that might be suitable hosts for salts.
5. The use of polymer electrolytes in rechargeable cells has advantages over conventional secondary batteries, particularly for applications where weight is critical such as micropower sources for portable electrical and electronic equipment.

In solid state batteries with polymer electrolytes, Li is used as the negative electrode and an ion-insertion compound such as titanium disulphide (TiS_2) is used as the positive electrode. When the battery is discharged, lithium is oxidized at the Li–polymer interface, transported as Li^+ through the polymer film and inserted into the TiS_2 at the positive electrode. The process is reversed when the cell is recharged. An ion exchange battery is a galvanic cell in which the electrolyte is an ion exchange membrane that is a partly hydrated solid.

9.2 POLYMERS IN ENERGY

Over recent years energy costs have risen rapidly and the supply of available petroleum feed stocks will be exhausted in a few centuries. Thus the development of new energy sources is a worldwide research subject. Connections between polymers and energy are pervasive and extremely complex. The contributions of polymers to energy can be categorized as (a) the production and conservation of energy and (b) chemical conversion and storage of solar energy.

The excellent electrical and thermal insulating properties of polymeric materials coupled with their ease of fabrication are essential in conventional power-generating equipment, which requires insulation that permits close windings and freedom from electrical breakdown, and in cables for power distribution. They also make an important contribution in energy conservation, e.g. in insulating materials in homes and other buildings, with enormous long-term savings in energy. The use of polymers in solar collectors, in order to use solar energy on economically large scales, has also received considerable attention. Although the use of polymers for collection and concentration of solar energy is economically advantageous because of their light weight, low cost and good mechanical strength, their durability is the major problem associated with this use. Since it is not necessary for the polymer to contain any reactive functional group in these applications, this subject is not discussed further here.

The uses of functionalized polymers in the conversion and storage of solar energy have recently received much attention in the quest to replace petroleum fuels. Sunlight, when it is converted into a usable form, represents a virtually inexhaustible source of permanent and clean energy with no net consumption of resources [95, 96].

9.2.1 Conversion of solar energy

The solar spectrum on the earth ranges from 250 to 2400 nm, with its maximum at 500 nm. The visible region between 400 and 800 nm occupies about half of the spectrum and the chemical conversion. of this irradiation is therefore important for the direct production of fuel which can easily be stored and transported.

For a fuel, an electron source is needed. Economically viable catalytic photosensitized water splitting as an electron source is the simplest of the chemical conversion systems of solar energy. Since the excited state of a molecule is a better electron acceptor or donor than its ground state, light absorption can drive a redox reaction non-spontaneously. For water oxidation, a redox potential of $E_0' = 2.33$ V is needed in the first step to abstract one electron from a water molecule:

$$H_2O \longrightarrow HO\cdot + e^- + H^+ \quad (E_0' = 2.33 \text{ V}) \tag{9.10}$$

When the intermediate is stabilized on a catalyst and four electrons of two molecules of water are oxidized without isolating the intermediate (four-electron process), the required redox potential is only 0.82 V:

$$2H_2O \longrightarrow O_2 + 4e^- + 4H^+ \quad (E_0' = 0.82 \text{ V}) \tag{9.11}$$

The potential of the site of O_2 evolution of the photosynthesis of around 0.82 V shows that a four-electron process occurs. In water photolysis as a model system for photochemical conversion of solar energy, the system consists of a photoreaction centre, two kinds of electron mediators and reduction as well as oxidation catalysts (C_1 and C_2) in which water should be oxidized at C_2 to give O_2 and protons should be reduced at C_1 to give H_2:

$$2H_2O \xrightarrow{\;C_2\;} O_2 + 4e^- + 4H^+ \quad (E_0' = 0.82 \text{ V}) \tag{9.12}$$

$$4e^- + 4H^+ \xrightarrow{\;C_1\;} 2H_2 \quad (E_0' = 0.41 \text{ V}) \tag{9.13}$$

In this system, the potential of C_1 should be lower than -0.41 V and that of C_2 higher than 0.82 V. For proton reduction, the two-electron process (eqn 9.13) is much more favourable than a stepwise reaction in which the first step

$$H^+ + e^- \longrightarrow H \cdot \tag{9.14}$$

requires -2.52 V. Thus a multi-electron process is preferable at both the catalyst sites of the water photolysis system. In the photochemical conversion model, the most serious problem is the undesired back electron transfer, when separated charges combine again to consume the energy, as well as side electron transfer. This problem of photoinduced charge separation can be prevented if the reactions are carried out in a heterogeneous conversion system using functionalized polymers. Noble metals are the most potent catalysts to realize a multi-electron catalytic reaction for water photolysis. A functionalized polymer containing a pendant viologen as electron mediator and platinum as the H_2-evolving catalyst [97] and a polymer combining viologen units with $Ru(bpy)_3^{2+}$ (bipyridine as compound **7**) as photoreaction centre [98] are interesting examples in the field of photochemical solar energy conversion.

9.2.2   Storage of solar energy

Since the solar irradiation of the earth is intermittent and unstable, depending on time, season, weather and region, storage of the converted energy is required if it is to be used on a large scale. The use of photochemical reactions to generate kinetically stable products of high energy content provides an exceedingly attractive fuel source. The photosensitization conversion of norbornadiene (NBD) **2** to an energy-rich quadricyclane (QC) **3**, coupled with reversion of QC to NBD, is the most promising photoisomerization system for storing energy from sunlight. The QC is thermodynamically unstable relative to NBD and a device based on this interconversion as a model system for solar energy storage requires two steps: (a) energy storage through the sensitized photolysis of NBD to QC in an endothermic reaction in visible light; (b) energy release through the catalysed reconversion of QC to NBD in an exothermic reaction ($\Delta H = 21$ kcal mol^{-1}):

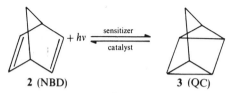

A different kind of sensitizer, which is a catalyst capable of absorbing radiation and transferring the energy to another molecule, is involved in the excitation of molecules in the first step. In general, sensitizers must absorb strongly in the region of available solar energy, be thermally and photochemically stable and effect the desired sensitization efficiently. A sensitizer excited by solar energy (*hv*) to the singlet state, which then passes into a longer-lived triplet state, can activate a molecule (M) and itself return to the ground state:

$$\text{Sens} + hv \longrightarrow {}^*\text{Sens} \longrightarrow {}^3\text{Sens} \tag{9.15}$$

$$^3\text{Sens} + \text{M} \longrightarrow {}^*\text{M} + \text{Sens} \tag{9.16}$$

However, an ideal catalyst for the reverse exothermic reaction in the second step must meet several requirements: (a) it should not produce undesirable side reactions; (b) it should be sufficiently active to evolve heat quantitatively in a rapid conversion; (c) it should have long-term stability; (d) the active species should not be leached by the reaction mixture.

The immobilization of the sensitizer and catalyst on polymeric supports is especially effective as it keeps them apart, because contamination of the materials with sensitizer or catalyst markedly lowers the efficiency of a solar energy storage system. Hence the isolation of the catalyst is attractive and important in order to keep the active catalyst away from the photochemical reactor where the conversion reaction of NBD to QC is taking place and to prevent the dispersion of the catalyst throughout the system. The use of functionalized polymers in energy storage has been directed toward the development of new polymeric sensitizers for the photochemical process [97–103] as well as the development of polymer-bound catalysts for the reversible valence isomerization of solar energy storage systems [104–106]. Various polymeric sensitizers, such as **4** [102], and polymeric catalysts, such as **5** [104], **6** [105] and **7** [106], have been shown to be active in NBD–QC solar energy storage systems:

5a $Z \equiv CO; X \equiv COOMe$
5b $Z \equiv NHCO; X \equiv COOMe$
5c $Z \equiv NHSO_2; X \equiv SO_3Me$

9.3 POLYMERS IN LITHOGRAPHIC PROCESSES

In semiconductor technology the manufacture of transistors and integrated circuits, involving microcircuit patterns on a silicon wafer, is required. A large variety of materials and processes are currently employed in the manufacture of hybrid microcircuits. In spite of the precautions taken by the electronics industry, it is often difficult to remove the last traces of processing materials completely by normal cleaning, and some may become electrostatically attached to the surface of the circuit, which can cause an electrical short and result in a missile or space craft failure.

One of the procedures used in making monolithic integrated circuits which overcome this problem is photolithography based on photoresists. However, in recent years electron beam lithography based on electron resists has been used as an alternative method for cases in which resolution is desired [107, 108]. The lithographic process involves coating of the substrate, e.g. silicon dioxide, with resist and its exposure to light through a mask (image boundaries). Images can be produced by solvent development followed by etching and stripping of the resist. Resists must (a) be capable of forming uniform pinhole-free films on a substrate by a simple process such as spinning, dip-coating or spraying, (b) be of a high degree of purity, i.e. contain only low levels of ions to achieve no electrical properties, (c) be easily removed by solvent dissolution, (d) be thermally stable up to 150°C to withstand the burn-in temperatures used for hybrid circuits, (e) cause no stresses on fine-wire bonds and no deleterious effects on active devices and (f) be compatible with the devices and wire bonds of high density circuits.

Usually polymers are the only materials which fulfil these requirements. The interaction of organic polymers with energetic electrons or ultraviolet light results in structurally changed molecules which may either be broken down to smaller fragments or link together to form larger molecules. Interaction that causes a break in the main polymer chain and results in irradiated material having a lower average molecular weight than unirradiated material allows 'positive-working resists' to be formulated. Clearly the successful removal of irradiation-degraded polymer without affecting the unirradiated material in the development process will occur if the original polymer has the highest possible molecular weight and is irradiated by moderate doses of electrons or ultraviolet light. Moreover, a polymer with a low glass transition temperature can easily be deformed and the resolution of fine patterns developed in a film of such a polymer may be impaired. Thus good positive resists should have a glass transition temperature above the highest temperature to which the resist will be subjected after development of the irradiated pattern in it.

For polymers which link together on irradiation, larger insoluble and infusible molecules are generally formed by a process known as crosslinking. This type of polymer forms the basis of 'negative-working resists' in which it is possible to dissolve and remove unirradiated material while irradiated material cannot be dissolved away after development. In principle, many polymers are available for use as negative resists; however, the choice may be limited by the sensitivity required, the solubility of the resist etc. [109, 110]. The requirement of a high glass transition temperature, which is necessary for positive-working resists, does not necessarily apply to negative resists because the resist material in this case is crosslinked during irradiation, which automatically increases its glass transition temperature.

9.3.1 Photoresists

Photolithography is generally the coating of a wafer surface with a polymer film containing photosensitive groups and its exposure to the light of a mercury ultraviolet lamp through a photomask. Depending on the type of photoresist, the exposed areas are either crosslinked (negative photoresists), and the non-exposed areas, i.e. uncrosslinked areas, are dissolved in a liquid developer, or degraded (positive photoresists), and the exposed areas are removed by a solvent under certain developer conditions (Fig. 9.2).

The ideal resist for ultraviolet lithography should (a) possess good sensitivity to 230–280 nm radiation with little or no absorption at longer wavelengths to eliminate the difficult task of filtering the long wavelength radiation present in

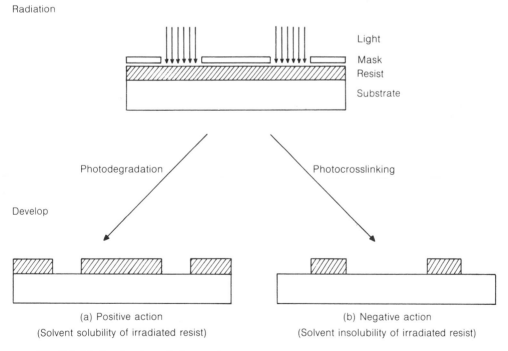

Fig. 9.2 Mode of action of photoresists: (a) positive action, (b) negative action.

Table 9.2 Photoresist polymers

Functional polymer	References

$\text{(P)}-Z-CO-CH\!=\!CH-Ph$
$\quad\backslash X$

(a) $Z \equiv O$; $X \equiv H$, OH, OAc	Refs 111, 112
	Refs 117–119
(b) $Z \equiv O(CH_2)_2O$; $X \equiv H$	Ref. 120
(c) $Z \equiv O(CH_2)_2O$; $X \equiv O(CH_2)_2OR$	Ref. 121
	Ref. 122

(d) $Z \equiv$ —⬡— CH_2O; $X \equiv$ Ref. 123

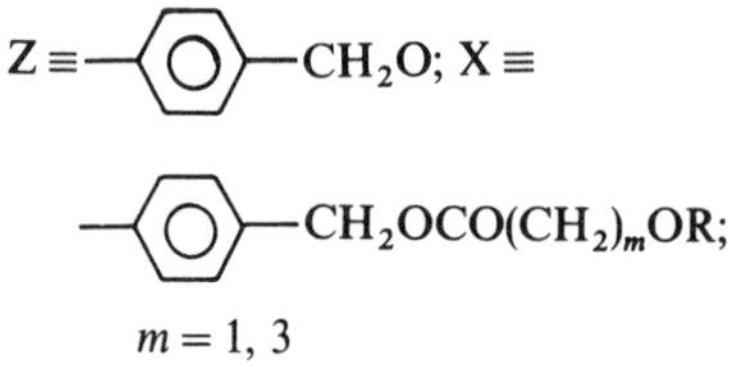

—⬡— $CH_2OCO(CH_2)_mOR$;

$m = 1, 3$

(e) $Z =$ —⬡— CH_2O; $X =$ —⬡— CH_2OR Ref. 123

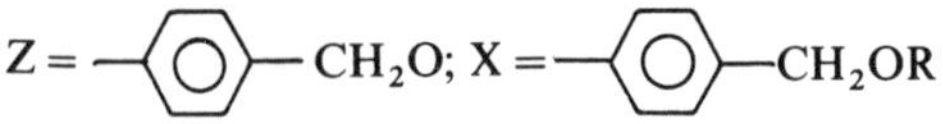

(f) $Z \equiv COO(CH_2)_2O$; $X \equiv CONH-R$	Ref. 124
(g) $Z \equiv COO(CH_2)_2O$; $X \equiv H$, COOEt	Ref. 125
(h) $Z \equiv COOCH_2CHO$; $X \equiv H$	Refs 122, 126

$\qquad\qquad\qquad\quad |$
$\qquad\qquad\quad CH_2OR$

(k) $Z = COO$—⬡—; $X \equiv COOCH_2CHOH$ Refs 114, 127

$\qquad\qquad\qquad\qquad\qquad\qquad |$
$\qquad\qquad\qquad\qquad\qquad CH_2OH$

(l) $Z \equiv$ —⬡N^+—$(CH_2)_2O$, $X = H$ Ref. 128

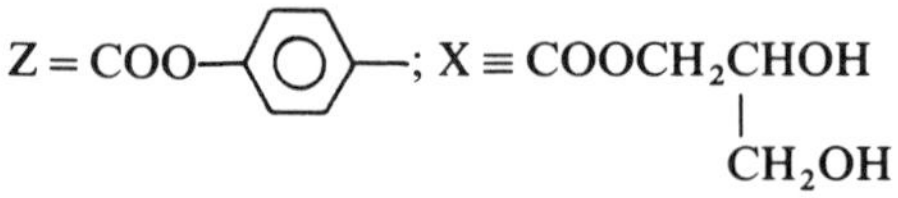

$R \equiv$ —⬡—NO_2,— naphthyl —NO_2

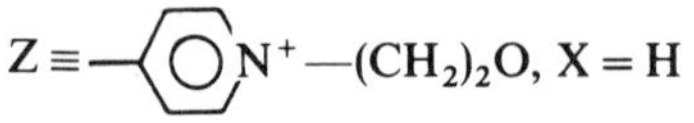

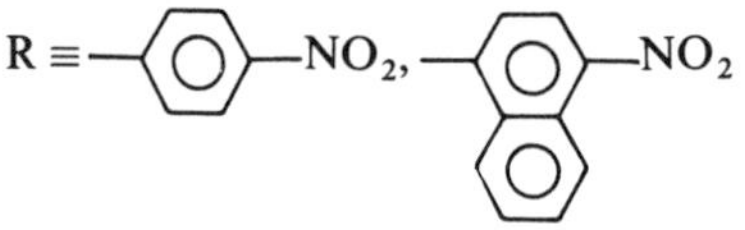

$\text{(P)}-Z-CO-C\!=\!CH-CH\!=\!CH-Ph$
$\qquad\qquad\quad |$
$\qquad\qquad\quad R$

(a) $Z \equiv O$; $X \equiv H$; $R \equiv H$	Ref. 129
(b) $Z \equiv NH$; $X \equiv H$; $R \equiv H$, CN	Refs 128, 130

(c) $Z \equiv$ —⬡—N^+—$(CH_2)_2O$; $X \equiv H$,   Refs 128, 130
$\qquad\qquad\qquad\qquad R \equiv H$, CN

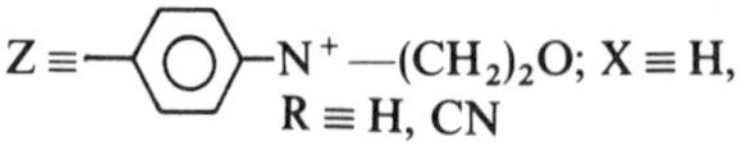

Table 9.2 (*Contd.*)

Functional polymer	References

(d) $Z \equiv \langle O \rangle N^+ - (CH_2)_2O$; $X \equiv COOMe$, Br^- $R \equiv H, Me$ — Ref. 130

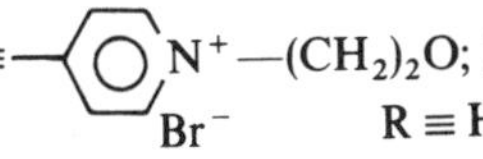

(e) $Z \equiv COO(CH_2)_2N^+ - (CH_2)_2O$; $R \equiv H, CN$ Me_2Br^- — Refs 128, 130

$\text{(P)}-COO-\langle O \rangle -CH{=}CH-COPh$ — Refs 114, 127

$\overset{|}{C}OOCH_2CHCH_2OH$ $\overset{|}{O}H$

$\text{(P)}-\langle O \rangle -CH{=}CH-Z$ — Ref. 131

$Z \equiv COOEt, COOH, CHO$

$\text{(P)}-OCO-\underset{}{\triangledown}\overset{R\ R'}{\underset{|\ \ |}{}}C{=}CH_2$ — Ref. 132

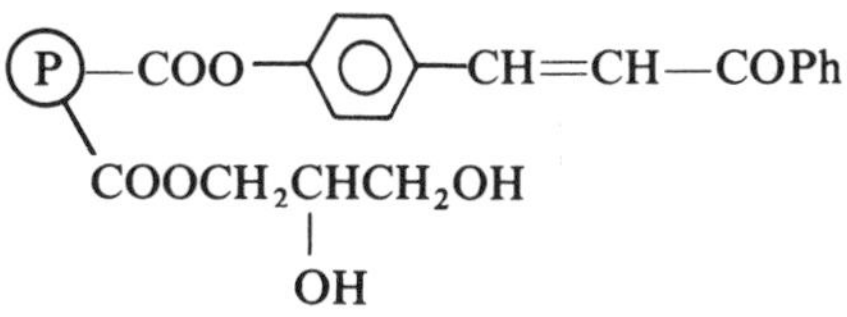

$R \equiv R' \equiv H, Me$

$\text{(P)}-OCO-CH{=}CH-\langle\!\langle O \rangle\!\rangle$ — Ref. 113

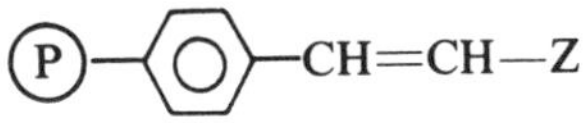

$\text{(P)}-\langle O \rangle N^+ -CH{=}CH-Ph$ — Ref. 115

$\text{(P)}-\langle O \rangle -CH_2-{}^+N\langle O \rangle -CH{=}CH_2$ — Ref. 133

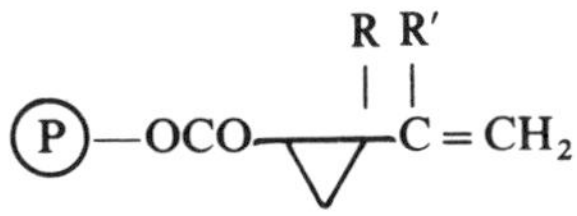

$\text{(P)}-\overset{Me}{\underset{+}{N}}\,-CH{=}CH-Ar\ MeSO_4^-$ — Ref. 134

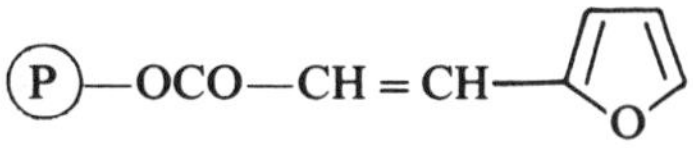

$\text{(P)}-OCO-C{=}CH-Ph$ — Ref. 135

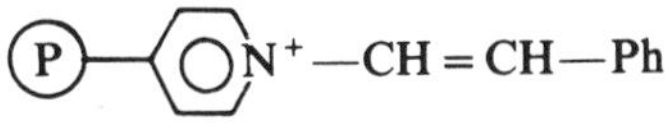

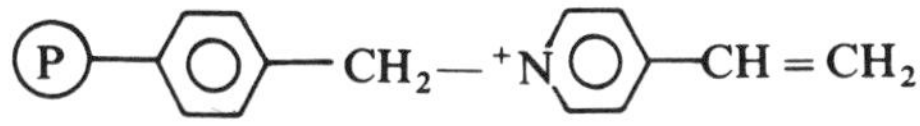

$\overset{|}{C}N$

$\text{(P)}-COOCH_2CHCH_2OCO(CH_2)_2\overset{Ph}{\underset{|}{C}}-Ph$ — Ref. 136

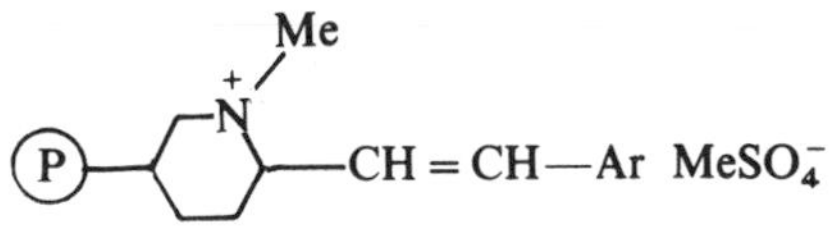

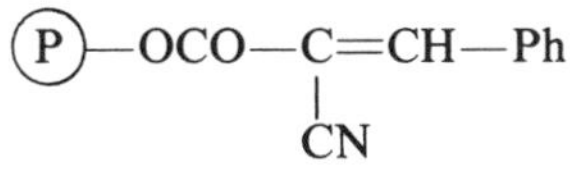

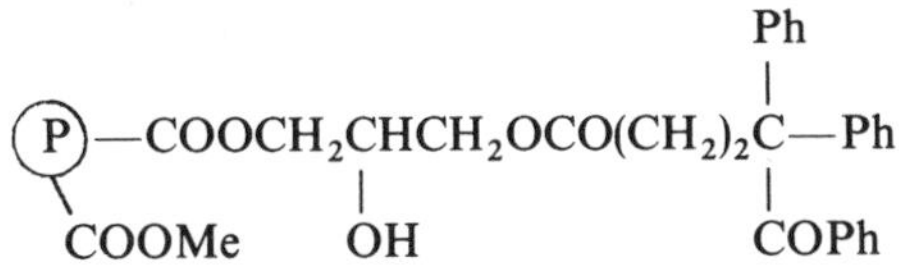

$\overset{|}{C}OOMe\quad \overset{|}{O}H\qquad\qquad \overset{|}{C}OPh$

Table 9.2 (*Contd.*)

Functional polymer	References

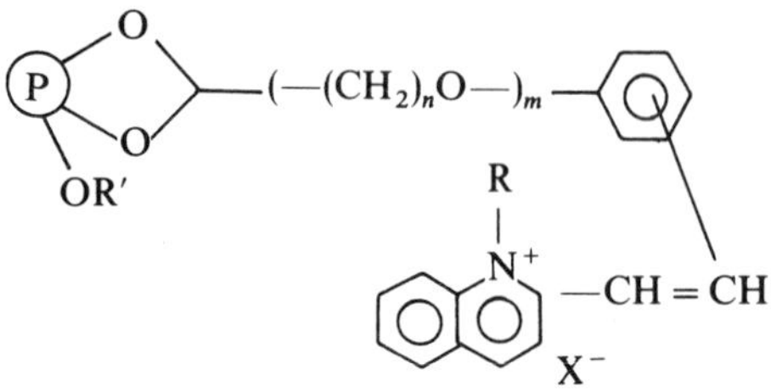

$R \equiv H, C_xH_{2x+1},; n = 1–6; R' \equiv H, Ac; m = 0, 1$ Ref. 137

$Z \equiv C(CN)_2$ Ref. 138
$Z \equiv O$ Refs 139, 138

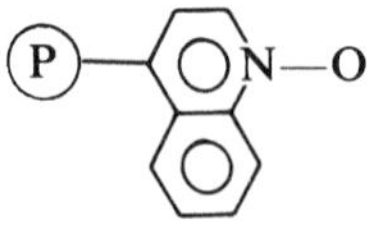

Ref. 138

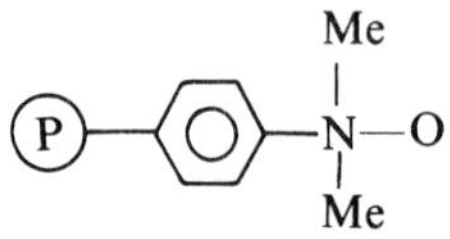

Refs 139, 138

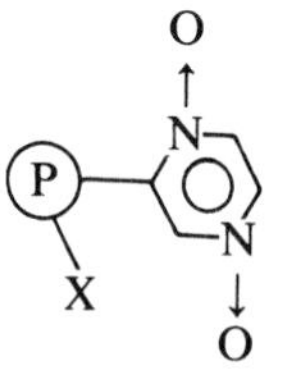

$X \equiv H, Ph, COOMe$ Ref. 140

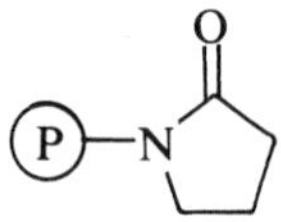

Ref. 141

$\widehat{P}$—NHCOOCMe$_3$ Ref. 142

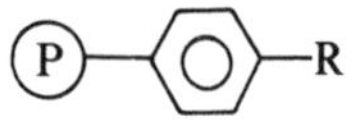

$R \equiv H$ Ref. 143
$R \equiv OH$ Ref. 144
$R \equiv CH_2Cl$ Ref. 145

Table 9.2 (*Contd.*)

Functional polymer	References

(P)—CO—R

   $R \equiv Me$ Ref. 146

   $R \equiv OMe$ Refs 143, 147

$$\begin{array}{c} R \\ | \\ +C-CH_2-SO_2\!+_n \\ | \\ (CH_2)_m-Me \end{array}$$

   $R \equiv H; m = 1$ Refs 144, 148, 149

   $R \equiv Me; m = 2$ Ref. 144

$+(CH_2)_4CO-CH_2SO_2\!+_n$ Ref. 144

$+CH_2C=C-CH=CHCOOCH_2CH=CH_2)_n$ Ref. 150

$$\begin{array}{cc} | & | \\ R & R \end{array}$$

   $R \equiv H, Me$

(PS)— $\equiv +CH_2-CH+_n$ (Si)— $\equiv$ silica

$$\begin{array}{c} | \\ \bigcirc \\ | \end{array}$$

(P)— $\equiv +CH_2-CH+_n$ (P)— $\equiv +CH_2-CH-CH_2-CH+_n$

$$\begin{array}{cccc} \quad\quad | & & \quad\quad | & | \\ \quad\quad X & & \quad\quad X & X \end{array}$$

$(P)^a$— $\equiv +CH_2-\overset{\displaystyle Me}{\underset{\displaystyle |}{C}}+_n$ $(P)^a$— $\equiv +CH_2-CH-CH_2-\overset{\displaystyle Me}{\underset{\displaystyle |}{C}}+_n$

$$\begin{array}{cc} \quad | & \quad | \\ \quad X & \quad X \end{array}$$

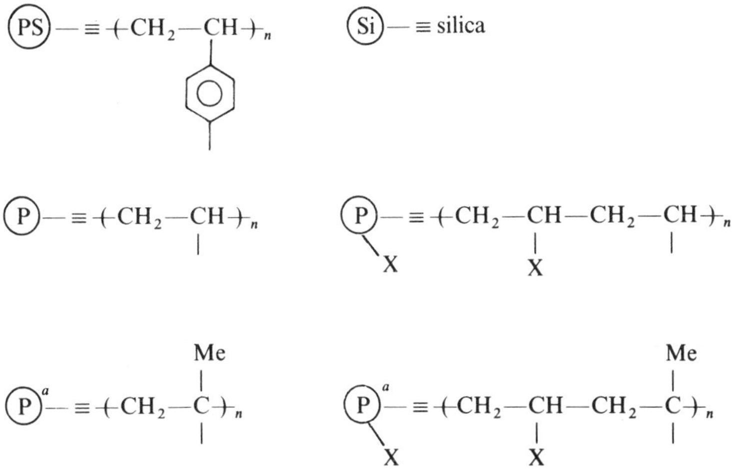

conventional sources, (b) be capable of high resolution, (c) have a reasonable exposure time and (d) be compatible with conventional microstructure fabrication processes.

Polymers with pendant cinnamoyl groups are well known as photosensitive polymers and have received considerable industrial application for the preparation of circuits because they form a thin lightly hardened film when exposed to light [111, 112]. Polymers with other pendant photosensitive moieties such as α-furylacrylic ester [113] **8** benzalacetophenone [114], styrylpyridinium [115] and α-phenylmaleimide [116] have been reported, as listed in Table 9.2.

$$\boxed{P}-OCO-CH{=}CH-\underset{}{\langle}O\rangle$$

8

Addition of a photosensitizer to a photosensitive polymer extends the range of spectral sensitivity and increases the apparent photographic speed. The correct choice of photosensitive–photosensitizer polymers results from consideration of the triplet-energy transfer of the sensitizer [151]. However, several photosensitizer photophysical properties are required before the triplet-energy transfer step will occur: (a) a large absorption coefficient of the sensitizer (excitation), (b) efficient intersystem crossing, (c) a small singlet–triplet gap (deactivation) and (d) efficient energy transfer from the triplet of the sensitizer to the chromophore (Fig. 9.3).

Polymers containing pendant photosensitive moieties have also been prepared either by radical copolymerization of 2-(cinnamoyloxy)ethylmethacrylate with photosensitizer monomers [152] or by chemical modification of chloromethylated polystyrene containing photosensitizer groups with salts of photosensitive compounds [153]. The photosensitivity, which is used for the

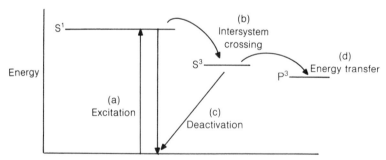

Fig. 9.3 Photophysical processes of photosensitization: (a) excitation of a sensitizer to a singlet state; (b) intersystem crossing from singlet to triplet state; (c) deactivation of singlet and triplet states of a sensitizer to ground state; (d) energy transfer from the triplet state of the sensitizer to the photosensitive chromophore.

determination of the insolubilization of the photoirradiated films, can be measured by the 'bismuth photoresist test' [139] or the 'photoresist test' [139].

9.3.2   Electron beam resists

Electron beam lithography techniques for defining patterns, in which the wafer is irradiated by electrons, have been used as an alternative to masks and mercury lamps in photolithographic procedures [154, 155]. The use of this method in the manufacture of integrated circuits offers the following advantages.

1. The beams have rather high electron energies (5–30 keV) so that diffraction effects are negligible. This allows resolution and accurate definition of details in the range 0.5–2 μm which is better than is possible with photolithographic procedures.
2. It is feasible to focus electron beams to very small diameters which can allow patterns to be made with detail much smaller than is possible by photolithography.
3. The ability to achieve finer detail can lead to an increase in the packing density of the components of integrated circuits with consequent reduction in their size and cost.
4. Smaller components will also reduce power requirements and increase the speed of operation.
5. The ease and precision with which electron beams can be deflected by electric or magnetic fields makes electron beam techniques relatively amenable to automation.
6. The energy of the electrons is much greater than that of ultraviolet photons which does not exceed a few electronvolts. This broadens the range of polymers from which a choice of resist can be made.

The radiation chemical yield is a measure of the intrinsic radiation sensitivity of the polymer and is expressed in terms of G values which are structure dependent. G(scission), or G(S), is equal to the number of main chain scissions produced per 100 eV of energy absorbed, and G(crosslinking), or G(X), is the number of crosslinks formed per 100 eV absorbed. For resist applications, G(S)/G(X) values greater than 4 indicate that the system is positive exposed relative to the unexposed regions. G(S)/G(X) ratios less than 4 indicate that crosslinking reactions will predominate, resulting in an insoluble gel and negative tone formation.

There are many polymeric materials which are available for use as negative electron resists, e.g. polymers of glycidyl acrylates, polystyrene, polysiloxanes and epoxidized polybutadiene. Poly(1-alkyl-vinylpyridinium halides) have also been shown to be good negative electron beam resists with sensitivities as high as 5 μC cm^{-2} in a 20 kV electron beam and line widths as low as 0.3 μm [156]. The electron beam sensitivities of the polymers increase with increasing polymer molecular weight and the size of the 1-alkyl group. The use of these polymeric salts eliminates the need to coat the substrate with a metal oxide and the natural electroconductivity of the polymer film prevents a buildup of electrostatic charge during exposure to an electron beam [156]. Polystyrene, as a weakly sensitive negative resist, has an excellent combination of properties which are difficult to achieve with positive resists, e.g. high dry-etching resistance because of its aromatic content and a high resolution of less than 1 μm. However, the sensitivity can be markedly enhanced by substitution on the ring with certain substituents, particularly halogen or halomethyl groups. Poly(chloromethylstyrene)s have been studied as negative electron resists that exhibit very high sensitivity while maintaining the other desirable characteristics of polystyrene [157–161].

Almost all polymers which give a good positive electron beam action should have a high $G(S)$ value and are of the general structure **9**:

$$-(-CH_2-\underset{\underset{R''}{|}}{\overset{\overset{R'}{|}}{C}}-)_n-$$

9

Here R' and R" are substituents other than hydrogen, e.g. polymethylmethacrylate (PMMA), poly(α-methylstyrene), polyisobutylene, poly(vinylidene chloride) or polytetrafluoroethylene. However, many of these are unsuitable because their glass transition temperatures are low or because they are not easily soluble which makes application of the film and development of the irradiation pattern difficult. The only materials which have so far found wide use are polysulphones and modified PMMA [148,162,163]. The combination of properties which include stability, sensitivity, contrast, adhesion and solubility has made the improved PMMA a good positive electron beam resist. Two directions have been taken towards improving the performance of PMMA without altering the processing characteristics: increasing the original molecular weight (e.g. by crosslinking) or inducing copolymerization with a comonomer of a 1,1-disubstituted vinyl type compound containing electron-withdrawing substituents such as halogen, cyano or carboxylic acid groups.

9.4 POLYMERIC NUCLEAR TRACK DETECTORS

One of the most widely used methods for the registration of nuclear particles in dielectric solids is the preferential etching of their latent tracks by appropriate chemical reagents. The use of thermosetting and highly crosslinked polymers as track detectors has recently been developed [164–166]. However, several important factors are required for the production of an ideal polymeric etch-track detector: (a) it should be optically highly transparent even after prolonged etching; (b) it should be completely amorphous; (c) it should be highly sensitive to radiation damage which can be preferentially attacked by a non-solvent reagent;

(d) it should possess a high molecular weight or good thermosetting properties; (e) it should etch homogeneously.

The important factors to be determined for detectors are (a) their bulk etch rates and (b) the relative track registration sensitivity to α particle and proton energies. The etching parameters of polymeric detectors, which are important in a comparison of their registration sensitivity, are (a) the bulk etch rate V_B (multiple beam interferometry is used to measure V_B), (b) the track etch rate V_T, (c) the etch rate ratio V_T/V_B and (d) the opaqueness after long etching periods. Since the etch rate ratio V_T/V_B as a function of Z/B indicates the sensitivity of detectors, accurate measurements of the bulk etch rate V_B and the track etch rate V_T are important for comparing the sensitivity of various polymeric nuclear track detectors.

The most important factors affecting the track registration sensitivity of an ideal detector include the following.

1. *Purity of casting compounds*: a high degree of purity in the casting compounds (monomer, initiator and other additives) is a necessary requirement for obtaining a good detector with optically smooth surfaces after etching. Hence, the compounds must be carefully filtered to remove all foreign contamination (larger than $0.1\ \mu m$) by using a nuclear filter.
2. *Vacuum outgassing of the casting compounds*: the presence of air in the monomer during the curing cycle may cause fogging and pitting on the upper surface of the finished foil, which may contribute to an increase in the density of more or less shallow etched pits. This effect can be avoided by keeping the casting compound under vacuum before use.
3. *Type and concentration of initiator and additives*: the uniformity of the rate of polymerization and the ultimate density of the converted double bonds depend on the type and concentration of the initiator and additives (plasticizers) employed. An initiator concentration providing the lowest value of V_B and the highest registration sensitivity (V_T/V_B) is desirable.
4. *Chemical structure*: the presence of aromatic rings in the structure of a polymeric detector leads to absorption of energy without destruction of the structure and acts to reduce the sensitivity of the polymer.

The development of the polymeric track detector CR-39, which is made from ethyleneglycol bis(allyl) carbonate, has recently been described [167, 168]. It has high sensitivity and uniformity with good etching properties. The three-dimensional structural network may be the reason for its high sensitivity, because the undamaged part of the polymer is very resistant to etching by NaOH solution. Radiation damage of CR-39 occurs at relatively low doses [169], there being no aromatic rings to soak up energy non-destructively. Etching proceeds by hydrolysis of the carbonate linkage, making it susceptible to chemical attack by strong bases and acids. Because of these unique characteristics, it has been used in many observations of ultra-heavy cosmic rays, in neutron dosimetry and in measurements of the uranium content of various materials. Some efforts have been made to improve the sensitivity of CR-39 by copolymerization with a small amount of diallyl chlorendate, which is effective for increasing the sensitivity and for decreasing the resistance of plastics to degradation by ultraviolet radiation. However, further improvements in the sensitivity of polymeric track detectors are required for the detection of high energy lithium, beryllium and boron nuclei in cosmic radiation. The composition and isotopic ratios of these nuclei at high energies will give new information about the origin of cosmic rays and their propagation in our galaxy.

A variety of polymers have been investigated for use as track detectors for the detection of charged particles. However, most of these polymers are more resistant to radiation damage than is CR-39 owing to the energy-absorbing properties of the aromatic rings.

9.5 POLYMERIC LIQUID CRYSTALS

Liquid crystals have become an essential part of displays in many electrical and electronic devices. The utilization of low molecular weight thermotropic liquid crystals in most cases is restricted by the necessity for special hermetic protective shells which maintain their constructional shapes and protect them from external influences. In the case of thermotropic liquid crystalline polymers there is no need for such sandwich-like construction. Polymeric liquid crystals which combine the properties of low molecular weight liquid crystals with those of polymers have recently received increased attention [170–187]. They are materials with a high degree of molecular orientation and excellent mechanical properties and they have the advantages of possessing a high modulus, high strength, high impact resistance, a high end-use temperature, a low melt viscosity, a low shear rate dependence and low density.

For a polymer to form a liquid crystalline phase, the molecule or a major portion of it should be stiff and rod-like in nature. A flexible chain polymer will not exhibit liquid crystalline behaviour while more rod-like polymers show only lyotropic mesomorphism. Semi-flexible polymers having intermediate chain extensions, i.e. incorporating both a rigid and a flexible segment, exhibit thermotropic mesomorphism. The liquid crystalline states (mesophases) are intermediate between the three dimensionally ordered crystalline state and the disordered isotropic fluid state, i.e. they are characterized by long-range orientationally ordered molecules. The intermediate states are classified with regard to their molecular order into nematic, smectic and cholesteric, and according to the molecular compositions involved into thermotropic and lyotropic. The classification, according to phase behaviour, into lyotropic or thermotropic liquid crystals depends upon whether the mesophase, i.e. the transition from a non-ordered isotropic state to an ordered anisotropic liquid crystalline state, is observed on variation of solvent content or variation of temperature respectively. Thus the crystalline order and hence the properties of the crystalline material, such as the thermal or mechanical properties, are strongly dependent on the existence of this long-range order.

The mesogenic group **10** must consist of at least two aromatic, cycloaliphatic or hetero-aromatic rings connected in the para-positions by a short rigid link (-X-) and they contain two terminal groups. The bridging groups, which maintain the linear arrangements of the aromatic rings, include imino $(-C=N-)$, azo $(-N=N-)$, azoxy $(-\overset{\uparrow}{\underset{}{N}}=N-)$, ester $(-COO-)$, trans-vinylene $(-CH=CH-)$ and p-phenylene groups.

R—⬡—X—⬡—R¹

10

The terminal groups include

$R \equiv alk, alk\text{-}O, alk\text{-}COO, alk\text{-}OCO, alk$ —⬡—COO,

alk —⬡—COO—

$R^1 = alk, CN, Cl, Br, F, NO_2$.

In the case of lyotropic mesomorphic polymers, liquid crystal characteristics are found in solution above a critical concentration; this results from very specific interactions between the amphiphilic solute and solvent which are concentration dependent as well as temperature dependent. These polymers are characterized by a high value of $(n/\eta)_\infty$ [170] where n is the flow birefringence determined from the equation

$$n = \lim_{g \to 0,\, C \to 0} \left(\frac{\Delta n}{g C \eta_0} \right)$$

Here Δn is the difference between the two main refractive indices in solution at concentration C and at the flow gradient g, η_0 is the viscosity of the solvent and η is the intrinsic viscosity of the solution. Hence the polymers are characterized by a high optical anisotropy of the molecule. Among the polymer molecules exhibiting lyotropic mesomorphism are synthetic polypeptides, polyalkylisocyanates, cellulose derivatives and p-aromatic polyamides. The liquid crystalline state in lyotropic systems has made possible the production of a new type of ultrahigh strength, high modulus polymer using wet spinning from ordered liquid crystalline solutions. Synthetic polyamides are examples of commercialized high modulus, high strength fibres spun from a liquid crystalline solution. There is a characteristic dramatic change in viscosity on increasing the concentration of the rod-like solute, whereby the viscosity increases rapidly until a critical con-centration is reached and a sharp decrease in viscosity then accompanies the formation of the lyotropic mesophase. The use of preliminary ordered liquid crystal melts should promote the most favourable conditions for subsequent crystallization of the polymer from the oriented state and the production of fibres and films with superior mechanical properties. The chemistry and physics of lyotropic systems have been discussed in the literature [188, 189].

In contrast, thermotropic systems show liquid crystalline behaviour in the melt, i.e. they melt to an ordered liquid and are temperature dependent with respect to mesophase behaviour. The temperature for transition to the thermotropic liquid crystal state depends primarily on the length-to-diameter ratio and interactions of the rigid molecule. Thermotropic mesogen polymers exhibit a mesophase in the melt state at temperatures above the crystalline solid state and before the formation of anisotropic melt. The first reported thermotropic liquid crystal polymer was a copolyester of polyethyleneterephthalate and p-hydroxybenzoic acid which has a stiff rod-like conformation and is the liquid-crystal-forming component [190].

Generally, there are two classes of thermotropic liquid crystal polymers in which liquid crystal groups are in either the side chains or the main chain of the polymer. In the case of crystallizable polymers, which are mainly those containing mesogenic groups in the main chain, the liquid crystalline state is observed from above the melting temperature T_m and up to the clearing temperature T_{Cl} when the melt displays anisotropy and may flow. Liquid crystal polymers in which the mesogenic moiety is incorporated into the polymer main chain are prepared by condensation polymerization and are of two types: (a) rigid segments joined by flexible segments; (b) totally rigid segments where there is no chain flexibility.

A considerable amount of work has been performed in this area of polymers in which the liquid crystal order is exhibited primarily by the mesogenic side groups. In this case the degree of mesomorphism is mainly determined by the length of the side chain rather than by the chain length as a whole, i.e. the order of the main chain is less well defined and the structure of the backbone is only of secondary importance. The low temperature limit for the existence of the liquid crystal state is T_g, above which a so-called segmental mobility, originating from the lability of distinct macromolecular segments, is present. This temperature is lower than the clearing temperature and in the T_g–T_{cl} interval the polymer in the form of either an elastomer or a viscous melt is in a liquid crystal state. The liquid crystal properties depend on several different structural factors, e.g. those of the mesogenic groups, the spacer arm and the backbone. These chemical structures influence the particular physical characteristics of the mesophase in liquid crystal polymers, including (a) dipolar effects, the planarity and rigidity of the mesogenic unit and its length-to-width ratio, (b) the clearing and glass transition temperatures, i.e. the mesophase temperature range, and (c) optical and dielectric anisotropy.

The direct linkage of rigid mesogenic groups to the polymer backbone restricts the transitional and rotational motions of the mesogenic moieties, which influences their interactions and prevents formation of the liquid crystal state. The systematic realization of liquid crystal side chain polymers requires a flexible spacer. Flexible spacers have been used to separate mesogenic groups placed in the main chain, and the backbone flexibility achieved by this approach has the effect of markedly improving solubility and lowering transition temperatures compared with those of rigid-rod polymers. An increase in the spacer length is associated with less hindered rotation of the mesogenic groups around their molecular axes and the main chain of the polymer does not influence the anisotropic packing of the side chains. The variation in the length of the flexible spacer leads to a change in the character and type of the mesophase as well as in its physical properties.

The effects of substituents introduced into mesogenic units are of interest in thermotropic polymers because of the high melting temperature of rigid-rod polymers, for which it is desirable to lower the melting temperature for processing. Such polymers are receiving increased interest for potential fibre applications because they can be processed by melt spinning techniques, thus avoiding the highly active solvents required in the solution spinning of fibres. In general, the influence caused by the substituents is complicated by both steric and polar effects. The steric effect results in a reduction in the thermal stability of the mesophase due to (a) a decrease in the coplanarity of adjacent units in the mesogenic group because of steric interactions between substituents and (b) a tendency for the substituents to force apart mesogenic groups in neighbouring polymer chains because of their space requirements. In contrast, substituents which impart an increased polarizability and stronger dipolar interactions between the mesogenic groups can have stronger intermolecular attractions, which would lead to higher thermal stabilities of both crystalline and liquid crystal phases, i.e. to higher melting points and clearing temperatures. The presence of substituents may also change the morphology of the mesophase, so that the substituted compounds may form only one mesophase, with the lowest degree of molecular order, while the unsubstituted compound may have more than one mesophase.

Other methods that can be used to reduce the transition temperature include (a) the addition of an element of dissymmetry to the main chain by copolymerizing mesogenic units of different shapes and (b) copolymerizing non-linear non-mesogenic units. The method of copolymerization of mesogenic and non-mesogenic monomers consists in varying the temperature range of mesophase existence. A non-mesogenic monomer is chosen so as to lower the T_g of a copolymer relative to the T_g of the liquid crystal polymer. This is achieved by increasing the distance between large mesogenic groups (mesogenic group dilution) as well as by changing the chemical nature of the polymeric backbone. The shift in T_g is usually accompanied by a lowering of the clearing temperature, i.e. mesophase thermostability is decreased.

The molecular structure of the liquid crystal phase, i.e. the organization of mesogenic groups, may be of three types, as shown in Fig. 9.4. The manner in

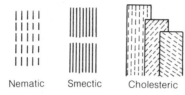

Nematic   Smectic   Cholesteric

Fig. 9.4 Organization of mesogenic groups.

which the macromolecules pack together in an ordered arrangement may therefore be changed appreciably by structural modification.

In the nematic phase there is order in the direction of the molecules and the long axes of the side groups are oriented parallel to the preferred direction of the common axis. Nematic liquid crystallinity is exhibited by certain compounds with relatively rigid polar rod-shaped molecules that tend to be oriented with their long axes parallel because of the anisotropy of their interactions. When such a compound is heated, the crystalline solid melts to a birefringent anisotropic liquid (nematic mesophase) in which adjoining molecules lie parallel to one another. At a higher temperature, the mesophase undergoes a transition to an isotropic liquid. Liquid crystal polymers of nematic type are mainly linear polymers containing mesogenic groups in their main chains. Nematic structure also occurs in homopolymers with short end substituents in mesogenic groups (Me, CN, OMe) and with a short spacer. In addition increasing the flexibility of the main chain promotes nematic phase formation, which allows for translational mobility of constituent molecules. For practical purposes there is a need for nematic mesophases at relatively low temperatures. Unfortunately, the molecular characteristics that are necessary for nematic mesomorphism also produce stable crystalline lattices. Accordingly, nematic compounds generally have high melting points. One way to lower crystal mesophase transition temperatures is to use substances with a relatively high degree of molecular dissymmetry.

In the smectic phase the molecules lie in planes with defined interlayer spacing and a long-range order exists in the preferential orientation of the side groups, in contrast with the nematic phase. The smectic phase is composed of molecular layers in which translational mobility is minimal. There are a variety of smectic phases that differ in (a) the ordering of molecules within the same layer, (b) the tilt of the average molecular axis with respect to the layer plane and (c) the positional correlation of molecules in different layers. In the smectic liquid phase of comb-like polymers the side mesogenic groups of neighbouring macromolecules are arranged either parallel or antiparallel to each other, as well as with partial overlapping of side groups (Fig. 9.5), so that the axes of side branches are perpendicular to the phase of smectic layers in which the main chains lie. For liquid crystal polymers of smectic type two packing possibilities exist: with orientation of side groups perpendicular to a layer plane or with a tilted orientation of side groups.

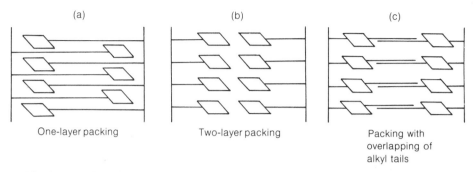

Fig. 9.5 Packing of side groups in oriented liquid crystal polymers.

Liquid crystal polymers with long flexible spacers form smectic layers independently of the chemical structure of backbones and mesogenic groups. The transformation from a nematic mesophase to a smectic mesophase is a result of a gradual decrease in flexibility of the main chain. This could be the consequence of a more defective packing of side groups due to incorporation within the backbone of units of different chemical nature. In polymers with short side groups, a smectic liquid crystal state is feasible only because of a rigid main chain. An increase in flexibility on incorporation of relatively small amounts of comonomer leads to the formation by the side groups of a nematic structure.

The cholesteric phase is described as twisted nematic, and the centres of gravity have no long-range order, i.e. it may actually be realized as a helical nematic structure. The cholesteric phase is locally similar to the nematic phase. However, the constituent molecules are chiral and give rise to asymmetric helical packing of the molecular sheets owing to the spontaneous twist resulting from molecular chirality. The main feature identifying a cholesteric mesophase in polymers is the presence of optical texture with selective reflection of circularly polarized light. One approach to creating the cholesteric type of polymeric liquid crystals is by the synthesis of copolymers that contain monomer units similar to nematic liquid crystals and monomer units with chiral mesogenic groups, such as cholesteryl esters. The cholesteric polymers exist only as copolymers, consisting of (a) nematogenic and chiral monomer units or (b) two chiral monomer units, where the polymer shows a cholesteric structure only at a defined composition. However, homopolymerization of chiral cholesteric monomers yields only smectic polymers. Some polymeric liquid crystals containing different mesogenic groups are listed in Table 9.3.

The main application of polymeric liquid crystals has been as a tool for producing high strength fibres from polymers which decompose without melting. The possibility of effectively regulating the orientation of polymer side groups by varying the electric field parameters, together with the possibility of fixing the oriented structure in a glassy state, leads to much use of liquid crystal systems for making polymeric materials with the necessary optical properties. The orient-

Table 9.3 Polymeric liquid crystals

$$\text{P}^a\!-\!\text{CO}-\text{Z}-\!\!\bigcirc\!\!-\!\text{OCO}-\!\!\bigcirc\!\!-\!\text{OR}$$
X

(a) $Z \equiv O$; $X \equiv H$, NO_2; $R \equiv C_3H_7$, $\hspace{2cm}$ Refs 191–196
    C_6H_{13}, C_9H_{19}, $C_{16}H_{33}$
(b) $Z \equiv O(CH_2)_{10}CO$; $X \equiv H$; $R \equiv n\text{-Bu}$ $\hspace{1cm}$ Ref. 191
(c) $Z \equiv NH(CH_2)_{11}COO$; $X \equiv H$; $R \equiv C_6H_{13}$ $\hspace{0.5cm}$ Refs 197–199

$$\text{P}\!-\!\text{Z}-\!\!\bigcirc\!\!-\!\text{COO}-\!\!\bigcirc\!\!-\!\text{R}$$
R'

(a) $Z \equiv COO$; $R' \equiv H$; $R \equiv -O-c\text{-}C_6H_{13}$ $\hspace{1.5cm}$ Ref. 200
(b) $Z \equiv COO(CH_2)_mO$; $R' \equiv H$; $R \equiv CN$, OBu, OMe $\hspace{0.3cm}$ Refs 201–216
(c) $Z \equiv COO(CH_2)_mCOO$; $R' \equiv H$; $R \equiv OBu$; $m = 0, 3, 10, 5$ $\hspace{0.2cm}$ Refs 217–226
(d) $Z \equiv (CH_2)_mO$; $R' \equiv Me$, H; $R \equiv CN$, C_3H_7 $\hspace{1cm}$ Refs 227, 228

$$\text{Si}\!-\!(CH_2)_m\!-\!O-\!\!\bigcirc\!\!-\!\text{COO}-\!\!\bigcirc\!\!-\!\text{R}$$ Refs 227, 229–233
R'

$m = 3, 4, 5, 6$; $R \equiv CN$, $OCH_2CHMeEt$, OMe; $R' \equiv Me$, H

$$\text{P}^a\!-\!\text{Z}-\!\text{O}-\!\!\bigcirc\!\!-\!\text{COOH}$$ Refs 234–238

$Z \equiv CO, CH_2, O(CH_2)_2$

$$\text{P}^a\!-\!\text{Z}-\!\text{O}-\!\!\bigcirc\!\!-\!\text{CH}=\text{N}-\!\!\bigcirc\!\!-\!\text{R}$$

(a) $Z \equiv CH_2$, $R \equiv n\text{-Bu}$, OBu $\hspace{3cm}$ Ref. 239
(b) $Z \equiv COO(CH_2)_m$, $R \equiv OH$, CN; $m - 2\text{-}11$ $\hspace{1cm}$ Refs 240–244
(c) $Z \equiv CO$; $R \equiv C_nH_{2n+1}$ $\hspace{3.5cm}$ Refs 245–247
    $Z \equiv CO$; $R \equiv OC_nH_{2n+1}$ $\hspace{3.3cm}$ Ref. 246
    $Z \equiv CO$; $R \equiv COOH$ or $COOC_nH_{2n+1}$ $\hspace{0.8cm}$ Refs 246, 248, 249
    $Z \equiv CO$; $R \equiv Br$, CN $\hspace{4cm}$ Ref. 245
    $Z \equiv CO$; $R \equiv CH{=}CHCOOC_nH_{2n+1}$ $\hspace{0.8cm}$ Ref. 250
    $Z \equiv CO$; $R \equiv SO_3Na$ $\hspace{3.8cm}$ Ref. 246
    $Z \equiv CO$; $R \equiv COMe$, NHCOMe, $N{=}NPh$ $\hspace{0.4cm}$ Ref. 248

$$\text{P}\!-\!\!\bigcirc\!\!-\!\text{N}=\text{CH}-\!\!\bigcirc\!\!-\!\text{R}$$
R'

(a) $R' \equiv H$; $R \equiv CN$, OC_nH_{2n+1} $\hspace{2.5cm}$ Refs 251–256
(b) $R' \equiv OH$, $R \equiv OC_nH_{2n+1}$ $\hspace{2.8cm}$ Ref. 257

Table 9.3 (*Contd.*)

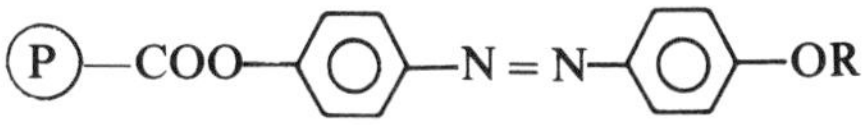

$(P)-COO-⟨C_6H_4⟩-N=N-⟨C_6H_4⟩-OR$ Ref. 258

$R \equiv Me, Et, n\text{-}Bu$

$(P)-COO-⟨C_6H_4⟩-N=N-⟨C_6H_4⟩-R$
with $\downarrow O$ on the N

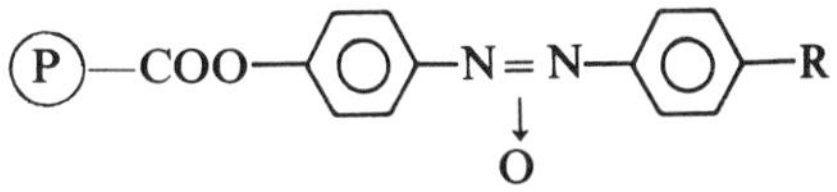

(a) $R \equiv H, Me, Et, n\text{-}Bu$ Refs 259, 260
(b) $R \equiv OMe, OEt, OBu\text{-}n$ Ref. 261

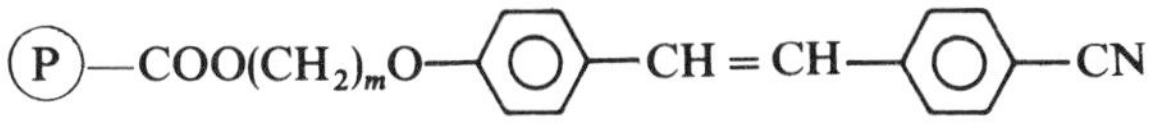

$(P)-COO(CH_2)_mO-⟨C_6H_4⟩-CH=CH-⟨C_6H_4⟩-CN$ Ref. 244

$m = 5, 6, 11$

$(P)^a-Z-⟨C_6H_4⟩-⟨C_6H_4⟩-R$

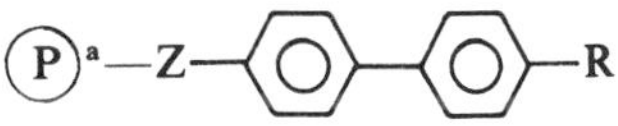

(a) $Z \equiv COO; R \equiv H$ Refs 262–267
(b) $Z \equiv COO(CH_2)_mO; R \equiv CN; m \equiv 2\text{–}11$ Refs 207, 244, 268–271
(c) $Z \equiv NH(CH_2)_{11}COO; R \equiv H$ Ref. 272
(d) $Z \equiv (CH_2)_mO; R \equiv CN; CH_2CHMeEt; m = 1, 6$ Refs 227, 228

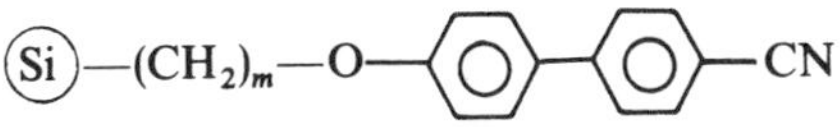

$(Si)-(CH_2)_m-O-⟨C_6H_4⟩-⟨C_6H_4⟩-CN$ Ref. 227

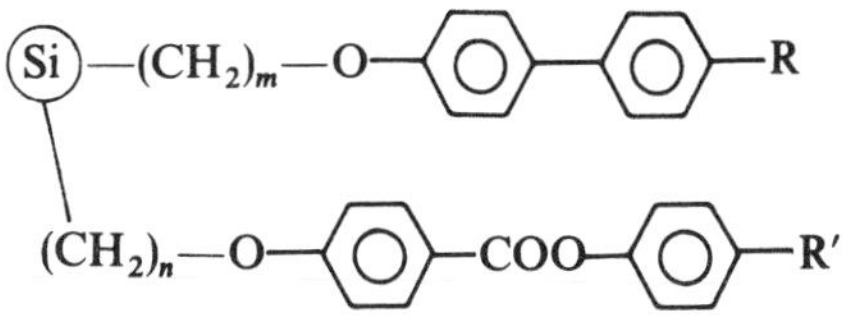

$m = 3\text{–}6$

$(Si)-(CH_2)_m-O-⟨C_6H_4⟩-⟨C_6H_4⟩-R$ Ref. 227
with side chain $(CH_2)_n-O-⟨C_6H_4⟩-COO-⟨C_6H_4⟩-R'$

$R' \equiv CN, C_3H_7; m = 3, 5; n = 4, 6; R \equiv CH_2CHMeEt, CN$

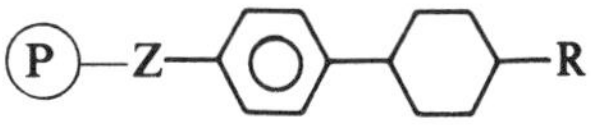

$(P)-Z-⟨C_6H_4⟩-⟨C_6H_{10}⟩-R$

(a) $Z \equiv COO; R \equiv H$ Refs 266, 267, 273
(b) $Z \equiv (CH_2)_6OCO; R \equiv C_3H_7$ Ref. 228

$(P)-O-(CH_2)_{15}-Me$ Ref. 274

$(P)^a-CONH(CH_2)_4-CHCOOR$ Ref. 275
with $|$ NHCOR' below the CH

$R \equiv Me, H; R' \equiv C_{13}H_{27}, C_{15}H_{31}, C_{17}H_{35}, C_{21}H_{43}$

Table 9.3 (*Contd.*)

(P)—COO$(CH_2)_6$—O—R COO$(CH_2)_6$—Z	Ref. 276
(a) Z ≡ 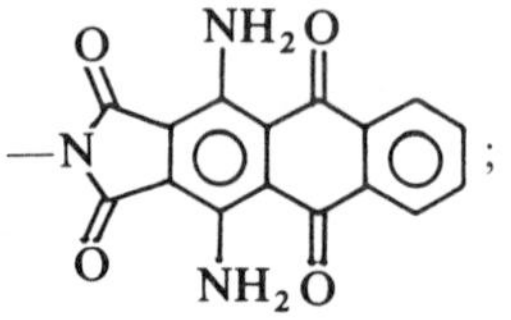 ; NH2 O O —N O NH2 O	
R ≡ 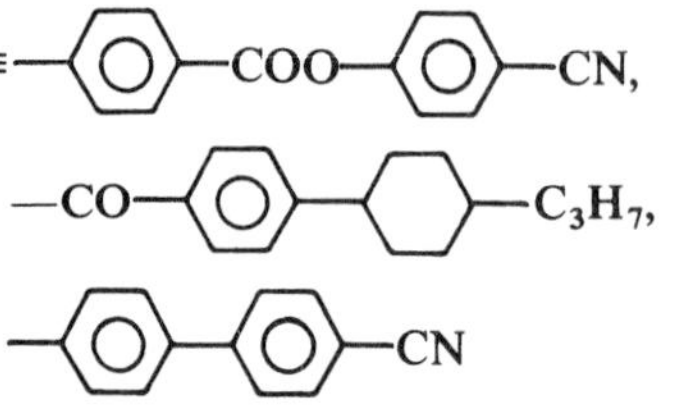 COO CN, —CO C_3H_7, CN	
(b) Z ≡ 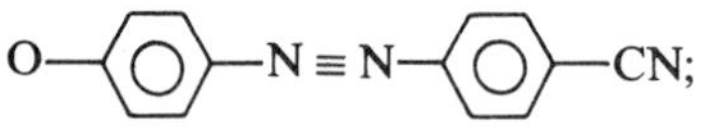 O N ≡ N CN;	
R ≡ 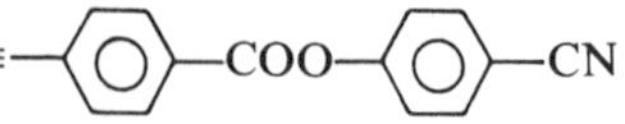 COO CN	
(P)[a]—Z—O— 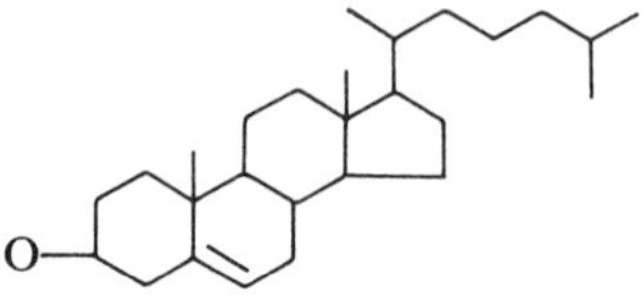	
(a) Z ≡ CO	Refs 238, 277–282
(b) Z ≡ —$(CH_2)_2$CO—	Ref. 238
(c) Z ≡ —COO$(CH_2)_m$CO—; $m = 2, 5, 10$	Refs 239, 260, 283–285
(d) Z ≡ —COO— 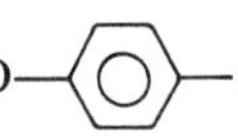 —CO	Refs 252, 256
(e) Z ≡ CO$(CH_2)_{10}$COO— 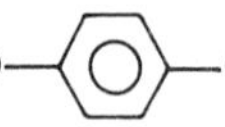 CH = CH CO—	Ref. 238
(f) Z ≡ CONH$(CH_2)_m$CO—; $m = 2, 5, 6, 8, 10, 11$	Ref. 286
(P)[a]—COO— 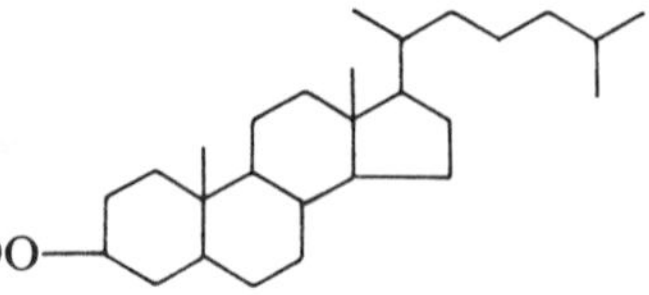	Refs 238, 252, 278, 279

ation in electric or magnetic fields is the basis of numerous technical applications of these polymers in electro-optical displays and temperature indicators. The covalent attachment of dye molecules and mesogenic units to the same polymer backbone is widely applied in this field of technology. The guest–host effect in these polymers reveals the possibility of obtaining regulated colour indicators. Mesogenic groups of a polymer that have been oriented in an external field (mechanical or electric) allow the dye molecules to orient and this causes a change in colour depending on the dye type and the parameters of the external field.

9.6 POLYMERIC SURFACTANTS

Surfactants are amphiphilic molecules characterized by aggregation behaviour which depends on the chemical structures, the nature of the media and the method of preparation. Opposing forces of repulsion between the polar head groups and of association between the hydrocarbon chains of the surfactants are responsible for the aggregation in water. Dipole–dipole interactions provide the driving force for association in polar solvents.

Vesicles derived from naturally occurring phospholipids (liposomes) are receiving interest as models for biological membranes. However, the use of these aggregates in mechanistic studies and practical applications, especially those based on long-term use, is seriously limited because they are thermodynamically unstable, having relatively short shelf lives. The need for enhanced stabilities and controllable permeabilities has led to the development of polymeric surfactant vesicles.

Recently, growing interest has been shown in the synthesis and characterization of polymerizable surfactants in order to produce membranes of enhanced stability and controllable size, rigidity and permeability for utilization in photochemical solar energy conversion and storage, catalysts for reactivity control, drug delivery and the fabrication and operation of solid state devices [287–290].

(PS)— ≡ $+CH_2—CH+_n$ with pendant phenyl-O group

(Si)— ≡ silica

(P)— ≡ $+CH_2—CH+_n$ | X

(P)— ≡ $+CH_2—CH—CH_2—CH+_n$ | X | X

(P)^a— ≡ $+CH_2—\underset{|}{\overset{Me}{C}}+_n$

(P)^a— ≡ $+CH_2—CH—CH_2—\underset{|}{\overset{Me}{C}}+_n$ | X | X

Polymeric surfactant aggregates combine the beneficial properties of stable and uniform polymer particles with the fluidities of the aggregate. The structure of these amphiphilic molecules is characterized by the presence of a hydrophilic head group and a hydrophobic alkyl of C_8–C_{18} chains, as well as the polymerizable groups. The hydrophilic group may be cationic (quaternary ammonium salts), anionic (phosphate (11, 12), sulphonate (13), carboxylate, ester, hydroxylate) or zwitterionic (14, 15):

$$-O-\overset{\overset{O}{\|}}{P}(OH)_2$$

11

$$\begin{matrix} -O \\ -O \end{matrix} \overset{\overset{O}{\|}}{P}-OH$$

12

$$>N-(CH_2)_2-SO_3H$$

13

$$-O-\overset{\overset{O}{\|}}{\underset{\underset{O^-}{|}}{P}}-O-(CH_2)_2-N^+Me_3/H_3$$

14

$$-O-\overset{\overset{O}{\|}}{\underset{\underset{O^-}{|}}{P}}-O-CH_2\underset{\underset{^+NH_3}{|}}{CH}-COO^-$$

15

The polymerizable group can be introduced into the surfactant molecules at the hydrophobic alkyl chain, in which case there is no influence on the head groups which preserve their physical properties such as charge and charge density, but the fluidity of the hydrophobic tails is changed. It can also be located at the hydrophilic head group of the surfactant, in which case the fluidity is not affected but there is no free choice of head groups. Molecules containing either diacetylene or acrylate residues have received most attention as polymerizable groups, although a few allyl, styryl and maleic acid based species have also been described (Table 9.4).

Polymerizable species have been used to produce organized surfactant aggregates, membranes (monolayer and multilayer films), micelles (spherical, rod-like, reversed) and vesicles (Fig. 9.6). Since the formation of monolayers at the gas–water interface is the oldest and simplest of the membrane models, considerable data exist on the ultraviolet polymerization of monolayers and organized multilayers [299, 319–321]. In contrast, there are only a few examples of the polymerization of microemulsions [322, 323] and bilayer vesicles [324]. Attention to the polymerization of aqueous micelles [325–328] and polymerized vesicles [296, 297, 314, 317, 329–331] has been reported. Micelles are organized systems usually formed in water by the aggregation of surfactant molecules when their concentration exceeds the so-called critical micelle concentration. The hydrophobic effect is the unique organizing force at the site of organization that is responsible for the assembly of surfactants. The shapes taken by micelles depend on the concentration of the surfactant: at relatively low concentration the aggregates may acquire a spherical shape, which elongates at higher con-

Table 9.4 Monomeric surfactants

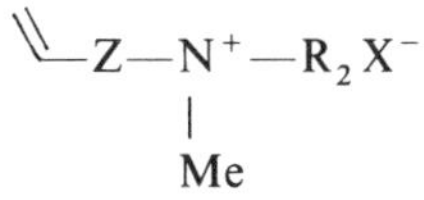

$$\text{\textbackslash\textbackslash} Z-N^+-R_2\,X^-$$
$$\overset{|}{\text{Me}}$$

Refs 291, 292–295

$$Z \equiv -CH_2-, \; -\langle\bigcirc\rangle-CH_2-;$$

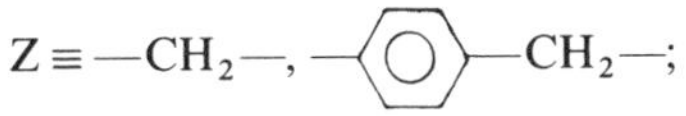

$$R \equiv -(CH_2)_2OCOC_{11}H_{23}, \; -C_{18}H_{37},$$
$$-(CH_2)_2OCOC_{15}H_{31}$$

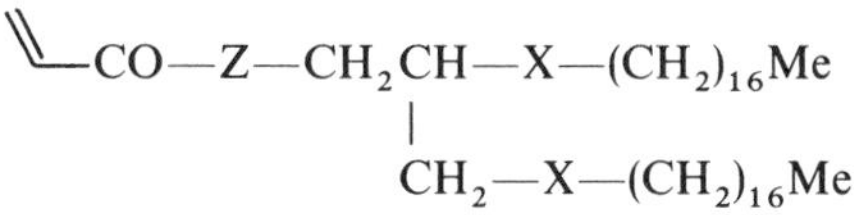

$$\text{\textbackslash\textbackslash}CO-Z-CH_2CH-X-(CH_2)_{16}Me$$
$$\overset{|}{\underset{CH_2-X-(CH_2)_{16}Me}{}}$$

(a) $Z \equiv -O-$; $X \equiv -OCO-$ Ref. 296
(b) $Z \equiv -NH(CH_2)_5COO-$, $X \equiv -OCH_2-$ Ref. 296
(c) $Z \equiv -NH-$; $X \equiv -COOCH_2-$ Ref. 297

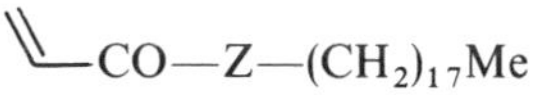

$$\text{\textbackslash\textbackslash}CO-Z-(CH_2)_{17}Me$$

(a) $Z \equiv -O-$ Ref. 298
(b) $Z \equiv -NH-$ Ref. 299

$$\text{\textbackslash\textbackslash}COO(CH_2)_2OCO-CH-Z-(CH_2)_9Me$$
$$\overset{|}{X}$$

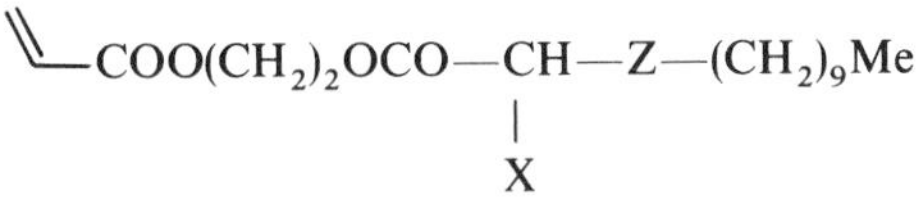

(a) $X \equiv -N^+Me_3Br^-$; $Z \equiv -(CH_2)_4-$ Ref. 300

(b) $X \equiv -{}^+N\langle\bigcirc\rangle \; Br^-$, $Z \equiv (CH_2)_4-$ Ref. 300

(c) $X \equiv -CH_2COOK$, $Z \equiv -CH=CH-$ Ref. 300

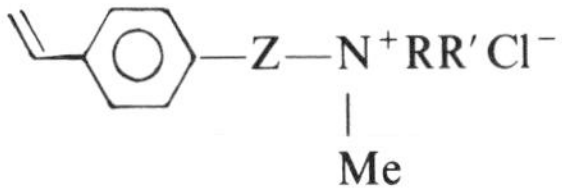

$$\text{\textbackslash\textbackslash}\langle\bigcirc\rangle-Z-N^+RR'Cl^-$$
$$\overset{|}{\text{Me}}$$

(a) $Z \equiv CH_2$; $R \equiv Me$; $R' \equiv C_{12}H_{25}$ Ref. 301
(b) $Z \equiv CH_2$; $R \equiv R'(CH_2)_2OCO(CH_2)_{14}Me$ Refs 302, 303
(c) $Z \equiv COO(CH_2)_2$; $R \equiv R' \equiv C_{18}H_{37}$ Ref. 304

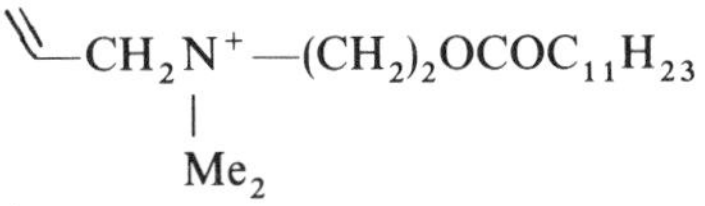

$$\text{\textbackslash\textbackslash}CH_2N^+-(CH_2)_2OCOC_{11}H_{23}$$
$$\overset{|}{Me_2}$$

Ref. 291

$$\text{\textbackslash\textbackslash}CONH-(CH_2)_3-N^+-CH_2CHOH$$
$$\overset{|}{Me_2} \qquad \overset{|}{CH_2OR'}$$
$$R' \equiv C_nH_{2n+1}; \; n = 8\text{–}20$$

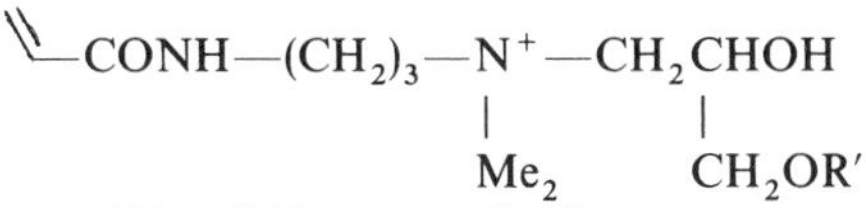

Ref. 305

Table 9.4 (*Contd.*)

$\diagdown$—CONH(CH$_2$)$_3$—N$^+$—((CH$_2$)$_{17}$Me)$_2$ $\quad\qquad\qquad\mid$ $\qquad\qquad\qquad$Me Br$^-$	Ref. 306
$\diagdown$—COO(CH$_2$CH$_2$O)$_m$—CO(CH$_2$)$_{16}$Me $m = 1, 2, 8, 9$	Ref. 300
$\diagdown$—COOCH(CH$_2$O(CH$_2$)$_n$—Me)$_2$	Ref. 307
$\diagdown$—CON(CH$_2$)$_n$—Me)$_2$ $n = 11, 17$	Ref. 307
$\diagdown\!\!\!\diagup$—COO$^-$ $^+$N—((CH$_2$)$_{17}$Me)$_2$ R$\diagup$ $\qquad\quad\mid$ $\qquad\qquad$Me$_2$	Refs 293, 308, 309
R ≡ H, Me	
$\diagdown$—Z—(CH$_2$)$_6$—N$^+$—(CH$_2$)$_m$—Me Br$^-$ $\qquad\qquad\qquad\mid$ $\qquad\qquad\qquad$R$_2$	
(a) Z ≡ —$\langle$O$\rangle$—NHCO(CH$_2$)$_4$; R ≡ H; $m = 15$	Ref. 296
(b) Z ≡ COO—(CH$_2$)$_5$; R ≡ Me; $m = 15$	Refs. 297, 310
(c) Z ≡ COO(CH$_2$)$_{11}$COO; R ≡ Me; $m = 17$	Refs 306, 311
$\diagdown$—COO(CH$_2$)$_m$—N$^+$—R Br$^-$ $\qquad\qquad\qquad\mid$ $\qquad\qquad\qquad$Me$_2$ $m = 11$, R ≡ (CH$_2$)$_{15}$	Ref. 310
$m = 2$; R ≡ CH$_2$COO—	Ref. 312
$m = 2$; R ≡ (CH$_2$)$_x$—Me; $x = 8, 11, 15$	Ref. 313
$m = 11$; R ≡ Me	Ref. 313

Table 9.4 (*Contd.*)

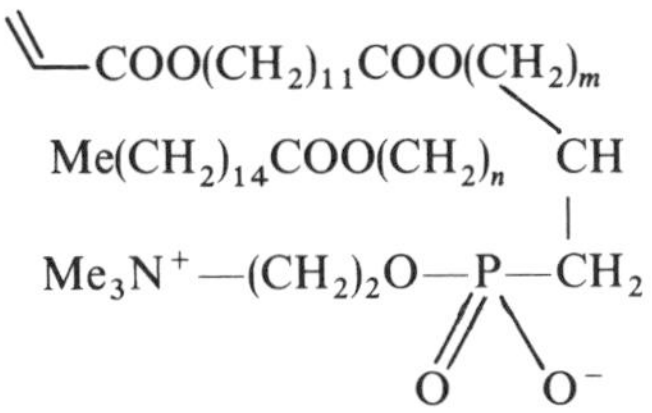

$$\text{\textbackslash}-COO(CH_2)_{11}COO(CH_2)_m$$

$$Me(CH_2)_{14}COO(CH_2)_n \quad CH$$

$$Me_3N^+-(CH_2)_2O-\underset{\underset{O^-}{\overset{O}{\shortparallel}}}{P}-CH_2$$

(a) $m = 0$; $n = 1$
(b) $m = 1$; $n = 0$

Ref. 314

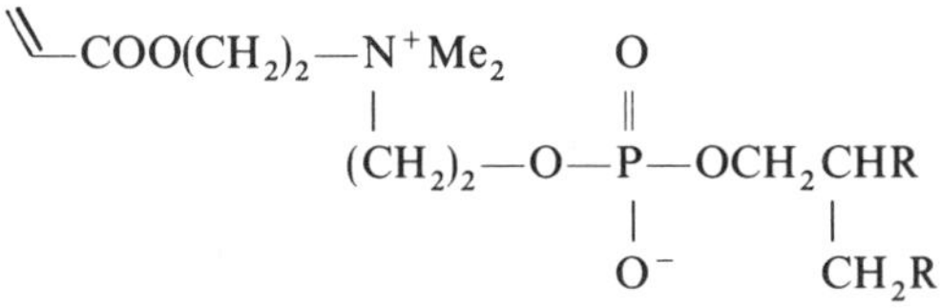

$$\text{\textbackslash}-COO(CH_2)_2-N^+Me_2 \quad O$$

$$(CH_2)_2-O-\underset{\underset{O^-}{\overset{\overset{O}{\shortparallel}}{P}}}{}-O\underset{\underset{CH_2R}{}}{C}H_2CHR$$

$$R \equiv OCO(CH_2)_{14}-Me$$

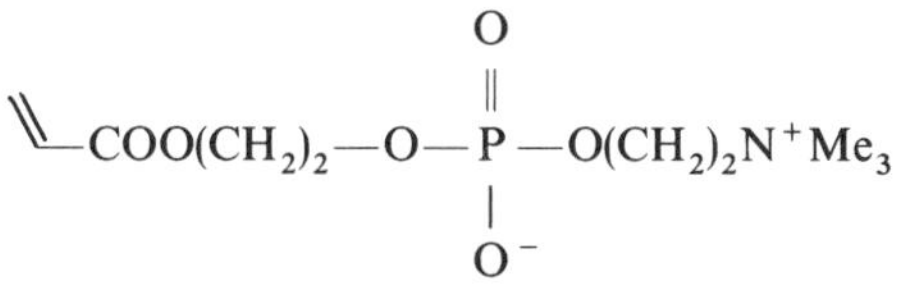

Ref. 315

$$\text{\textbackslash}-COO(CH_2)_2-O-\underset{\underset{O^-}{\overset{\overset{O}{\shortparallel}}{P}}}{}-O(CH_2)_2N^+Me_3$$

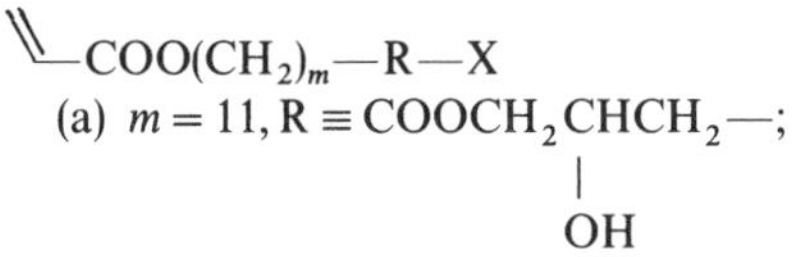

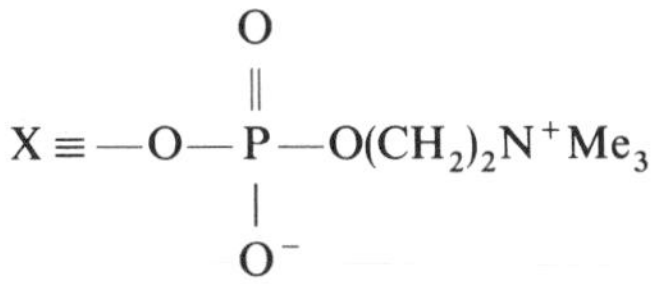

Ref. 316

$$\text{\textbackslash}-COO(CH_2)_m-R-X$$
(a) $m = 11$, $R \equiv COOCH_2CHCH_2-$;
$$\underset{OH}{|}$$

Refs 314, 317

$$X \equiv -O-\underset{\underset{O^-}{\overset{\overset{O}{\shortparallel}}{P}}}{}-O(CH_2)_2N^+Me_3$$

(b) $m = 2, 10$; $R \equiv OCO(CH_2)_{10}-$, $-CH_2-$;

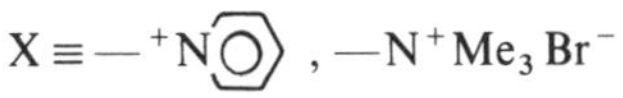

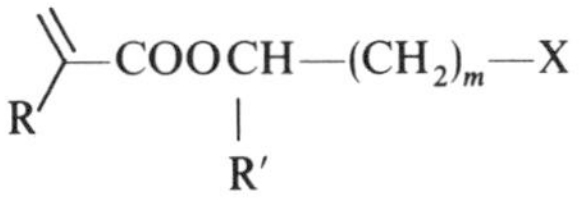

$$X \equiv -\,^+N\!\!\bigcirc\!\!\!\!\diagdown \ , -N^+Me_3 \ Br^-$$

Ref. 300

$$\underset{R}{\overset{}{}}\text{\textbackslash}-COOCH-(CH_2)_m-X$$
$$\underset{R'}{|}$$
$m = 10, 14$, $R \equiv H$, Me; $R' \equiv H$, $-(CH_2)_5Me$;
$$X \equiv COOK, N^+Me_3, -\,^+N\!\!\bigcirc\!\!\!\!\diagdown$$

Ref. 300

Table 9.4 (*Contd.*)

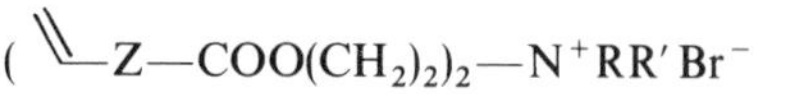

$(\diagdown\!\!-Z-COO(CH_2)_2)_2-N^+RR'\,Br^-$

  (a) $Z \equiv (CH_2)_8;\ R \equiv H,\ Me;$ Ref. 291

      $R' \equiv Me,\ (CH_2)_2OH,\ (CH_2)_2SO_3{}^-$

  (b) $Z \equiv CONH(CH_2)_{10};\ R \equiv R' = Me$ Refs, 314, 296

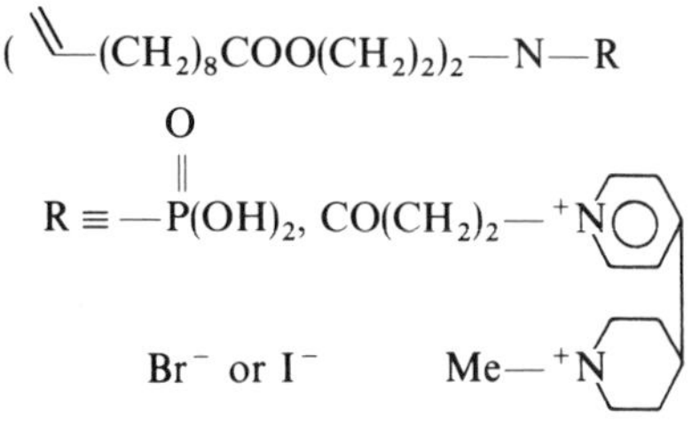

$(\diagdown\!\!-(CH_2)_8COO(CH_2)_2)_2-N-R$ Refs 291, 293

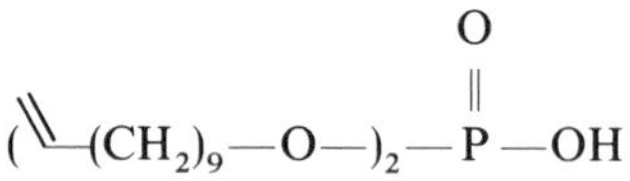

$R \equiv -P(OH)_2,\ CO(CH_2)_2-{}^+\overset{O}{\underset{\parallel}{N}}\bigcirc$

    $Br^-\ or\ I^-$ $Me-{}^+N$

$(\diagdown\!\!-(CH_2)_9-O-)_2-\overset{O}{\underset{\parallel}{P}}-OH$ Refs, 309, 318

$\text{(PS)}- \equiv +CH_2-CH+_n$ $\text{(Si)}- \equiv silica$

$\text{(P)}- \equiv +CH_2-CH+_n$ $\text{(P)}- \equiv +CH_2-CH-CH_2-CH+_n$
                   X X

$\text{(P)}^a- \equiv +CH_2-\overset{Me}{\underset{}{C}}+_n$ $\text{(P)}^a- \equiv +CH_2-CH-CH_2-\overset{Me}{\underset{}{C}}+_n$
                                X X

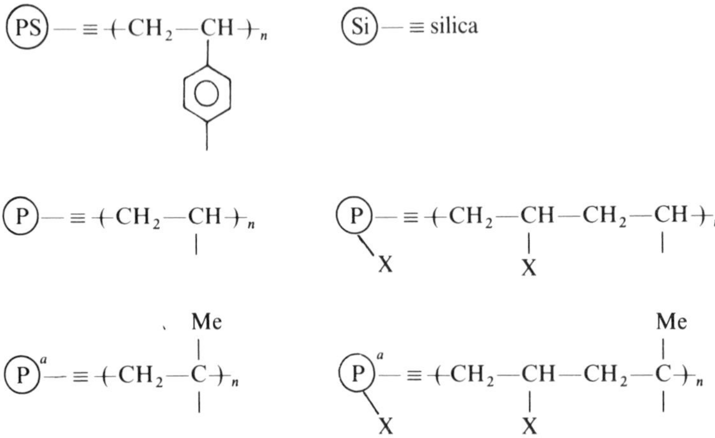

centrations and becomes rod-like. The vesicles, which are the nearest approach to biomembranes, are closed spherical structures having an aliphatic double chain, an aqueous interior and one or several lipid double layers of different sizes depending on the method of formation. There are two methods for the formation of vesicles; by slow injection of a surfactant solution in alcohol or ether into thermostated water, or by the ultrasonic dispersal of the surfactant crystal suspension in water. However, in the ultrasonic method, the type of sonicator, the applied power and the time and temperature of sonication need to be specified.

    Polymerization of vesicles formed by sonicating the surfactant has been effectively carried out by direct irradiation with ultraviolet light or by heating the

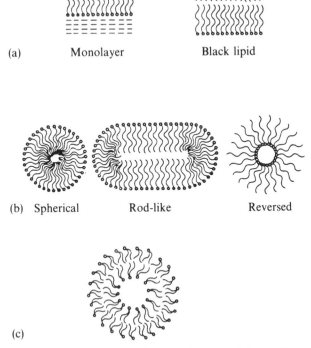

(a)   Monolayer   Black lipid

(b) Spherical   Rod-like   Reversed

(c)

Fig. 9.6 Organized structures formed from surfactants: (a) membrane; (b) micelle; (c) vesicle.

surfactant with a free radical initiator. The extent of polymerization can be monitored by following the disappearance of the polymerizable double bonds either by nuclear magnetic resonance or ultraviolet absorption spectroscopy [291, 321].

Polymerization of vinyl vesicles in a bimolecular lipid membrane has recently been described in the literature. Depending on the position of the double bonds, vesicles can be linked either across their bilayers or across their head groups. Vesicles having double bonds in their head groups can be polymerized at their inner or outer surfaces, or at both their outer surfaces or at both their outer and their inner surfaces. Irradiation by light results in the complete loss of vinyl protons. Conversely, polymerization by external addition of an initiator to already sonicated vesicles causes only 60% loss of the vinyl protons.

Polymerized surfactant vesicles are now receiving intense interest and have been proposed for use as stable models for biomembranes and in a variety of different applications. They have been proposed to act as antitumour agents on a cellular level [332], in analogy with the action of the immune system of mammals against tumour cells [290], and to provide chemical membrane dissymmetry [333] which may lead to enhanced utility in photochemical energy transfer [288], e.g. in devices for solar energy conversion and in artificial photosynthesis.

9.7 POLYMERIC STABILIZERS

The growing use of polymers in place of traditional materials such as metals, glass, wood and stone in building and construction applications has intensified the search for improving durability where resistance to degradation in the outdoor environment or to catastrophic combustion is important.

Polymer degradation can be utilized positively in several areas such as (a) recycling for the conversion of polymers to a useful chemical after their initial use so that they do not become an ecological waste problem, e.g. the hydrolysis of waste packing materials or automotive foams to monomers, (b) agrochemical or drug delivery systems based on the chemical erosion of a polymer and (c) controlled degradation for determining the structure of the natural polymer, e.g. protein, or for obtaining valuable low molecular weight substances from them, e.g. glucose from cellulose.

However, polymers under weathering conditions are susceptible to deterioration, which restricts their use in outdoor applications. The deterioration of the chemical and physical properties of the polymers from the initial optimum may result in a changed appearance, e.g. discoloration or a transparent material's becoming opaque, or degradation of the mechanical properties which make them technologically not useful. The flaking of paint and perishing of rubber are regrettably well-known examples of the deterioration suffered by most organic polymers after prolonged exposure to air and light. Hence the protection of polymers against atmospheric ageing and degradation is a prerequisite for their successful technological development and application. The most important destructive agents for polymers are environmental factors as well as aggressive conditions of processing or service, especially in the presence of other constituents of the polymer itself added either for a specific purpose or present as an impurity. The applications of functionalized polymers as stabilizers in the fields of chain-breaking antioxidants and ultraviolet and flame retardants have recently received considerable interest [334–343], as shown in Table 9.5.

9.7.1 Polymeric antioxidants

The basic oxidative chain reaction that occurs in polymers subjected to weathering conditions or high temperature processing is a sequence of reactions leading to the formation of hydroperoxide groups. The first step in the oxidation cycle is the generation of a free radical at some point on the polymer hydrocarbon chain:

$$\text{polymer} \longrightarrow R^{\bullet} \tag{9.17}$$

Table 9.5 Monomeric and polymeric stabilizers

Monomeric and polymeric antioxidants

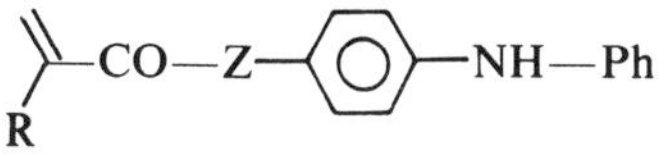

$$\underset{R}{\diagdown}\!\!\!\diagup\!\!-CO-Z-\!\!\diagcirc\!\!-NH-Ph$$

(a) $Z \equiv NH$; $R \equiv Me$	Refs 344–351
(b) $Z \equiv O$; $R \equiv H$, Me	Ref. 352
(c) $Z \equiv SO_2NH$; $R \equiv H$	Ref. 353
(d) $Z \equiv CH_2SO_2NH$; $R \equiv H$	Ref. 353
(e) $Z \equiv OCH_2CHCH_2N-$; $R \equiv Me$	Ref. 354

$$\overset{|}{OH}\quad\overset{|}{R'}$$

$$\diagcirc\!\!-Z-\!\!\diagcirc\!\!-NH-Ph$$

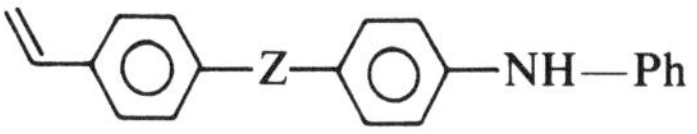

(a) $Z \equiv CH_2NH-$	Ref. 353
(b) $Z \equiv CH_2O-$	Ref. 353
(c) $Z \equiv SO_2NH-$	Ref. 353
(d) $Z \equiv NH-CS-NH-$	Ref. 355

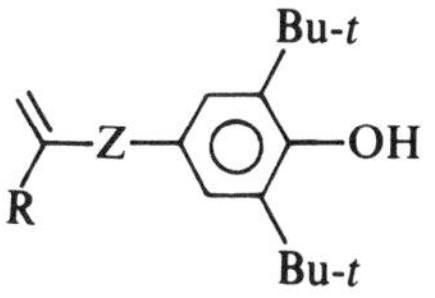

Bu-t

$$\underset{R}{\diagdown}\!\!\!\diagup\!\!-Z-\!\!\diagcirc\!\!-OH$$

Bu-t

(a) $Z \equiv$ nothing, $R \equiv H$	Ref. 356
(b) $Z \equiv COO$; $R \equiv H$, Me	Refs 357, 358
(c) $Z \equiv COOCH_2$; $R \equiv H$, Me	Refs 359, 360
(d) $Z \equiv COOCH_2CMe_2CH_2$; $R \equiv H$, Me	Ref. 361
(e) $Z \equiv COO(CH_2)_2OCH_2$; $R \equiv H$, Me	Ref. 362
(f) $Z \equiv COO(CH_2CH_2O)_mCO$; $R \equiv H$	Ref. 363
(g) $Z \equiv COS(CH_2)_m$; $R \equiv H$, Me; $m = 1, 2, 3$	Ref. 364
(h) $Z \equiv CONH$; $R \equiv H$	Ref. 364
(k) $Z \equiv CONHCH_2OCH_2$; $R \equiv H$, Me	Ref. 360
(l) $Z \equiv OCO(CH_2)_2$, $R \equiv H$	Refs 365, 366
(m) $Z \equiv O(CH_2)_2OCO$; $R \equiv H$	Refs 367, 368

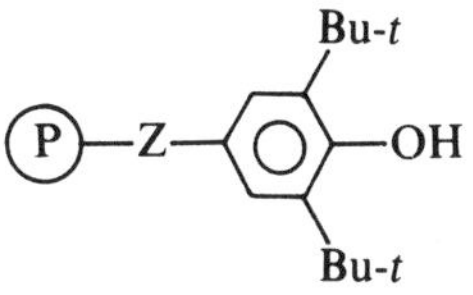

Bu-t

$$\text{(P)}\!-Z-\!\!\diagcirc\!\!-OH$$

Bu-t

(a) $Z \equiv -CH_2-$; $\text{(P)} \equiv \text{(P)}^a$	Ref. 369

Table 9.5 (*Contd.*)

(b) $Z \equiv -O-$ [t-Bu substituted aromatic ring] $-CH_2$; (P) $\equiv$ (P)[b,c]   Ref. 370

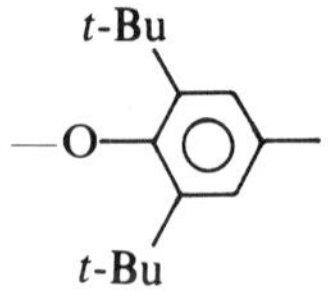

(c) $Z \equiv COO(CH_2)_2NCO(CH_2)_2$;   Ref. 371
$\qquad$ |
$\qquad$ Me

(P)=(P)[a]

[structure with XO, R′, Z, R, R′, OX substituents]

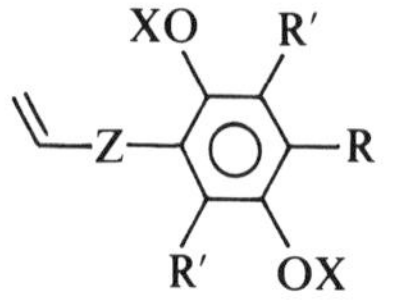

(a) $Z \equiv$ nothing; $X \equiv H$, Ac; $R \equiv R' \equiv$ alk   Ref. 372
(b) $Z \equiv CONHCH_2$; $X \equiv H$; $R \equiv R' \equiv$ Me   Ref. 373

(c) $Z \equiv -$[benzene ring]$-$; $X \equiv H$; $R \equiv$ Me; R'—H   Ref. 374

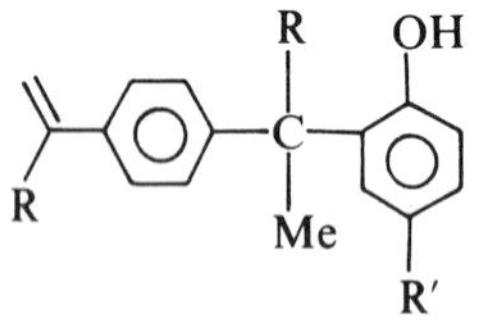

[structure with R, OH, C, Me, R′ substituents]   Ref. 375

$R \equiv H$, Me; $R' \equiv$ Me

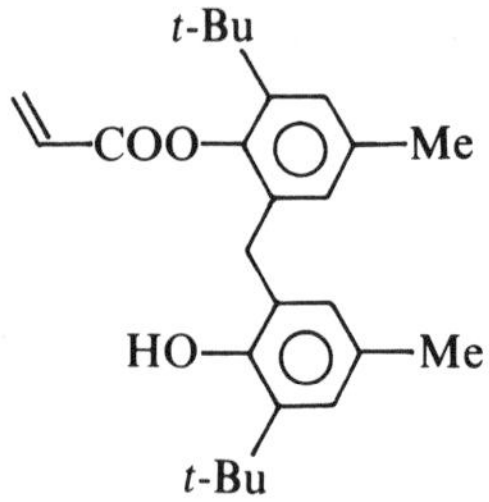

[structure: t-Bu, COO, Me, HO, Me, t-Bu substituted rings]   Ref. 376

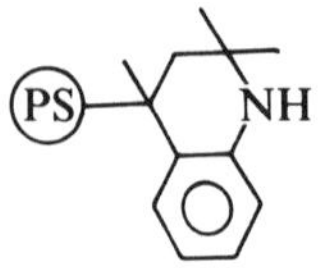

(PS)[structure with NH]   Ref. 377

Table 9.5 (*Contd.*)

Monomeric ultraviolet absorbers

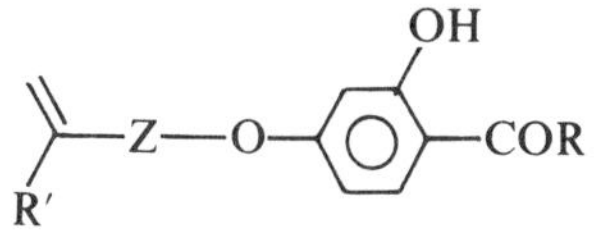

R′ ≡ H, Me

(a) Z ≡ CO; R ≡ Ph   Refs 378–380
(b) Z ≡ CO; R ≡ Me   Ref. 378
(c) Z ≡ COOCH$_2$CHCH$_2$; R ≡ Ph   Refs 381–383
      |
     OH

(d) Z ≡ (CH$_2$)$_n$OCH$_2$CHCH$_2$, R ≡ Ph; n = 1, 2   Refs 384, 385
         |
        OH

(e) Z ≡ COO(CH$_2$)$_2$; R ≡ Ph   Ref. 386

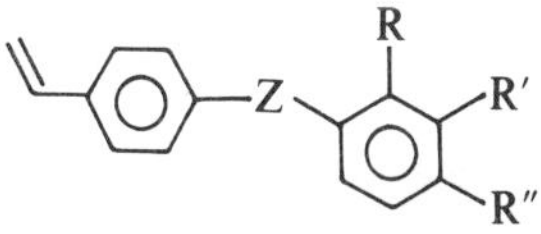

(a) Z ≡ CH$_2$O; R ≡ H; R′ ≡ OH; R″ ≡ COPh   Refs 387, 388
(b) Z ≡ CH$_2$O; R ≡ COPh; R′ ≡ R″ ≡ H   Ref. 387
(c) Z ≡ CO; R ≡ R″ ≡ OH; R′ ≡ H   Refs 389, 390
(d) Z ≡ CO; R ≡ OMe; R′ ≡ R″ ≡ H   Ref. 391

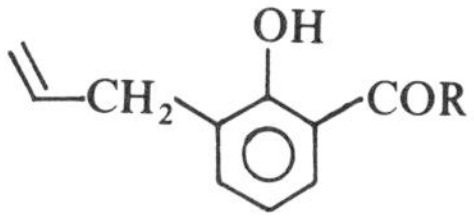

R ≡ Me, Ph   Ref. 392

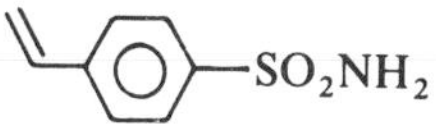

Ref. 393

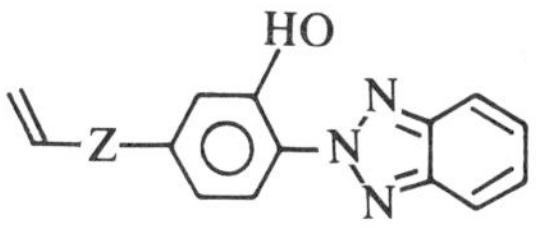

Refs 388, 394, 395

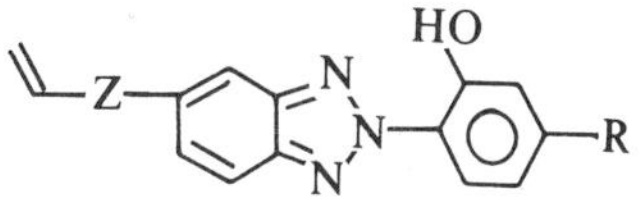

Refs 388, 394, 395

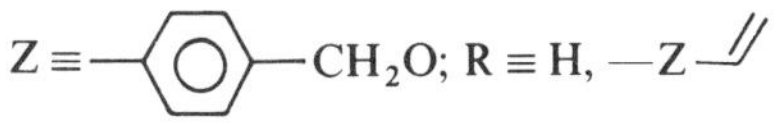

Z ≡ —⟨O⟩—CH$_2$O; R ≡ H, —Z—

Table 9.5 (*Contd.*)

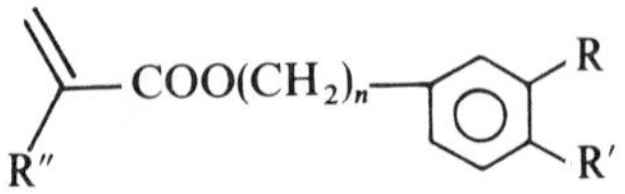 (a) $n = 1$; $R \equiv COOPh$; $R' \equiv OH$	Ref. 396
(b) $n = 0$, $R \equiv COOPh$; $R' \equiv OH$	Refs 396–398
(c) $n = 0$; $R \equiv OH$; $R' \equiv COOMe$	Ref. 399

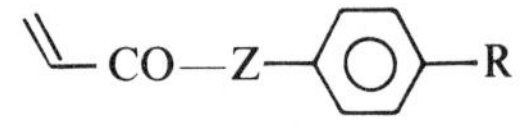

Refs 400–402

(a) $Z \equiv NH$; $R \equiv COOEt$
(b) $Z \equiv O$; $R \equiv NMe_2$

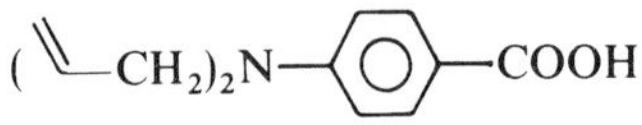

Ref. 403

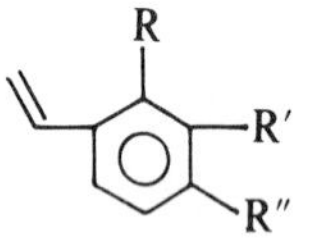 (a) $R \equiv H$; $R' \equiv COOC_nH_{2n+1}$; $R'' \equiv OC_nH_{2n+1}$	Ref. 404
(b) $R \equiv OC_nH_{2n+1}$; $R' \equiv COOC_nH_{2n+1}$; $R'' \equiv H$	Ref. 405
(c) $R \equiv H$; $R' \equiv OC_nH_{2n+1}$; $R'' \equiv COOC_nH_{2n+1}$	Ref. 406

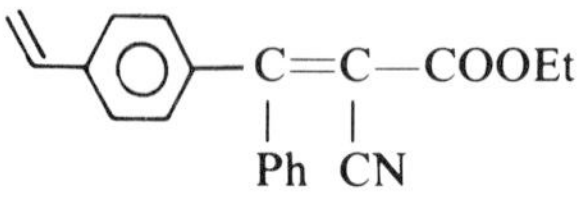

Ref. 407

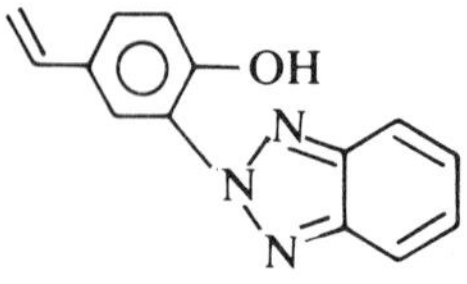

Ref. 408, 409, 410

Monomeric and polymeric flame retardant

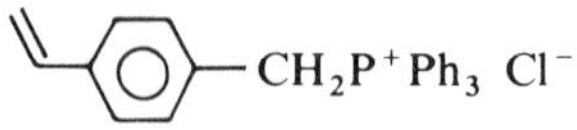

Ref. 411

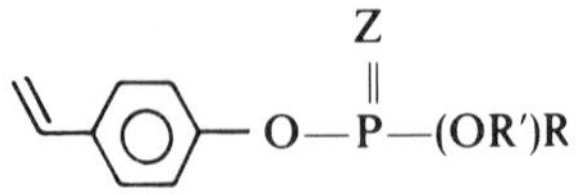

Ref. 412, 413

$Z \equiv S, O$; $R' \equiv Et$, Bu; $R \equiv OEt$, Cl

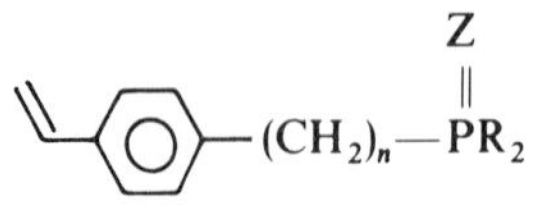

(a) $n \equiv 0$; $Z \equiv S, O$; $R \equiv Ph$

Ref. 414

Table 9.5 (*Contd.*)

(b) $n = 1$; $Z \equiv O$; $R \equiv NH_2$ Ref. 415
(c) $n = 1$; $Z \equiv O$; $R \equiv OC_nH_{2n+1}$ Ref. 416

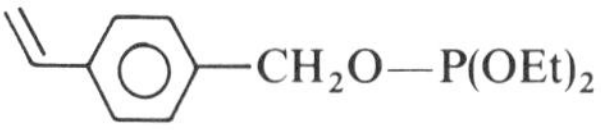

$\langle\!\langle$—$\bigcirc$—CH_2O—$P(OEt)_2$ Ref. 416

(P)—$\bigcirc$N$^+$—H H$_2$PO$_4{}^-$ Ref. 417

$\langle\!\langle$—$\bigcirc$—SO_2NH_2 Ref. 393

$$\langle\!\langle\text{—CH}_2\text{OCH}_2)_2\overset{\displaystyle O}{\overset{\displaystyle \|}{P}}\text{—OH}$$ Ref. 418

$$\langle\!\langle\text{—CH}_2\text{OCH}_2\text{—}\overset{\displaystyle O}{\overset{\displaystyle \|}{\underset{\displaystyle BrCH_2CH_2OCH_2}{P}}}\text{—OH}$$ Ref. 418

$$\langle\!\langle\text{—}\overset{\displaystyle O}{\overset{\displaystyle \|}{P}}\text{—(OCH}_2\text{CH}_2\text{Cl)}_2$$ Ref. 419

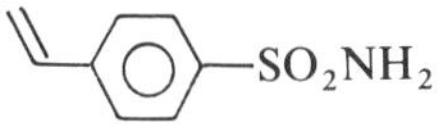

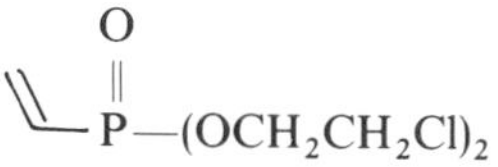

(PS)— $\equiv$ $\{CH_2$—$CH\}_n$ (Si)— $\equiv$ silica

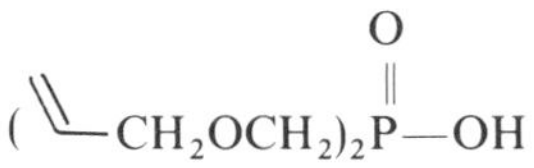

        $\bigcirc$

(P)— $\equiv$ $\{CH_2$—$CH\}_n$ (P)— $\equiv$ $\{CH_2$—CH—CH_2—$CH\}_n$
        | X   X

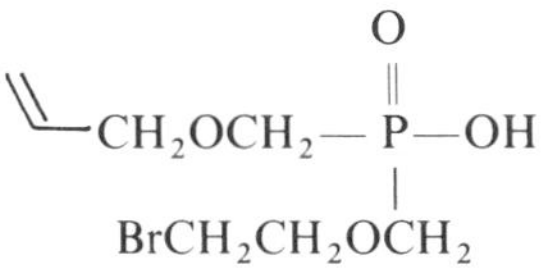

(P)^a— $\equiv$ $\{CH_2$—$\overset{\displaystyle Me}{\underset{\displaystyle |}{C}}\}_n$ (P)^a— $\equiv$ $\{CH_2$—CH—CH_2—$\overset{\displaystyle Me}{\underset{\displaystyle |}{C}}\}_n$
        X   X

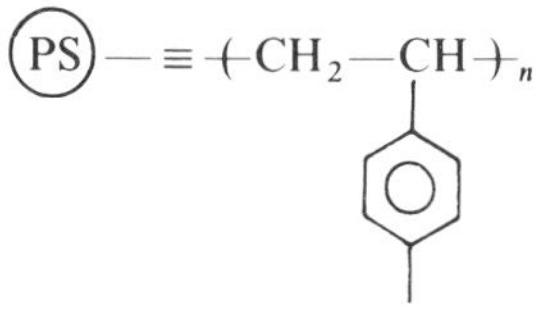

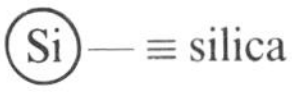

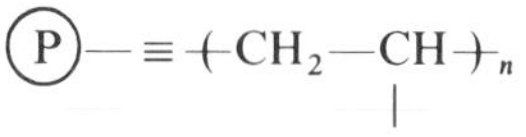

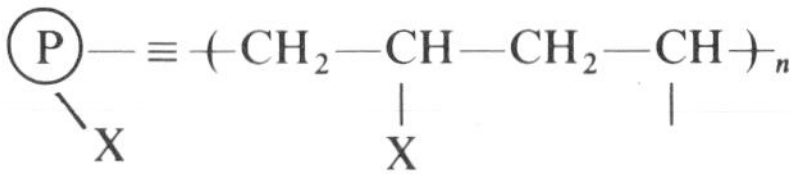

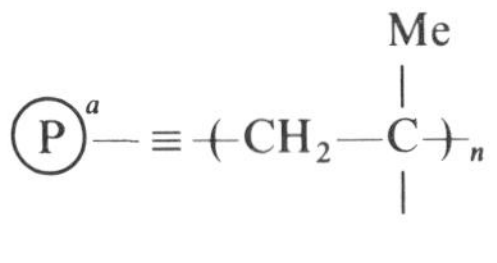

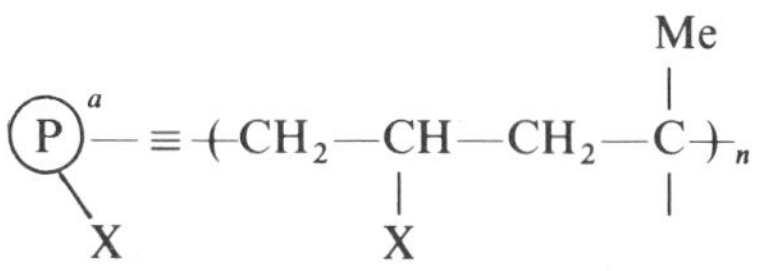

A hydrocarbon radical can arise from a variety of applied stresses, such as heating or mechanical damage when the polymer is stretched, flexed or milled. Once generated, the radical can undergo further reactions that result in breaking or crosslinking of the polymer chain, or it can combine with another radical to form inactive products. The radical can also react with a ground-state diradical oxygen molecule to give a peroxy radical

$$R^{\cdot} + O_2 \longrightarrow RO_2^{\cdot} \tag{9.18}$$

which then undergoes a hydrogen abstraction reaction:

$$RO_2^{\cdot} + R{-}H \longrightarrow R{-}OOH + R^{\cdot} \tag{9.19}$$

The auto-oxidation reaction is therefore autocatalytic in character as a result of both the chain reaction behaviour and the formation of large amounts of labile hydroperoxide groups which are potential free radical initiators.

The loss of desirable mechanical, physical, chemical and electrical properties brought about by oxidation is due to the change in the molecular weight distribution of the polymer via decomposition or crosslinking. Chain scission, which causes a decrease in chain length, has been shown to occur by decomposition of the hydroperoxide groups formed [420]:

$$\tag{9.20}$$

Crosslinking occurs as a result of either a combination reaction between two of the parent radicals involved in the oxidation mechanism or the attack of one of the radicals on an unsaturated point of the chain.

Thus antioxidants are needed for the fabrication and long-term use of most polymers to prevent the initiation or interrupt the propagation step of the auto-oxidation reaction. However, the protection given by conventional antioxidants against deterioration is often short lived under some service conditions owing to their loss by volatilization or by leaching processes and thus a relatively high concentration of antioxidants is required.

Recently, many attempts have been made to increase the inhibition efficiency of small molecule antioxidants by producing high molecular weight antioxidants or by chemically combining them with the polymers that they are protecting. Polymeric antioxidants have great potential because they do not readily migrate out of the polymer system and they give solvent- and detergent-resistant polymers. The most important classes of antioxidants which have been incorporated into polymers are phenol and aniline derivatives [421].

9.7.2 Polymeric ultraviolet stabilizers

Polymers in outdoor applications are usually exposed to damaging sunlight radiation and hence they are susceptible to photodeterioration. The most damaging part of the solar radiation is at approximately 33 000 cm^{-1} (300 nm) in the near ultraviolet while the shorter wavelength radiations, below 300 nm, are absorbed by the ozone layer of the outer atmosphere which protects the earth's surface from this harmful radiation. The spectral composition and intensity of solar radiation at the earth's surface varies with the time of day, the location and the weather conditions owing to changes in the thickness of the ozone layer.

Photoinitiation may be attributed to the chromophoric groups present as impurities or created during processing. Absorbed ultraviolet radiation leads to excitation of the polymer and the polymer returns to the electronic ground state with re-emission of the energy in a non-harmful form. Photochemical reactions are chain cleaving and lead to the formation of functional groups, which may be starting points for degradation, or to auto-oxidation in the presence of oxygen. Thus it is of considerable practical importance that polymers should be protected against the effects of solar radiation to increase their durability [422–434]. However, the solubility of conventional stabilizer in the matrix is often less than the minimum effective concentration, leading to loss of significant amounts of absorbers from polymers by physical means, i.e. through exudation, volatilization or solvent extraction during fabrication and end use. Moreover, blended-in ultraviolet-absorbing additives tend to aggregate in spots within the film and therefore cannot provide full protection. These problems are particularly acute in the stabilization of thin films and coatings. Recent work with stabilizers grafted onto polymers has shown that the disadvantages decreased with an increase in stabilizer molecular weight. Polymeric ultraviolet absorbers have been prepared through polymerization of the vinyl group directly attached to the absorbers. For example, vinylbenzyl monomers having o-hydroxybenzophenone substituents have been copolymerized with styrene and other monomers to give a polymer composition (0.5%–1%) that has good ultraviolet stability in plastic articles, films, lacquers, oil-base varnishes etc.

Polymeric ultraviolet stabilizers are more resistant to solvent extraction, volatilization during high temperature processing and exudation from plastics than low molecular weight stabilizers. Thus, over a long period of time or in the presence of solvents, polymeric stabilizers appear to be more permanent and superior to conventional additives.

9.7.3 Polymeric flame retardants

A rapid increase has occurred in the replacement of conventional materials with synthetic polymers; very large amounts are used in construction and decorative finishing as paint and other coatings of buildings. The increasing use of polymers has increased fire hazards, resulting in damage to structures and death or incapacitation of occupants of buildings. Thus the flammability characteristics of polymers are a barrier to their increased use in buildings because (a) they make a major contribution to the rate of fire growth, i.e. increase the fire load, (b) they generate dense smoke and toxic fumes as a result of burning which creates problems for fire fighters by limiting visibility and necessitating the use of breathing apparatus and (c) they rapidly lose their properties as a result of combustion.

Thus the problems involved in reducing the fire hazards associated with the use of synthetic polymeric materials are very complex from both an economic and a social point of view. The nature of the combustion process, which occurs when a fire starts, may be represented as shown in Fig. 9.7. The initial stage of a fire occurs when a source of ignition decomposes the polymeric materials to inflammable volatile products. Thus, for continuous burning to occur, (a) the application of heat must be sufficient to decompose the polymer, (b) the temperature must be sufficient to ignite the products of decomposition and (c) the amount of heat transferred from the flame back to the polymer must be sufficient to maintain the cycle.

Hence, to reduce flame spread once ignition has taken place, i.e. to make the polymer fire retardant, it is necessary to break the cycle. This may be done in one of three general ways: (a) by modifying the polymer so that the initial mode of the thermal degradation process is changed, resulting in the evolution of less flammable products, (b) by inhibiting burning or by quenching the flame or (c) by reducing the feed-back of heat from the flame to the decomposing polymer so that the transferred heat is not sufficient to cause the evolution of flammable gases.

Polymers especially designed to resist burning and thus to reduce the fire hazard in their use have been reported. Low flammability can be achieved by making the polymer rich in ring structure and low in readily oxidizable side groups. For example, aromatic polyimides 16 [435], that contain no hydrogen atoms, have excellent fire resistance but are expensive and therefore are limited to special uses such as aerospace applications [436].

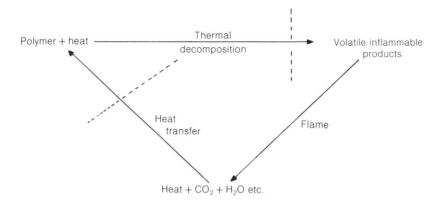

Fig. 9.7 Combustion cycle.

16

In the production of fire-retardant grades of commercial polymeric materials use can be made of appropriate additives which have a dual purpose: to ensure that the polymeric material does not present a fire hazard and to enable the hazards from the other flammable materials to be reduced. Many halogen- and phosphorus-containing compounds are outstanding in this respect. The combustion of gaseous fuels is a high temperature process which proceeds via a free radical mechanism. The use of halogens as flame retardants is based on the theory that they function in the gas phase as radical traps. The ideal flame retarder must be inexpensive, colourless, easily blended, compatible, heat and light stable and effectively permanent, and must have no deleterious effects upon the properties of the base polymer.

Unfortunately, conventional flame-retardant additives usually induce the formation of greater volumes of more toxic fumes and volatilize before they perform their action. Recently, flame-retardant substances chemically bonded to polymeric materials, monomeric flame retardants, have been introduced to overcome the problems of the volatility of conventional retardants [437–439].

9.8 POLYMERIC CORROSION INHIBITORS

Recent years have seen a considerable increase in the size of merchant ships, particularly tankers and bulk carriers. The costs of these vessels and their operation (interest, capital repayment, insurance etc.) are correspondingly high. These charges continue whether the ship is in service or out of service undergoing maintenance or repairs, and hence it is necessary to extend the effective life of the protective paint systems and to minimize the time out of service when maintenance does become necessary.

At present, the factor limiting the intervals between dry dockings is the performance of the antifouling composition applied to the underwater hull to prevent the attachment and growth of marine plants and animals.

Coatings applied to protect steel hulls against corrosion are capable of providing even longer periods of maintenance-free service [440]. Corrosion is an undesirable electrochemical process and is defined as the destructive attack of a metal by the environment; significant factors are (a) the oxygen concentrations in the water or the atmosphere, (b) the temperature, (c) the concentrations of various salts in solution in contact with the metal and (d) the pH of the electrolyte.

The most common origin of corrosion is the basic thermodynamic tendency for metals to react, as expressed in terms of the free energy of reaction G. A negative free energy of formation of a metal oxide indicates a tendency for the metal to react, i.e. the oxide is stable (e.g. Fe_2O_3),

$$2Fe + \tfrac{3}{2}O_2 \longrightarrow Fe_2O_3 \quad \Delta G = -741\,kJ\,mol^{-1}(-177.1\,kcal\,mol^{-1}) \qquad (9.21)$$

but a positive G indicates that the metal is stable (e.g. Au_2O_3). However, there are different types of corrosion: (a) galvanic cells in which corrosion operates between two dissimilar metals, e.g. if an iron pipe is connected to a copper pipe in a moist environment such as the soil; (b) salt concentration cells which are common where a metal is in contact with two concentrations of the same salt, e.g. copper with $CuSO_4$, and in which corrosion is accelerated at the site of low salt concentration; (c) oxygen concentration cells which are differential aeration cells in which corrosion is accelerated at the low oxygen concentration site by an electrochemical process because the high concentration site is cathodic. Potential differences on the surface of steel arise from a variety of causes, including (a) the presence or absence of millscale (the oxide film formed on steel when it is hot rolled), (b) local differences in the composition of the steel and (c) local variations in heat treatment of the steel. An electrochemical corrosion cell is set up, at a break in the millscale on steel immersed in seawater, in which the potential difference is between the scale-free areas that become anodic and the scale-covered areas that become cathodic. Millscale is a major cause of corrosion in seawater, there being a potential difference of about 0.3 V between the scale-free and the scale-covered areas. The anodic and cathodic reactions are as follows.

$$\text{Anode: } 4Fe \longrightarrow 4Fe^{2+} + 8e \qquad (9.22)$$

$$\text{Cathode: } 8e + 4H_2O + 2O_2 \longrightarrow 8OH^- \qquad (9.23)$$

The initial anodic (eqn 9.22) and cathodic (eqn 9.23) products—ferrous ions and hydroxyl ions—diffuse from the surface and react to form ferrous hydroxide

$$4F^{2+} + 8OH^- \longrightarrow 4Fe(OH)_2 \qquad (9.24)$$

which in turn is oxidized by dissolved oxygen to hydrated ferric oxide, rust:

$$4Fe(OH)_2 + O_2 \longrightarrow 2Fe_2O_3 \cdot H_2O + 2H_2O \qquad (9.25)$$

It should be noted that (a) the ferrous hydroxide is not precipitated in contact with either the anodic or cathodic areas but at some intermediate point, and hence does not stifle the corrosion reaction, and (b) conditions at cathodic areas become alkaline, necessitating that protective paints have good alkali resistance.

Inhibitors are chemical substances that, when added in small amounts to the environment in which the metal would corrode, will retard or entirely prevent this corrosion. An inhibitor must be adsorbed to the metal surface for it to be effective, but the type of interaction bond (chemisorption, electrostatic or π-bond (delocalized electron) adsorption varies with the chemical configuration of the molecule.

Corrosion can be prevented by stopping the flow of corrosion currents in the cells, which is possible by (a) anodic inhibition, in which the anode reaction (9.22) is suppressed by paints that supply electrons to the metal surface or reinforce the naturally occurring oxide film on the surface, (b) cathodic inhibition (although suppression of the cathodic reaction (9.23) is in general not practicable because paint films of normal thickness are so permeable to water and oxygen that the reaction proceeds almost unhindered) or (c) resistance inhibition, in which the movement of ions in the electrolyte surrounding the metal is impeded by interposing a high resistance in the electrolytic path of the cells; this is the most general mechanism of protection by paints. The resistance of a paint film is affected by (a) the presence of electrolytes in the film, and hence water-soluble impurities in pigments must be kept to a minimum, (b) the presence of electrolytes beneath the film, and hence surface cleanliness prior to paint application is important, (c) penetration of the film by water and electrolytes from outside, highly crosslinked films being in general more resistant to such penetration, and (d) its thickness—hence the practical importance of applying thick films. On the basis of these considerations it is clear that the essential requirements for protective paint coatings on the underwater hulls of ships are that (a) they should be applied to a clean steel surface free from millscale, rust or other contaminants, (b) they should provide a high electrical resistance between the metal and seawater and (c) they should withstand alkaline conditions.

Poly(4-vinylpyridine) and its derivatives including quaternary alkyl halide salts and betaines, such as those prepared from sodium chloroacetate, function as effective inhibitors for preventing the corrosion of metals, including iron, aluminium, magnesium and brass, in non-oxidizing acid solutions [441–443]. Sodium poly(styrenesulphonate) also forms a protective coating on phosphated mild steel that prevents rusting [444]. Polyvinylimidazoles 17 and 18 provide effective corrosion protection for copper at elevated temperatures through complex formation between copper and nitrogen atoms which inhibits oxygen adsorption on the copper surface [445]:

17   18

9.9 POLYMERIC FLOCCULATING AGENTS

When a suspension of solid particles in a medium is allowed to stand, the large particles settle out first and very fine material settles very slowly or not at all. This resistance to settling by the very fine particles relates to the size and density of the particles, to their surface properties and to the composition of the suspending medium.

In H_2O suspensions the solid surfaces tend to have a net electrical charge, balanced by ions of opposite charge in the solution in the vicinity of the particle surface. The primary sources of this surface charge are (a) ionization of surface groups, (b) isomorphic substitution in the solid lattice and (c) preferential adsorption of ions or ionizable species from the suspending medium. This situation of a charge localized on the particle surface with a diffuse distribution of counter-ions extending into the liquid medium is referred to as an ionic double layer.

The surface charge generates a repulsive force between particles, tending to keep them apart. For large particles, gravitational forces are dominant and make them unstable with respect to suspension, but for individual particles of colloidal dimensions the electrostatic and microhydrodynamic forces dominate and make their suspensions relatively stable. Flocculation is the coagulation of the colloidal particles into larger aggregates (floc) for which gravitational forces dominate and suspension stability is lost. The chemical agent that enhances the process is termed a flocculating agent or a flocculant. Flocculating agents enhance aggregation either by modifying the energy of interaction between a pair of particles or by introducing new interaction terms, thereby destabilizing the suspension and effecting solid–liquid separation.

The stability of dispersions in polymer solutions depends on (a) the polymer–particle interaction, (b) the polymer–solvent interaction and (c) the solvent–particle interaction. However, the temperature may affect all these interactions, and hence the degree of adsorption of the polymer on the particle may also be changed by temperature. When a polymer which has less adsorption energy than the solvent is added to the system, the dispersion flocculates. The energy of this type of flocculation is attributed to solvent–particle interaction. The solvent molecules in the solvation layer on the particle may be desorbed during a collision of particles and dilution of the polymer by this desorption of the solvent brings about an attraction between the particles. The attractive energy caused by the heat of dilution is large enough to overcome the electrical repulsion due to the electrical charge on the particles.

The stability of a dispersion is quantitatively expressed by the free energy change ΔG of the system on the approach of two particles. If ΔG becomes negative, stabilization will result, and if ΔG becomes positive, flocculation will result. In the presence of a polymer, whether the polymer acts as a dispersant for stabilization of dispersions or as a flocculant, ΔG is a complex function of various parameters such as molecular weight, molecular structure, concentration, electrical charge, degree of adsorption, conformation in the adsorbed layer, solvency and the dielectric properties of the medium. The energy of interaction between a pair of particles (a factor in the formation of aggregates) is formulated as arising from two components: (a) a term due to overlap of double layers leading to repulsion and (b) a term due to van der Waals attraction. Since dispersion forces are short range, the particles must be close together before a significant attractive force develops. The distance of closest approach for colloidal particles is strongly influenced by the magnitude of the double layer, which therefore influences the stability of a suspension. The flocculation by polymers that bring about destabilization of aqueous suspensions is usually explained by two general methods: (a) polymer charge neutralization and (b) polymer bridging.

(a) Polymer charge patch

Addition of polyelectrolyte of the opposite sign to the dispersed particles can cause flocculation due to charge neutralization because of adsorption of the polymer onto the particle surface via electrostatic bonds and consequent charge neutralization on the particle; this charge neutralization results in a lowering of electrical repulsion and thereby leads to flocculation of the dispersion. When a cationic polymer is adsorbed onto the surface of a particle, it neutralizes the negative charge and provides excess cationic charge to compensate for other negative charge sites on the surface. Thus the polymer adsorbs onto a particle surface in patches of molecular size, forming positive patches surrounded by regions of negative surface sites. This mosaic of positive and negative surface areas then allows a direct electrostatic attraction between particles.

(b) Polymer bridging

An ionic polymer or a polyelectrolyte of the same sign as the particles can cause flocculation by bridging. Water-soluble polymers can bond to a particle surface by a variety of mechanisms, including electrostatic bonds, the hydrophobic bonds, van der Waals bonds and a group of rather specific physicochemical interactions ranging from hydrogen bonding to covalent bonding.

When very long polymer molecules are adsorbed onto the surface of particles, they tend to form loops that extend some distance from the surface into the aqueous phase. These loops and ends may attach to another particle to form a bridge between the two particles.

An interesting industrial application of polymeric flocculating agents is the filtration and decoloration of various types of slurries, such as bentonite, corn starch and uranium ore [446]. Flocculation by polymers is also important in waste water treatment and in the improvement of soil structure because improved sedimentation and filtration properties frequently result.

In decoloration, which can be followed by a significant improvement in light transmission, normally soluble colour bodies are rendered insoluble by association with the flocculating polymer. For example, poly-(vinylbenzyltrimethylammonium chloride) (19) has been used for the decoloration of blackstrap molasses, in which the colour bodies were precipitated as a brown floc [447, 448]:

19

Table 9.6 Polymeric flocculating agents

Functionalized polymer	Application	Reference
(PS)—$CH_2N^+Me_3 X^-$		
$X^- \equiv Cl^-$	Decolorization of blackstrap molasses	447, 448
	Filtration to remove bacteria and viruses from solutions	449
	Flocculation of slurries such as bentonite, corn starch, uranium ore	446
	Flocculation of negatively charged colloids	450
	Filtration of sewage sludges	451
$X^- \equiv Cl^- \cdot SO_2$	Flocculation of kaolin slurry	452
(PS)—$CH_2S^+Me_2 Cl^-$	Flocculation of taconite slurry	453
(PS)—CH_2SO_3Na		

(PS)—R
$\quad$ ⟍X

$X \equiv H; R \equiv$ —⟨O⟩—CH_2—	Flocculation of uranium ore slurry	454
$X \equiv H; R \equiv$ —⟨O⟩—	As flocculant	455–461
	As hardener for gelatin	454
	As emulsifier	462, 463
$X \equiv CONH_2; R \equiv$ —⟨O⟩—	As flocculant	464, 465
(P)—⟨O⟩N^+—RX^-	As flocculant	466–469

(PS) — $\equiv$ ⟮CH_2—CH⟯$_n$ (Si) — $\equiv$ silica
$\qquad\qquad$ ⟨O⟩

(P) — $\equiv$ ⟮CH_2—CH⟯$_n$ (P) — $\equiv$ ⟮CH_2—CH—CH_2—CH⟯$_n$
$\qquad\qquad\qquad$ | $\qquad\qquad\qquad\qquad\qquad$ | $\qquad\qquad$ |
$\qquad\qquad\qquad$ X $\qquad\qquad\qquad\qquad\qquad$ X $\qquad\qquad$ X

$\qquad\qquad\qquad$ Me $\qquad\qquad\qquad\qquad\qquad\qquad\qquad\qquad$ Me
$\qquad\qquad\qquad$ | $\qquad\qquad\qquad\qquad\qquad\qquad\qquad\qquad\qquad$ |
(P)^a — $\equiv$ ⟮CH_2—C⟯$_n$ (P)^a — $\equiv$ ⟮CH_2—CH—CH_2—C⟯$_n$
$\qquad\qquad\qquad$ | $\qquad\qquad\qquad\qquad\qquad\qquad\qquad\qquad$ | $\qquad\qquad$ |
$\qquad\qquad\qquad$ X $\qquad\qquad\qquad\qquad\qquad\qquad\qquad$ X $\qquad\qquad$ X

Another important application of this polymer is the treatment of filter aid, such as diatomaceous earth, to yield a material which has excellent ability to remove bacteria and viruses from a solution [449]. Some other functionalized polymers which have been applied as flocculating agents are listed in Table 9.6.

9.10   IONOMERS

Ionomers are synthetic organic polymers that have an ion content of up to 10–15 mol.% and are thus generally insoluble in water [470–477]. In contrast, polyelectrolytes have much higher ion contents and are insoluble in organic solvents. Thus an ionomer can be defined as an ionized copolymer whose major component is a non-ionic backbone, usually a hydrocarbon, and whose minor component consists of ionic comonomers with associated counter-ions such as carboxylate, sulphonate, phosphonate and quaternary ammonium salts, e.g. **20** [478–481]:

$$-[-\overset{\overset{\displaystyle Me}{|}}{\underset{\underset{\displaystyle R}{|}}{N^+}}-(CH_2)_z-\overset{\overset{\displaystyle Me}{|}}{\underset{\underset{\displaystyle R}{|}}{N^+}}-(CH_2)_y-]_n-$$

$$X^- \qquad\qquad X^-$$

20a $R \equiv$ dodecyl, $z = y = 3$ or 5

20b $R \equiv$ octadecyl, $z = y = 5$

The ionic groups may be introduced either by modifying a non-ionic polymer through appropriate chemical techniques or by copolymerization with the major components.

The reason for the great interest and important industrial potential of ionomers lies in the often profound changes in structure and properties caused by the introduction of ionic groups into non-ionic organic polymers. For example, Nafion (**21**), as an ionomer, has found extensive application as a membrane in electrochemical processes such as fuel cells, electrodialysis, spent acid regeneration and selectively permeable separations in chemical processing.

$$+CF_2-CF_2\text{+}+CF_2-CF\text{+}_n-$$
$$|$$
$$(OCF_2-CF\text{+}_m-OCF_2CF_2-SO_3^-\,Na^+$$
$$|$$
$$CF_3$$

21

The extent to which the properties are altered depends on a number of factors such as the dielectric constant of the backbone, the position and type of ionic group, the counter-ion type, the ion concentration and the degree of neutralization. The major effects of the ions on polymers are on their static and dynamic mechanical properties and on their melt rheology: an increase in the ion content of a polymer raises its T_g, melt strength and melt viscosity, and the modulus at temperatures beyond the glassy plateau, and broadens the transition regions. Owing to the great difference in polarity between the ionic functionality and the hydrophobic repeat units of the polymer backbone, the hydrophilic ionic groups lead to microphase separation, which results in the coexistence of two types of ionic aggregates.

1. Multiplets are small aggregates with tight isolated groups which contain no organic material and are dispersed in the matrix. Consisting of a few ions or ion pairs, they act as moderately strong temporary ionic crosslinks.
2. Clusters are larger aggregates, more loosely interacting, and contain a relatively large number of ion pairs and some organic material. They act as crosslinks and as a strongly reinforcing filler.

Cluster formation has the most dramatic effect on polymer properties, the greatest changes being observed in polymers with a low dielectric constant and containing small ions, since such a situation provides the greatest incompatibility between the ionic and non-ionic regions of the system. The position of the ions on the chains and the nature of the ionic group have an influence on the state of aggregation.

As a result of changes in physical properties caused by ionic aggregation, ionomers are finding broad application. The chemical modification of asphaltic bitumen by treatment with maleic anhydride or sulphur trioxide–trimethylamine complex, followed by reaction with suitable oxides or bases, yields asphalt ionomers. The resulting chemically modified products when mixed with aggregate fillers give composites that retain a very high fraction of their strength when wet, in contrast with unmodified asphalt which loses most of its strength upon exposure to water. These asphalt ionomers have been suggested as road-paving materials [482].

9.11 POLYMERS IN BUILDING

9.11.1 Polymers in stone preservation

Natural building stone is often associated with a sense of permanence and durability. However, some types of stone are subject to relatively rapid decay that may result in the deterioration of carved work, i.e. lead to the loss of highly artistic work. Thus it is important to preserve ancient monuments by the treatment of stone to prevent their decay.

The causes of stone decay are either physical or chemical. Physical processes involve two types of damage.

1. Frost damage is due to the expansion of water upon freezing and occurs only in those features of a building that are frequently frozen while wet. The frost

susceptibility of a stone is governed by its pore size distribution, because the pore structure governs the natural degree of saturation and the magnitude of the stresses that may be generated upon freezing.

2. Crystallization damage is due to the crystallization of salts within the pores of a stone that causes a powdering or blistering of the surface, or leads to the formation of deep cavities. Sulphur dioxide in the atmosphere leads to the formation of $CaSO_4$ from limestones and to both $CaSO_4$ and $MgSO_4$ from magnesian limestones. Other common sources of salts include the soil, seawater and unsuitable cleaning materials. The resistance of a limestone to this type of damage is dependent on its pore size distribution, and the durability decreases as the proportion of fine pores increases.

Chemical attack occurs when the chemical reaction of $CaCO_3$ with SO_2 and CO_2 dissolved in rain-water causes rapid deterioration of calcareous sandstones, which consist of grains of silica bound by a matrix of $CaCO_3$.

Since water is involved in almost every type of stone decay, water-repellent surface treatments have been used as promising preservations methods. Many water repellents have been described for the preservation of stone [483], e.g. the deposition of barium salts within the pores of calcareous stones because $BaSO_4$ causes less crystallization damage than $CaSO_4$ in view of its lower solubility [484, 485]. Impregnation of stone with molten wax immobilizes soluble salts but it may increase the rate at which the stone picks up dirt and it is also difficult to achieve practical adequate penetration of the wax. Such treatments have no long-term preservation effect and may even accelerate decay as a result of absorption of rain or groundwater at some unprotected point followed by its evaporation from behind the water-repellent layer; any salts in solution then crystallize there and lead to spalling of the treated surface.

However, the important factors that govern the selection of a treatment are cost and effectiveness, i.e. the relative lifetimes of the various treatments and the quantity of material that is required to treat a given area, which depends on the porosity of the stone. An impregnation treatment should consolidate friable stone and prevent further deterioration caused by salt crystallization, either by making the salts inaccessible to water or by making the stone more resistant to crystallization damage. Increased resistance to crystallization damage could be achieved by an increase in the stone's tensile strength or by a modification of its pore structure. In order to achieve adequate penetration, the treatment should have a high surface tension, a low contact angle and very low viscosity at the time of application. Young's equation states that

$$\cos \theta = \frac{\gamma_{SA} - \gamma_{SL}}{\gamma_{LA}} \tag{9.26}$$

where θ is the contact angle, γ_{SA} and γ_{LA} are the surface free energies of the solid and the liquid, and γ_{SL} is the interfacial energy. For wetting θ must be minimized;

for perfect wetting $\theta = 0$. Hence, for good wetting a large value of γ_{LA} is required and $\gamma_{SA} - \gamma_{SL}$ should be small but must be positive.

The choice of treatment may be further restricted by consideration of such properties as flammability, toxicity, vapour pressure, water miscibility and elastic modulus upon curing. The viscosity requirement is usually the governing factor and is met by one of two approaches. The first possibility is that of dissolving a resin in a solvent of low viscosity, but this approach suffers from a number of disadvantages: (a) the resin may migrate back to the surface as the solvent evaporates [486]; (b) large polymer molecules may be too large to enter the smallest pores of the stone. The second approach is the impregnation of monomers followed by *in situ* polymerization of functionalized vinyl monomers, alkoxysilanes or epoxides. For example, the radical polymerization of methacrylate esters with trimethylolpropane trimethacrylate as crosslinker has been investigated [487, 488]. However, the disadvantages of this are the inhibition of polymerization by oxygen and the increase in viscosity as the polymerization commences on mixing, which inhibits penetration.

Curing polymerizations of alkoxysilanes (**22**), which depend on hydrolysis and loss of alcohol by evaporation, have received most attention [489].

$(R'O)_3Si{-}R$
22a $R' \equiv Et; R \equiv OEt$
22b $R' \equiv Et; R \equiv Me$
22c $R' \equiv Me; R \equiv Me$

However, the main limitation of this system is that the loss of alcohol makes it impossible to achieve a complete pore filling; this can be solved by replacing the methyl groups by some larger groups. Compounds containing two epoxide groups and having a low viscosity, such as 1:2-, 3:4-diepoxybutane, diglycidyl ether, 1,4-butanediol diglycidyl ether, diluted with tetramethoxysilane to decrease the viscosity, have cured with 1,8-diamine-*p*-methane [490]. The drawback of this system is the white efflorescence that may be produced by reaction of the hardener with CO_2.

9.11.2 Polymeric cement additives

Many of the undesirable properties of products made of Portland cement result from the porosity of the cement matrix. The porosity of the cement may be as high as 25% in total; pore diameters may range from gel-pores (1–10 nm) and capillary pores (10–100 nm) to macroscopic pores (> 100 nm) arising from entrapped air and poor particle packing. Thus an interconnecting pore system within the cement matrix allows the ingress of water or aqueous solutions containing sulphate and chloride ions. Such penetration promotes attack on the cement matrix itself or on the bond between the cement matrix and the aggregate, and the mechanical properties of the composite structure deteriorate. Several other problems related to moisture movement in concrete are as follows: (a) expansion, shrinkage and cracking; (b) efflorescence, staining and mildew; (c) lowered resistance to freezing and thawing; (d) increased thermal and electrical conductivity; (e) chemical attack; (f) damage to contents from leakage; (g) damage to finished walls; (h) corrosion of the reinforcing steel; (i) damage to structures from settlement. Thus water resistance is an important factor in concrete and masonry construction for the safety, health and comfort of the occupants of buildings. Organic polymeric materials have recently been incorporated into Portland cement for achieving water resistance to improve the mechanical properties and durability of concrete. The incorporation of polymers into cement results in concrete products with (a) increased compressive and tensile strength, (b) improved fracture toughness, (c) enhanced adhesion to steel or old concrete surfaces, (d) increased resistance to cracking because of crack bridging by the polymer network, (e) increased mix fluidity which allows reduction of water requirements, (f) reduced porosity with reduced permeability to water and aqueous solutions, and hence lower water penetration, (g) increased freeze–thaw resistance, (h) enhanced durability in extreme environments and (i) reduced creep deformation.

Polymeric materials may be used in two ways to modify the properties of cement and concrete products [491–494]: (a) the preformed concrete structure is impregnated with liquid monomers and then polymerized *in situ* by irradiation or thermocatalytically (using initiator and heat) [495], e.g. polymerization of a mixture of methylmethacrylate, isodecylmethacrylate (or isobutylmethacrylate), trimethylolpropanetrimethacrylate and benzoyl peroxide [487], (b) the polymer is incorporated into the mix of cement, aggregate and water, forming an intimate part of the hardened structure which is called polymer Portland cement concrete [496, 497]. However, the polymer must be capable of resisting the high alkalinity of the concrete mix and must not interfere with hydration reactions.

9.12   POLYMERS IN ENHANCED OIL PRODUCTION

Primary recovery is the normal depletion of wells from crude oil which results from the driving forces provided by dissolved gases or associated liquids that allow the oil to be brought to the surface. However, secondary oil recovery is carried out by injecting fluids, such as water or immiscible gas, into a portion of the wells to provide an increase in oil production.

Enhanced oil recovery (tertiary oil recovery) is the recovery of residual oil which remains in the reservoir after the primary and secondary methods because of capillary, viscous and gravity forces. It is achieved by a variety of methods that depend on the reduction of oil viscosity through heat or the injection of interfacially active fluids or fluids that act as solvents. Micellar polymer flooding is one of the methods available for improving oil recovery [498–500]. The basic principle of this process for displacing residual oil from a reservoir is to reduce the high interfacial tension of the trapped reservoir oil by injecting a surfactant solution to achieve a miscible displacement. Mobilization of the residual oil is due to the changes in fluid properties caused by retention of the chemical and salinity/hardness changes that result from ion exchange with the mineral surfaces. Polymers such as polyacrylamide and its derivatives are added to the micellar fluid to increase its viscosity. The viscosity of a polymer solution depends on the salinity, i.e. the viscosity increases as the salinity decreases.

9.13 POLYMERS FOR IMPROVING FIBRE PROPERTIES

Although the use of synthetic fibres has expanded and their commercial importance has grown, many of these polymeric materials have some disadvantages which reduce customer acceptability. The drawbacks include (a) poor dyeability due to the unreactive structure, (b) the collecting of static electricity on fabrics and (c) the obviousness of deposited dirt (soil).

9.13.1 Polymeric dye carriers

For a fibre to be dyeable, some physical or chemical interaction must take place between the dye and the fibre to achieve complete dye penetration of the fibres so that the dye is not removed by subsequent processing, e.g. washing. For example, polymeric fibres, such as polyacrylonitrile and polypropylene, have little affinity for dyestuffs owing to their unreactive structure. Functionalized polymers can be used to provide active sites to improve the dyeability of fibres. This improvement in dyeability can be achieved by using other comonomers selected to confer affinity for dyestuffs on the resulting copolymers by copolymerization, grafting or blending techniques. The most common modified fibres contain either acidic or basic groups which can be dyed with conventional anionic or cationic dyestuffs respectively. Both types of dyestuffs become anchored within the fibre by salt-like bonds.

Copolymerization of the fibre-forming monomer acrylonitrile with *p*-dimethylaminomethylstyrene (2%–10%) [501] or vinylbenzylmethyl ether of diethylene glycol [502] gives fibres with a relatively hydrophilic composition which are strongly receptive to acid dye. In another modification process, a graft copolymer formed by, for example, the reaction of a vinylbenzylmethyl ether of diethylene glycol with a copolymer of an *N*-vinylpyrrolidone and *N*-vinyl-5-methyloxazolidinone (70:30) is used to confer dyeability on polyacrylonitrile. The dye-receptive polymer is mechanically trapped between the fibrils of the strand while it is subjected to irreversible drying. The product has excellent dyeability with acid and direct dyes [503]. The graft copolymerization of vinylbenzyl chloride on polypropylene fibres by ionizing radiation gives a product which, directly or after treatment with a nucleophile, has excellent receptivity to acid and direct dyes [504].

An interesting third method of achieving dyeability consists of physical impregnation of the fibre with a suitable functionalized monomer followed by polymerization *in situ* by X-radiation. For example, molten polypropylene, from which the fibre in spun, may be admixed with polyvinylpyridine or a copolymer of *N*-vinylpyrrolidone and dimethylaminoethylmethacrylate and the fibres can then be dyed with conventional acid dyes.

9.13.2   Polymeric antistatic agents

Static electricity is a surface phenomenon caused by the transfer of electrons across surfaces brought in contact with each other. In general this exchange between the two surfaces is not symmetrical, one acquiring an excess of electrons and the other a shortage of electrons, i.e. a deficiency. After separation, part of the charge is discharged into air and part remains on the surface. Thus, the material becomes electrically charged with an equal magnitude of charges of the opposite sign. In a conducting substrate the electrons move freely and the excess of electrons is dissipated and eliminated by backflow. However, in an insulator, which is non-conductive, the electrons are not mobile and the phenomenon of static electricity becomes apparent.

The density of static charge accumulated by an insulator is limited owing to leakage through air and therefore the attraction and repulsion of charged bodies are also limited. Hence the quantities of static electricity are generally small. Some characteristics that depend on the chemical composition and structure of the surfaces have a significant influence on static charge generation: (a) static charges are generated when two surfaces are brought into contact and separated again; (b) electron transfer is possible between metals and insulators as well as between two insulators, which is attributed to the presence of impurities and to surface asperities; (c) rubbing usually increases charge generation; (d) materials vary in their susceptibility, extent and rate of charge generation and decay, and in the sign of charge.

Generation of static electricity is one of the most important problems encountered in the textile industry, both during the processing of textile materials and during their subsequent use. Synthetic fibres are much more susceptible to static charge problems than are natural fibres, because they are better electrical insulators. The problem of static electricity is also encountered in other industries, such as the paper, photography, printing and powder resin industries. Electrostatic charge tends to accumulate on the surface of plastic articles causing fabrication problems and attracting dust and dirt particles to the fabricated article surface during its subsequent use.

A reduction in the accumulation of static charge is possible either (a) by reducing the rate of generation, e.g., by neutralizing the charge generation via blending two kinds of fibre with a tendency to attain opposite charges or (b) by increasing the rate of dissipation of the electrostatic charge. Static charge on materials dissipates or decays through two processes: (a) surface and volume conductivity of charges through the substrate and (b) loss of charge to the air. The electrical conductivity of the material can be improved by increasing its electrolytic (ionic) or electronic conductivity. The electrolytic conductivity can be increased (a) by increasing the moisture content of the surrounding atmosphere, (b) by application of antistatic agents or (c) by chemically modifying the material. Chemical modification or grafting of functional groups is especially used for textile fibres where the hygroscopic properties are improved by the functional groups which attract moisture.

Antistatic agents are substances that are added to reduce the material's propensity to accumulate electrostatic charge. They must not interfere with processing of the product or impair colour, odour, appearance or performance properties of the substrate; they must be non-toxic and non-flammable. It is necessary, especially when flammable materials are involved, to eliminate spark discharge, because the discharge of electrons at high intensity may cause explosion. Antistatic agents can function either by reducing the generation of charge or by increasing the rate of charge dissipation. They are substances of high electrical conductivity of the following types: amines, amides, quaternary ammonium salts, esters of fatty acids, sulphonic acids, sulphonates, polyethers, polyalcohols and phosphoric acid derivatives. The degree of efficiency of an antistatic agent depends upon (a) the degree of its hygroscopicity, (b) the distribution of moisture and (c) its ability to supply mobile ions to the aqueous layer it forms with absorbed water.

Simple antistatic agents are non-durable because they are water soluble and easily removed by washing. Hence functionalized polymers carrying ionizable groups, i.e. polyelectrolytes, have been used as durable antistatic agents.

The build-up of electric charge on polymeric fibres attracts dust and dirt particles and thus the fibres become soiled. Thus, the problems of antistatic and antisoiling treatments are coupled. A successful solution for preventing the build-up of electrostatic charge and soiling is copolymerization with a monomeric quaternary salt, e.g. methacrylatoethyldiethylmethylammonium methosulphate (23), and polyoxyethylene compounds (24) as an antistat. The resultant copolymers are less soluble than the original, and these agents are not easily removed by washing and are active for quite long periods.

$$\diagdown\!\!\!\!\diagup\text{--COO---(CH}_2)_2\text{---N}^+\text{MeEt}_2\,\text{MeSO}_4{}^-$$
23

$$\diagdown\!\!\!\!\diagup\text{--CO}\text{+OCH}_2\text{CH}_2)_m\text{---OH}$$
24 $m = 3, 10$

Success in prevention of soiling in wear is based on (a) coating fabrics with low energy surfaces which will avoid adherence of soil, e.g. with fluorochemicals such as **25**, which have low surface tension and are not readily wetted even by mineral and vegetable oils and (b) application of hydrophilic polymeric finishes to the fabric to repel oils and greasy soil and ensure ready wetting of the fabric surface by aqueous detergents.

$$\diagdown\!\!\!\!\diagup\text{--COO---(CF}_2)_7\text{---CF}_3$$
25

REFERENCES

1. Hoegl, H., Süs, O. and Neugebauer, W. (1958) US Patent 3 037 861.
2. Gutmann, F. and Lyons, L.E. (eds) (1967) *Organic Semiconduction*, Wiley, New York.
3. Katon, J.E. (ed.) (1970) *Organic Semiconducting Polymers*, Marcel Dekker, New York.
4. Mort, J. and Pai, D.M. (eds) (1976) *Photoconductivity and Related Phenomena*, Elsevier, New York.
5. Pearson, J.M. (1977) *Pure Appl. Chem.*, **48**, 463.
6. Hafano, M. and Tanikawa, K. (1978) *Prog. Org. Coatings*, **6**, 65.
7. Stolka, M. and Pai, D.M. (1978) *Adv. Polym. Sci.*, **29**, 1.
8. Gibson, H.W. (1974) *Macromolecules*, **7**, 711; (1984) *Polymer*, **25**, 3; (1984) *Polym. Sci. Technol.*, **25**, 381.
9. Percec, V., Natansohn, A., Tocaciu, D.C. and Simionescu, C.I. (1981) *Polym. Bull.*, **5**, 247.
10. Natansohn, A., (1983) *Polym. Bull.*, **9**, 67.
11. Biswas, M., and Das, K. (1982) *Polymer*, **23**, 1713.
12. Hoegl, H., Süs, O. and Neugebauer, W. (1956) Ger. Patent 1068115.
13. Chiellini, E., Solaro, R. and Ledwith, A. (1978) *Makromol. Chem.*, **179**, 1929.
14. Limburg, W.W. (1975) Ger. Patents 2 430 748, 2 430 783; *Chem. Abstr.*, **82**, 178209-F; *Chem. Abstr.*, **83**, 351679-u; (1975) US Patent 3 884 689.
15. Griffiths, C.H., Okumura, K. and Van Laeken, A. (1977) *J. Polym. Sci., Polym. Phys. Ed.*, **15**, 1677.
16. Inami, A., Morimoto, K. and Murakami, Y. (1969) US Patent 3421891.
17. Okamoto, K., Oda, N., Itaya, A. and Kusobayashi, S. (1976) *Bull. Chem. Soc. Jpn*, **49**, 1415.
18. Morimoto, K. and Murahami, Y. (1970) Jpn. Patent 7 015 508; *Chem. Abstr.*, **73**, 56615-c.
19. Morimoto, K. and Monobe, A. (1970) Jpn. Patent 7 015 509; *Chem. Abstr.*, **73**, 56615-b.
20. Morimoto, K., Inami, A. and Monobe, A. (1967) Jpn. Patent XX, 049.
21. Inami, A. and Morimoto, K. (1958) Ger. Patent 1 264 954.
22. Pochan, J.M. and Gibson, H.W. (1982) *J. Polym. Sci., Polym. Phys. Ed.*, **20**, 2059.
23. Watorai, S. and Seoka, Y. (1975) US Patent 3895945.
24. Chiellini, E., Solaro, R., Colella, O. and Ledwith, A. (1978) *Eur. Polym. J.*, **14**, 489.
25. Chiellini, E., Solaro, R., Galli, G. and Ledwith, A. (1980) *Macromolecules*, **13**, 1654.
26. Limburg, W.W. and Seanor, D.A. (1975) US Patent 3 877 936.
27. Chang, D.M., Gromelski, S., Rupp, R. and Mulvaney, J. (1977) *J. Polym. Sci., Polym. Chem. Ed.*, **45**, 571.
28. Gibson, H.W. and Bailey, F.C. (1977) *Macromolecules*, **10**, 602.
29. Kobayashi, T., Suzuki, K., Murakam, H., Nishiide, K., Yamanouchi, T. and Kinjo, K. (1973) Jpn. Patent 7 359 843; (1973) *Chem. Abstr.*, **79**, 131380-c; (1976) *Macromolecules*, **9**, 10; (1974) *J. Polym. Sci., Polym. Chem. Ed.*, **12**, 2141.
30. Okamoto, K., Itaya, A. and Kusabayashi, S. (1975) *Polymer*, **7**, 622; (1976) *J. Polym. Sci., Polym. Phys. Ed.*, **14**, 869.
31. Ito, H., Tazuka, S. and Ohkavara, M. (1976) Jpn. Patent 76 101 534; (1977) *Chem. Abstr.*, **86**, 163615.
32. Yoshikawa, M., Nomori, H. and Hatano, M. (1978) *Makromol. Chem.*, **179**, 2397.
33. Tanikawa, K., Okuno, Z., Iwaoka, T. and Hatano, M. (1977) *Makromol. Chem.*, **178**, 1779.
34. Oshima, R. and Kumanotani, J. (1979) *Polym. Prepr., Am. Chem. Soc., Div. Polym. Chem.*, **20** (1), 522; (1985) *J. Polym. Sci., Polym. Chem. Ed.*, **23**, 911.
35. Galli, G., Solaro, R., Chiellini, E. and Ledwith, A. (1981) *Polymer*, **22**, 1088.

36. Gibson, H.W., Olin, G.R. and Pochan, J.M. (1981) *J. Chem. Soc., Perkin Trans. 2*, 1267.
37. Hatano, M. and Enomoto, T. (1974) Jpn. Patent 7 475 688; (1975) *Chem. Abstr.*, **83**, 59940-y.
38. Natansohn, A. (1983) *Polym. Prepr., Am. Chem. Soc., Div. Polym. Chem.*, **24** (2), 358; (1984) *J. Polym. Sci., Polym. Chem. Ed.*, **22**, 3161; (1984) *J. Polym. Sci., Polym. Lett. Ed.*, **22**, 579.
39. Yanus, J.F. and Pearson, J.M. (1974) *Macromolecules*, **7**, 716.
40. Sirotkina, E.E. (1974) Ger. Patent 2 320 855; US Patent 4 038 468.
41. Gipstein, E. and Hewett, W.A. (1971) US Patent 3 554 741.
42. Hoegl, H. and Schlesinger, H. (1962) Ger. Patent 1 131 988; (1963) US Patent 3 307 940.
43. Gipstein, E. and Hewett, W.A. (1971) Ger Patent 2 104 557.
44. Merrill, S.H. and Brantly, T.B. (1973) US Patent 3 779 750.
45. Nakaya, T., Kodera, H. and Imoto, M. (1979) *Polym. Prepr., Am. Chem. Soc., Div. Polym. Chem.*, **20** (1), 520.
46. Inami, A., Morimoto, K. and Hayashi, Y. (1964) *Bull. Chem. Soc. Jpn*, **37**, 842.
47. Fox, C.J. (1966) US Patent 3 265 496.
48. Mitsubishi Co. (1975) Jpn. Patent 7 579 638.
49. Histatake, O., Honjo, S. and Watana, O. (1970) Ger Patent 2 007 962.
50. Asahi Glass Co. (1983) Jpn. Patent Tokyo Koho JP 58 160 359; (1984) *Chem. Abstr.*, **100**, 69854-g.
51. Goodman, C.H.L. (1973) In *Electronic and Structural Properties of Amorphous Semiconductors*, (eds P.G. LeComber and J. Mort), Academic Press, New York, p. 549.
52. Newmann, A.A. (1964) *Br. J. Photogr.*, (9), 784.
53. Dessauer, J.H. and Clark, H.E. (1965) *Xerography and Related Processes*, Focal Press, London.
54. Weigl, J.W. (1977) *Angew. Chem., Int. Ed. Engl.*, **16**, 374.
55. Winkelmann, D.J. (1978) *Electrostatics*, **4**, 193.
56. Hoegl, H., Süs, O. and Neugebauer, W. (1957) Ger. Patents 1 068 115, 1 111 935.
57. Weiser, G. (1972) *J. Appl. Phys.*, **43**, 5028.
58. Gibson, H.W. and Bailey, F.C. (1972) *J. Polym. Sci., Polym. Chem. Ed.*, **10**, 3017.
59. Gibson, H.W., Bailey, F.C., Mincer, J.L. and Gunther, W.H.H. (1979) *J. Polym. Sci., Polym. Chem. Ed.*, **17**, 2961.
60. Gibson, H.W. and Gunther, W.H.H. (1978) US Patent 4 070 296.
61. MacDiarmid, A.G. and Heeger, A.J. (1980) *Synth. Met.*, **1**, 101.
62. Wegner, G. (1981) *Angew. Chem., Int. Ed. Engl.*, **20**, 361.
63. Street, G.B. and Clarke, T.C. (1981) *IBM J. Res. Dev.*, **25**, 51.
64. Baughman, R.H., Bredas, J.L., Chance, R.R., Elsenbaumer, R.E. and Shacklette, L.W. (1982) *Chem. Rev.*, **82**, 209.
65. Duke, C.B. and Gibson, H.W. (1982) In *Encyclopedia of Chemical Technology*, vol. 18, Wiley, New York, pp. 735–93.
66. Simionescu, C.J. and Percec, V. (1982) *Prog. Polym. Sci.*, **8**, 133.
67. Etemad, S. and Heeger, A. (1982) *Ann. Rev. Phys. Chem.*, **33**, 443.
68. Bloor, D. and Movaghar, B. (1983) *IEEE Proc.*, **130**, 225.
69. Masuda, T. and Higashimura, T. (1984) *Acc. Chem. Res.*, **17**, 51.
70. Baughman, R.H. (1984) In *Contemporay Topics in Polymer Science*, vol. 5, (ed. E.J. Vandenberg), Plenum, New York, p. 321.
71. Greene, R.L. and Street, G.B. (1984) *Science*, **226**, 651.
72. Seymour, R.B. (ed.) (1981) *Conductive Polymers*, Plenum, New York.
73. Bryce, M. and Murphy, L. (1984) *Nature*, **309**, 119.
74. Bredas, J.L. and Street, G.B. (1985) *Acc. Chem. Res.*, **18**, 309.

75. Feast, W.J. (1985) *Chem. Ind.*, (8), 263.
76. Frommer, J.E. (1986) *Acc. Chem. Res.*, **19**, 2.
77. Frommer, J.E. and Chance, R.R. 'Electrically Conductive Polymers'. In *Encyclopedia of Polymer Science and Engineering*, vol. 5 (eds M. Grayson and I. Kroschwitz), Wiley, New York, pp. 462–507.
78. Chien, J.C.W., Wnek, G.E., Karasz, F.E. and Hirsch, J.A. (1981) *Macromolecules*, **14**, 479.
79. Cukor, P., Krugler, J.I. and Rubner, M.F. (1981) *Makromol. Chem.*, **182**, 165.
80. Rubner, M. and Delts, W. (1982) *J. Polym. Sci., Polym. Chem. Ed.*, **20**, 2043.
81. Diaz, A.F., Castillo, J., Kanazawa, K.K. and Logan, J.A. (1982) *J. Electroanal. Chem.*, **133**, 233.
82. Chiang, C.K., Druy, M.A., Gau, S.C., Heeger, A.J., Lewis, E.J. MacDiarmid, A.G., Park, Y.K. and Shirakawa, H.J. (1978) *J. Am. Chem. Soc.*, **100**, 1013.
83. Chiang, C.K., Gau, S.C., Fincher, C.R., Park, Y.W., MacDiarmid, A.G. and Heeger, A.J. (1978) *Appl. Phys. Lett.*, **33**, 18.
84. Murray, R.W. (1980) *Acc. Chem. Res.*, **13**, 135.
85. Faulkner, L.R. (1984) *Chem. Eng. News*, **62**, 28.
86. Zumbrunnen, H.R. and Anson, F.C. (1983) *J. Electroanal. Chem.*, **152**, 111.
87. Anson, F.C., Ohsaka, T. and Saveant, J.M. (1983) *J. Am. Chem. Soc.*, **105**, 4883.
88. Nigrey, R.J., MacDiarmid, A.G. and Heeger, A.J. (1979) *J. Am. Chem. Chem. Commun.*, 594; (1982) *Mol. Cryst. Liq. Cryst.*, **83**, 309.
89. Shacklette, L.W., Elsenbaumer, R.L., Chance, R.R., Sowa, J.M., Ivory, D.M., Miller, G.G. and Baughman, R.H. (1982) *J. Chem. Soc. Chem. Commun.*, 361.
90. Schneider, A.A., Harney, D.E. and Harney, M.J. (1980) *J. Power Sources*, **5**, 15.
91. Yamamoto, T. and Kuroda, S.I. (1983) *J. Electroanal. Chem.*, **158**, 1.
92. Moutet, J.C. (1984) *J. Electroanal. Chem.*, **161**, 181.
93. Kamat, P.V., Bashear, R. and Fox, M.A. (1985) *Macromolecules*, **18**, 1366.
94. Hooper, A. and Tofield, B.C. (1984) *J. Power Sources*, **11**, 33.
95. Gebelein, C.G., Williams, D.J. and Deanin, R.D. (eds) (1983) *Polymers in Solar Energy Utilization*, Am. Chem. Soc. Symp. Ser. 220, Washington, DC.
96. Kaneko, M. and Yamada, A. (1984) *Adv. Polym. Sci.*, **55**, 1.
97. Nishijima, T., Nagamura, T. and Matsuo, T. (1981) *J. Polym. Sci., Polym. Lett. Ed.*, **19**, 65.
98. Matsuo, T., Sakamoto, T., Takuma, K., Sakurai, K. and Ohsaka, T. (1981) *J. Phys. Chem.*, **85**, 1277.
99. Wöhrle, D. (1983) *Adv. Polym. Sci.*, **50**, 1.
100. Hautala, R.H., Little, J. and Sweet, E. (1977) *Sol. Energy*, **19**, 503.
101. Grätzel, M. (1980) *Ber. Bunsenges Phys. Chem.*, **84**, 981.
102. Hautala, R.R. and Little, J. (1980) In *Interfacial Photoprocesses: Applications to Energy Conversion and Synthesis* (ed. M.S. Wrighton), Advances in Chemistry Series 184, p. 1.
103. Schwendiman, D.P. and Kutal, C. (1977) *J. Am. Chem. Soc.*, **99**, 5677.
104. King, R.B. and Sweet, E.M. (1979) *J. Org. Chem.*, **44**, 385.
105. King, R.B. and Hanes, R.M. (1979) *J. Org. Chem.*, **44**, 1092.
106. Card, R.J. and Neckers, D.C. (1978) *J. Org. Chem.*, **43**, 2958.
107. Davidson, T. (ed.) (1984) *Polymers in Electronics*, Am. Chem. Soc. Symp. Ser. 242, Washington, DC.
108. Feit, E.D. and Wilkins, C. (eds) (1982) *Polymer Materials for Electronic Applications*, Am. Chem. Soc. Symp. Ser. 184, Washington, DC.
109. Hirai, T., Hatano, Y. and Nonogaki, S. (1971) *J. Electrochem. Soc.*, **118**, 669.
110. Aoc, H., Yatsui, Y. and Hayashida, T. (1970) *Microelectron. Reliab.*, **9**, 267.
111. Fedorov, Y.I., Voskobivnik, G.A., Ryabov, A.V. and Lebedev, V.P. (1968) *Vysokomol. Soedin.*, **10**, 611.

112. Nakamura, K. and Kikuchi, S. (1968) *Bull. Chem. Soc. Jpn.*, **41**, 1977.
113. Tsuda, M. (1969) *J. Polym. Sci., A-1*, **7**, 259.
114. Unruh, C.C. and Smith, A.C. (1960) *J. Appl. Polym. Sci.*, **3**, 310.
115. Borden, D.G. and Williams, J.L.R. (1977) *Makromol. Chem.*, **178**, 3035.
116. Ichimura, K., Watanabe, S. and Ochi, H. (1976) *J. Polym. Sci., Polym. Lett. Ed.*, **14**, 207.
117. Minsk, L.M., Smith, J.G., Van Densen, W. and Wright, J.R. (1959) *J. Appl. Polym. Sci.*, **2**, 302, 308.
118. Kato, M., Ichijyo, T., Ishii, K. and Hasegawa, M. (1971) *J. Polym. Sci., A-1*, **9**, 2109.
119. Kawai, W. (1980) *Kobunshi Ronbunshu*, **37**, 303.
120. Watanabe, S., Kato, M. and Kosakai, S. (1984) *J. Polym. Sci., Polym. Chem. Ed.*, **22**, 2801.
121. Nishikubo, T., Lizawa, T., Takahashi, E. and Udagawa, A (1984) *Makromol. Chem. Rapid Commun.*, **5**, 131.
122. Nishikubo, T. and Lizawa, T. (1984) *Polym. Prepr., Am. Chem. Soc., Div. Polym., Chem.*, **25**(1), 315.
123. Nishikubo, T., Takahashi, E., Lizawa, T. and Hasegawa, M. (1984) *Nippon Kagaku Kaishi*, (2), 306; (1984) *Chem. Abstr.*, **100**, 175471-q.
124. Nishikubo, T., Lizawa, T. and Yamada, M. (1981) *J. Polym. Sci., Polym. Chem. Ed.*, **19**, 177.
125. Osada, C., Satomura, M. and Ono, H. (1972) Ger. Patent 2 164 625; (1972) *Chem. Abstr.*, **77**, 107730-e.
126. Nishikubo, T., Lizawa, T., Takahashi, E. and Nono, F. (1984) *Polym. J.*, **16**, 371.
127. Matsushita Electric Ind. Co. (1984) Jpn. Patent Tokkyo Koho JP 5 915 418; (1984) *Chem. Abstr.*, **100**, 210966-d.
128. Roucoux, C., Loucheux, C. and Lablache-Combier, A. (1981) *J. Appl. Polym. Sci.*, **26**, 1221.
129. Tanaka, H., Tsuda, M. and Nakanishi, H. (1972) *J. Polym. Sci., A-1*, **10**, 1729.
130. M'Bon, G., Roucoux, C., Lablache-Combier, A. and Loucheux, C. (1984) *J. Appl. Polym. Sci.*, **29**, 651.
131. Farrall, M.J. (1984) *Polym. Bull.*, **11**, 191.
132. Guliev, A.M., Gasanova, S.S., Ramazanov, G.A., Musina, E.A., Mazzhukhin, D.D., Selivanov, G.K., Densikin, V.V., Alyev, A.T. and Agaev, U.K. (1983) USSR Patent 1 058 972; (1984) *Chem. Abstr.*, **100**, 104781-e.
133. J. Gevaert-Agfa, N.V. (1966) Neth. Appl. 6 604 633; (1966) *Chem. Abstr.*, **65**, 16302.
134. Leubner, G.W., Williams, J.L.R. and Unruh, C.C. (1957) US Patent 2 811 510.
135. Nishikubo, T., Ichijyo, T. and Takaoka, T. (1984) *J. Appl. Polym. Sci.*, **18**, 2009.
136. Shim, J.S., Park, N.G., Kim, U.Y. and Ahn, K.D. (1984) *Pollimo*, **8**, 34; (1984) *Chem. Abstr.*, **100**, 210503-a.
137. Matsushita Electric Ind. Co. (1983) Jpn. Patent Tokkyo Koho JP 58 132 006; (1984) *Chem. Abstr.*, **100**, 104105-n.
138. Cottart, J.J., Loucheux, C. and Lablache-Combier, A. (1981) *J. Appl. Polym. Sci.*, **26**, 1233.
139. Decout, J.D., Lablache-Combier, A. and Loucheux, C. (1979) *Photogr. Sci. Eng.*, **23**, 309; (1980) *J. Polym. Sci., Polym. Chem. Ed.*, **18**, 2371, 2391.
140. Deledalle, P., Lablache-Combier, A. and Loucheux, C. (1984) *J. Appl. Polym. Sci.*, **29**, 125.
141. Jellinek, H.H.G. and Wang, L.C. (1968) *J. Macromol. Sci.*, **A-2**, 781, 1353.
142. Monahan, A.R. (1968) *Macromolecules*, **1**, 408.
143. Decout, J.L., Lablache-Combier, A. and Loucheux, C. (1980) *Photogr. Sci. Eng.*, **24**, 255.
144. Hiraoka, H. and Welsh, L.W. (1983) *Proc. 10th Int. Conf. on Electron and Ion Beam Science and Technology*, Electrochem. Soc. Proc. 83-2, p. 171; (1983) *Am. Chem. Soc.*,

Div. Org. Coatings, Appl. Polym. Sci. Papers, **48**, 48; (1983) In Polymers in Electronics, Am. Chem. Soc. Symp. Ser. 242, Washington, DC.

145. Choong, H.S. and Kahn, F.J. (1983) J. Vac. Sci. Technol., **B-1**, 1066.
146. Tsuda, M., Olkawa, S., Nakamura, Y. and Nakane, H. (1979) Photogr. Sci. Eng., **23**, 290.
147. Lin, B.J. (1975) J. Vac. Sci. Technol., **12**, 1317.
148. Haller, I., Hatzakis, M. and Srinivasan, R. (1968). IBM J. Res. Dev., **12**, 251.
149. Bowden, M.J. and Thompson, L.F. (1973) J. Appl. Polym. Sci., **17**, 3211.
150. Guliev, J.A.M., Ramazanov, G.A., Gasanova, S.S., Alyev, A.T. and Nefedov, O.M. (1983) USSR Patent 1 073 241; (1984) Chem. Abstr., **100**, 176207-v.
151. Curme, H.G., Natale, C.C. and Kelley, D.J. (1967) J. Phys. Chem., **71**, 767.
152. Nishikubo, T., Iizawa, T., Yamada, M. and Tsuchiya, K. (1983) J. Polym. Sci., Polym. Chem. Ed., **21**, 2025.
153. Iizawa, T., Nishikubo, T., Takahashi, E. and Hasegawa, M. (1983) Kobunshi Ronbunshu, **40**, 425.
154. Roberts, E.D. (1975) Philips Tech. Rev., **35**(2–3), 41.
155. Jensen, J.E. (1984) Solid State Technol., (6), 145.
156. Lee, K.I., Jopson, H., Cukor, P. and Shaver, D. (1982) Proc. SPIE Int. Soc. Opt. Eng., **333**, 15; (1982) Chem. Abstr., **97**, 118140-b; (1982) Ger. Patent 321 3771; (1983) Chem. Abstr., **98**, 81524-x.
157. Choong, H.S. and Kahn, F.J. (1981) J. Vac. Sci. Technol., **19**, 1121.
158. Imamura, S., Tamamura, T., Harada, K. and Sugawara, S. (1982) J. Appl. Polym. Sci., **27**, 937.
159. Imamura, S. (1979) J. Electrochem. Soc., **126**, 1628.
160. Tamamura, T., Sukegawa, K. and Sugawara, S. (1982) J. Electrochem. Soc., **129**, 1831.
161. Feit, E.D. and Stillwagon, L.E. (1980) Polym. Eng. Sci., **20**, 1058.
162. Ku, H.Y. and Scala, L.C. (1969) J. Electrochem. Soc., **116**, 980.
163. Wolf, E.D., Bauer, L.O., Bower, R.W., Garvin, H.L. and Buckley, C.R. (1970) IEEE Trans., **ED-17**, 446.
164. Salamon, M.H., Drach, J., Guo, S.L., Price, P.B. and Tarle, G. (1984) Nucl. Instr. Meth., **224**, 217.
165. Salamon, M.H., Price, P.B., Tinckell, M., Guo, S.L. and Tarle, G. (1985) Nucl. Instr. Meth., **B-6**, 504.
166. Tarle, G., Ahlen, S.P. and Price, P.B. (1981) Nature, **293**, 556.
167. Price, P.B. and Tarle, G. (1985) Nucl. Instr. Meth. Phys. Res., **B-6**, 513.
168. Fujii, M., Nishimura, J. and Kobayashi, T. (1984) Nucl. Instr. Meth. Phys. Res., **226**, 496.
169. O'Sullivan, D., Price, P.B., Kinoshita, K. and Willson, C.G. (1982) J. Electrochem. Soc., **129**, 811.
170. Blumstein, A. (ed.) (1978) Liquid Crystalline Order in Polymers, Academic Press, New York.
171. Plate, N.A. and Shibaev, V.P. (1980) J. Polym. Sci., Polym. Symp., **67**, 1; (1984) Makromol. Chem., Suppl., **6**, 2.
172. Griffin, A.C. (1984) Polym. Mater. Sci. Eng., **50**, 139.
173. Ciferri, A., Krigbaum, W.R. and Meyer, R.B. (eds) (1982) Polymer Liquid Crystals, Academic Press, New York.
174. Plate, N.A., Tal'roze, R.V. and Shibaev, V.P. (1984) Pure Appl. Chem., **56**, 403; (1984) Makromol. Chem., Suppl., **8**, 47.
175. Sandman, D.J. and Cukor, P. (1984) Mol. Cryst. Liq. Cryst., **105** (1–4), 1.
176. Zugenmaier, P. (1984) Makromol. Chem., Suppl., **6**, 31; (1986) Makromol. Chem., Macromol. Symp., **2**, 33.
177. Ober, C.K., Jin, J.I. and Lenz, R.W. (1984) Adv. Polym. Sci., **59**, 103.

178. Economy, J. (1984) *J. Macromol. Sci.*, **A-21**, 1705.
179. Finkelmann, H. and Rehage, G. (1984) *Adv. Polym. Sci.*, **60–61**, 99.
180. Shibaev, V.P. and Plate, N.A. (1984) *Adv. Polym. Sci.*, **60–61**, 173; (1985) *Pure Appl. Chem.*, **57**, 1589.
181. Lipatov, Y.S., Tsukruk, V.V. and Shilov, V.V. (1984) *J. Macromol. Sci., Rev. Macromol. Chem. Phys.*, **C-24**, 173.
182. Lenz, R.W. (1985) *J. Polym. Sci., Polym. Symp.*, **72**, 1; (1985) *Polym. J.*, **17**, 105.
183. Krigbaum, W.R. (1985) *J. Appl. Polym. Sci., Appl. Polym. Symp.*, **41**, 105.
184. Plate, N.A., Freidon, Y.S. and Shibaev, V.P. (1985) *Pure Appl. Chem.*, **57**, 1715.
185. Cser, F., Horvath, J., Nyitrai, K. and Hardy, G. (1985) *Isr. J. Chem.*, **25**, 252.
186. Blumstein, A. (ed.) (1985) *Polymeric Liquid Crystals, Polym. Sci. Technol.*, vol. 28, Plenum, New York.
187. Chapoy, L.L. (ed.) (1985) *Recent Advances in Liquid Crystalline Polymers*, Applied Science, London.
188. Ciferri, A. and Valent, B. (1979) In *Ultra High Modulus Polymers* (eds A. Ciferri and I.M. Ward), Applied Science, London, p. 203.
189. Papkov, S.P. (1984) *Adv. Polym. Sci.*, **59**, 75.
190. Jackson, W.J. and Kuhfuss, H.F. (1976) *J. Polym. Sci., Polym. Chem. Ed.*, **14**, 2043.
191. Tsvetkov, V.N. and Shtennikova, I.N. (1983) *Polym. Prepr., Am. Chem. Soc., Div. Polym. Chem.*, **24**(2), 280.
192. Tsvetkov, V.N., Shtennikova, I.N., Rjumtsev, E.I., Kolbina, G.F., Konstantinov, I.I., Amerik, Y.B. and Krentsel, B.A. (1969) *Vysokomol. Soedin. Ser A*, **11**, 2528.
193. Tsvetkov, V.N., Rjumtsev, E.I., Shtennikova, I.N., Korneeva, E.V., Krentsel, B.A. and Amerik, Y.B. (1973) *Eur. Polym. J.*, **9**, 481.
194. Tsukruk, V.V., Shilov, V.V., Lipatov, Y.S., Konstantinov, J.J. and Amerik, Y.B. (1982) *Acta Polym.*, **33**, 63.
195. Konstantinov, I.I., Amerik, Y.B. and Sitnov, A.A. (1984) *Eur. Polym. J.*, **20**, 1127.
196. Tsukruk, V.V., Shilov, V.V. and Lipatov, Y.S. (1985) *Kristallogr.*, **30**, 198; (1985) *Chem. Abstr.*, **103**, 6952-d.
197. Gudkov, V.A. (1984) *Kristallogr.*, **29**, 529; (1984) *Chem. Abstr.*, **101**, 111632-f.
198. Shibaev, V.P. and Plate, N.A. (1980) *Makromol. Chem.*, **181**, 1393.
199. Gudkov, V.A., Chistaykov, I.G., Vainshtein, B.K. and Shibaev, V.P. (1982) *Kristallogr.*, **27**, 537.
200. Lorkowski, H.J. and Reuther, F. (1976) *Plast. Kautschuk*, **2**, 81.
201. Ringsdorf, H., Schmidt, H.W., Strobl, G. and Zental, R. (1983) *Polym. Prepr., Am. Chem. Soc., Div. Polym. Chem.*, **24**(2), 308.
202. Finkelmann, H. and Wendorff, H.J. (1983) *Polym. Prepr., Am. Chem. Soc., Div. Polym. Chem.*, **24**(2), 284.
203. Frenzel, J. and Rehage, G. (1980) *Makromol. Chem. Rapid Commun.*, **1**, 129.
204. Finkelmann, H., Ringsdorf, H. and Wendorff, J. (1978) *Makromol. Chem.*, **179**, 273.
205. Ringsdorf, H. and Schneller, A. (1981) *Br. Polym. J.*, **13**, 43.
206. Ringsdorf, H. and Zentel, R. (1982) *Makromol. Chem.*, **183**, 1245.
207. Finkelmann, H., Happ, M., Portugal, M. and Ringsdorf, H. (1978) *Makromol. Chem.*, **179**, 2541.
208. Kelker, H. and Wirzing, U. (1979) *Mol. Cryst. Liq. Cryst.*, **49**, 175.
209. Springer, J. and Weigelt, F.W. (1983) *Makromol. Chem.*, **184**, 1489, 2635.
210. Ringsdorf, H., Schmidt, H.W. and Schneller, A. (1982) *Makromol. Chem. Rapid Commun.*, **3**, 745.
211. Boeffel, C., Hisgen, B., Pschorn, U., Ringsdorf, H. and Spiess, H.W. (1983) *Isr. J. Chem.*, **23**, 388.

212. Portugal, M., Ringsdorf, H. and Zentel, R. (1982) *Makromol. Chem.*, **183**, 2311.
213. Ringsdorf, H. and Schneller, A. (1982) *Makromol. Chem. Rapid Commun.*, **3**, 557.
214. Paleos, C.M., Filippakis, S.E. and Leonidopoulou, G.M. (1981) *J. Polym. Sci., Polym. Chem. Ed.*, **19**, 1427.
215. Wassmer, K., Ohmes, E., Kothe, G., Portugal, M. and Ringsdorf, H. (1982) *Makromol. Chem. Rapid Commun.*, **3**, 281.
216. Geib, H., Hisgen, B., Pschorn, U., Rungsdorf, H. and Spiess, H. (1982) *J. Am. Chem. Soc.*, **104**, 917.
217. Freidzon, Y.S., Kostromin, S.G., Bioko, N.I., Shibaev, V.P. and Plate, N.A. (1983) *Polym. Prepr., Am. Chem. Soc., Div. Polym. Chem.*, **24**(2), 279.
218. Shibaev, V.P., Moiseenko, V.M. and Plate, N.A. (1980) *Makromol. Chem.*, **181**, 1381.
219. Keller, P. (1984) *Macromolecules*, **17**, 2937.
220. Shilov, U.V., Tsukruk, V.V. and Lipatov, Y.S. (1984) *J. Polym. Sci., Polym. Phys. Ed.*, **22**, 41.
221. Mousa, A.M., Freidzon, Y.S., Shibaev, V.P. and Plate, N.A. (1982) *Polym. Bull.*, **6**, 485.
222. Molchanov, Y.V., Privalov, A.F., Amerik, Y.B., Grebneva, V.G. and Konstatinov, I.I. (1985) *Vysokomol. Soedin. Ser. A*, **27**, 2206; (1986) *Chem. Abstr.*, **104**, 69367-h.
223. Tsukruk, V., Shilov, V. and Lipatov, Y. (1986) *Macromolecules*, **19**, 1308.
224. Freidzon, Y.S., Boiko, N.I., Shibaev, V.P. and Plate, N.A. (1986) *Eur. Polym. J.*, **22**, 13.
225. Decobert, G., Soyer, F., Dubois, J.C. and Davidson, P. (1985) *Polym. Bull.*, **14**, 179, 549.
226. Horvath, J., Nyitrai, K., Cser, F. and Hardy, G. (1985) *Eur. Poly. J.*, **21**, 251.
227. Gemmell, P.A., Gray, H.W. and Lacey, D. (1983) *Polym. Prepr., Am. Chem. Soc., Div. Polym. Chem.*, **24** (2), 253.
228. Ringsdorf, H., Schmidt, H.W., Baur, G. and Kiefer, R. (1983) *Polym. Prepr., Am. Chem. Soc., Div. Polym. Chem.*, **24** (2), 306.
229. Kock, H.J., Finkelmann, H., Gleim, W. and Rehage, G. (1983) *Polym. Prepr., Am. Chem. Soc., Div. Polym. Chem.*, **24** (2), 300.
230. Finkelmann, H. and Rehage, G. (1983) *Polym. Prepr., Am. Chem. Soc., Div. Polym. Chem.*, **24** (2), 277; (1980) *Makromol. Chem. Rapid Commun.*, **1**, 31, 733.
231. Finkelmann, H., Kock, H. and Rehage, G. (1981) *Makromol. Chem. Rapid Commun.*, **2**, 317.
232. Zugenmaier, P. and Mügge, J. (1984) *Makromol. Chem. Rapid Commun.*, **5**, 11.
233. Janini, G.M., Laub, R.J. and Shaw, T.J. (1985) *Makromol. Chem. Rapid Commun.*, **6**, 57.
234. Amerik, Y.B. and Krentsel, B.A. (1967) *J. Polym. Sci.*, **C-16**, 1383.
235. Blumstein, A., Kitagawa, N. and Blumstein, R.B. (1971) *Mol. Cryst. Liq. Cryst.*, **12**, 215.
236. Blumstein, A., Billard, J. and Blumstein, R. (1974) *Mol. Cryst. Liq. Cryst.*, **25**, 83.
237. Blumstein, A., Clough, S.B., Patel, L., Blumstein, R.B. and Hsu, E.C. (1976) *Macromolecules*, **9**, 243.
238. Tanaka, Y. (1976) *Org. Synth. Chem. J.*, **34**, 2.
239. Steiger, E.L. and Dietrich, H.J. (1972) *Mol. Cryst. Liq. Cryst.*, **16**, 279.
240. Tal'roze, R.V., Shibaev, V.P., Sinitzyn, V.V. and Plate, N.A. (1983) *Polym. Prepr., Am. Chem. Soc., Div. Polym. Chem.*, **24** (2), 309.
241. Kostromin, S.G., Sinitzyn, V.V., Tal'roze, R.V. and Shibaev, V.P. (1984) *Vysokomol. Soedin, Ser. A*, **26**, 335; *Chem. Abstr.*, **100**, 157235-m.
242. Kostromin, S.G., Sinitzyn, V.V., Tal'roze, R.V., Shibaeva V.P. and Plate, N.A. (1982) *Makromol. Chem. Rapid Commun.*, **3**, 809.
243. Tal'roze, R.V., Sinitzyn, V.V., Shibaev, V.P. and Plate, N.A. (1982) *Mol. Cryst. Liq. Cryst.*, **80**, 211.

244. Tal'roze, R.V., Sinitzyn, V.V., Korobeinikova, I.A., Shibaev, V.P. and Plate, N.A. (1984) *Dokl. Akad. Nauk SSSR*, **274**, 1149; (1984) *Chem. Abstr.*, **101**, 7884-d.
245. Strzelecki, L. and Liebert, L. (1973) *Bull. Soc. Chim. Fr.*, 597.
246. Perplies, E., Ringsdorf, H. and Wendorff, J.H. (1974) *Makromol. Chem.*, **175**, 553; (1974) *Ber. Bunsenges. Phys. Chem.*, **78**, 92; (1977) In *Polymerization of Organized Systems* (ed. H.G. Elias), Gordon and Breach, New York, p. 149.
247. Paleos, C.M. and Voliotirs, S. (1979) *Isr. J. Chem.*, **18**, 192.
248. Strzelecki, L. and Liebert, L. (1973) *Bull. Soc. Chim. Fr.*, 605.
249. Hsu, E.C., Lim, L.K., Blumstein, R.B. and Blumstein, A. (1976) *Mol. Cryst. Liq. Cryst.*, **33**, 35.
250. Strzelecki, L., Liebert, L. and Keller, P. (1975) *Bull. Soc. Chim. Fr.*, 2750.
251. Blumstein, A., Blumstein, R.B., Clough, S.B. and Hsu, E.C. (1975) *Macromolecules*, **8**, 73.
252. Hsu, E.C. and Blumstein, A. (1977) *J. Polym. Sci., Polym. Lett. Ed.*, **15**, 129.
253. Paleos, C.M., Laronge, T.M. and Labes, M.M. (1968) *J. Chem. Soc. Chem. Commun.*, 1115.
254. Hsu, E.C., Clough, S.B. and Blumstein, A. (1977) *J. Polym. Sci., Polym. Lett. Ed.*, **15**, 545.
255. Blumstein, A. (1977) *Macromolecules*, **10**, 872.
256. Clough,, S.B., Blumstein, A. and Vries, A.D. (1977) *Polym. Prepr., Am. Chem. Soc., Div. Polym. Chem.*, **18**, 1.
257. Paleos, C.M. and Labes, M.M. (1970) *Mol. Cryst. Liq. Cryst.*, **11**, 385.
258. Lecoin, D., Hochapjfel, A. and Viovy, R. (1975) *Mol. Cryst. Liq. Cryst.*, **31**, 233.
259. Cser, F., Nyitral, K., Seyfried, E. and Hardy, G. (1976) *Magy. Kem. Polym. J. (Hung.)*, **82**, 207; (1977) *Eur. Polym. J.*, **13**, 678.
260. Nyitrai, K., Cser, F., Bui, D.N. and Hardy, G. (1976) *Magy. Kem. Polym. J. (Hung.)*, **82**, 210.
261. Hochapfel, A., Lecoin, D. and Viory, R. (1976) *Mol. Cryst. Liq. Cryst.*, **37**, 109.
262. Baccaredda, M., Magagnini, P.L., Pizzirani, G. and Giusti, P. (1971) *J. Polym. Sci.*, **B-9**, 303.
263. Ceccarelli, G., Frosini, V., Magagnini, P.L. and Newman, B.A. (1975) *J. Polym. Sci., Polym. Lett. Ed.*, **13**, 109.
264. Paleos, C.M., Leonidopoulou, G.M., Filippakis, S.E., Malliaris, A. and Dias, P. (1982) *J. Polym. Sci., Polym. Chem. Ed.*, **20**, 2267.
265. Piskunov, M.V., Kostromin, S.G., Stroganov, L.B., Shibaev, V.P. and Plate, N.A. (1982) *Makromol. Chem. Rapid Commun.*, **3**, 443.
266. Magagnini, P.L. (1981) *Makromol. Chem., Suppl.*, **4**, 223.
267. Frosini, V., Levita, G., Lupinacci, D. and Magagnini, P. (1981) *Mol. Cryst. Liq. Cryst.*, **66**, 341.
268. Bressi, B., Frosini, V., Lupinacci, D. and Magagnini, P.L. (1980) *Makromol. Chem. Rapid Commun.*, **1**, 183.
269. Shibaev, V.P., Kostromin, S.G. and Plate, N.A. (1982) *Eur. Polym. J.*, **18**, 651.
270. Hahn, B., Wendorff, I.H., Portugal, M. and Ringsdorf, H. (1981) *Colloid Polym. Sci.*, **259**, 875.
271. Gemmell, P.A., Gray, G.W., Lacey, D., Alimoglu, A.K. and Ledwith, A. (1985) *Polymer*, **26**, 615.
272. Shibaev, V.P., Moissenko, V.M., Freidon, Y.S. and Plate, N.A. (1980) *Eur. Polym. J.*, **16**, 272.
273. Magagnini, P.L., Marchetti, A., Matera, F., Pizzirani, G. and Turchi, G. (1974) *Eur. Polym. J.*, **10**, 585.
274. Hardy, G., Nyitrai, K., Cser, F., Selik, G.C. and Nagy, I. (1969) *Eur. Polym. J.*, **5**, 133.
275. Shibaev, V.P., Tal'roze, R.V., Karakhanova, F.I., Kharitonov, A.V. and Plate, N.A. (1975) *Dokl. Akad. Nauk SSSR*, **225**, 632.

276. Ringsdorf, H. and Schmidt, H.W. (1983) *Polym. Prepr., Am. Chem. Soc., Div. Polym. Chem.*, **24** (2), 306; (1984) *Makromol. Chem.*, **185**, 1327.
277. Hardy, G., Cser, F., Kallo, A., Nyitrai, K., Bodor, G. and Lengyl, M. (1970) *Acta Chim. Acad. Sci. (Hung.)*, **65**, 287.
278. Toth, W.J. and Tobolsky, A.V. (1970)*J. Polym. Sci.*, **B-8**, 289.
279. deVisser, A.C., DeGroot, K., Feyen, J. and Banties, A. (1970) *J. Polym. Sci.*, **B-8**, 805; (1971) *J. Polym. Sci.*, **A-19**, 1893; (1972) *J. Polym. Sci.*, **B-10**, 851.
280. Tanaka, Y., Shiozaki, H. and Shimura, Y. (1973) *Chem. Abstr.*, **78**, 136543.
281. Tanaka, Y., Kabaya, S., Shimura, Y., Okada, A., Kuribata, Y. and Sakakibara, Y. (1972) *J. Polym. Sci.*, **B-10**, 261.
282. Saeki, H., Limura, K. and Takeda, M. (1972) *Polym. J.*, **3**, 414.
283. Nyitrai, K., Cser, F., Lengyel, M., Seyfried, E. and Hardy, G. (1976) *Magy. Kem. Polym. J. (Hung.)*, **82**, 195.
284. Freidzon, Y.S., Tropsha, E.G., Shibaev, V.P. and Plate, N.A. (1985) *Makromol. Chem. Rapid Commun.*, **6**, 625.
285. Finkelmann, H., Ringsdorf, H., Siol, W. and Wendorff, J. (1978) *Makromol. Chem.*, **179**, 829.
286. Shibaev, V.B., Friedzon, J.S. and Plate, N.A. (1976) *Dokl. Akad. Nauk SSSR*, **227**, 1412; (1979) *J. Polym. Sci., Polym. Chem. Ed.*, **17**, 1655.
287. Fendler, J.H. and Tundo, P. (1984) *Acc. Chem. Res.*, **17**, 3.
288. Fendler, J.H. (1982) *Pure Appl. Chem.*, **54**, 1809.
289. Fendler, J.H. (1984) *Science*, **223**, 888; *Chem. Eng. News*, **62**, 25.
290. Bader, H., Dorn, K., Hupfer, B. and Ringsdorf, H. (1985) *Adv. Polym. Sci.*, **64**, 1.
291. Tundo, P., Kippenberger, D.J., Klahn, P.L., Prieto, N.E., Jao, T.C. and Fendler, J.H. (1982) *J. Am. Chem. Soc.*, **104**, 456.
292. Fendler, J.H. (1981) *J. Photochem.*, **17**, 303.
293. Tundo, P., Kippenberger, D.J., Politi, M.J., Klahn, P. and Fendler, J.H. (1982) *J. Am. Chem. Soc.*, **104**, 5352.
294. Tundo, P., Kurichara, K., Kippenberger, D.J., Politi, M. and Fendler, J.H. (1982) *Angew. Chem., Int. Ed. Engl.*, **21**, 81.
295. Kippenberger, D., Rosenquist, K., Odberg, L., Tundo, P. and Fendler, J.H. (1983)*J. Am. Chem. Soc.*, **105**, 1129.
296. O'Brien, D.F., Whitesides, T.H. and Klingbiel, R.T. (1981) *J. Polym. Sci., Polym. Lett. Ed.*, **19**, 95.
297. Regen, S.C., Czech, B. and Singh, A. (1980) *J. Am. Chem. Soc.*, **102**, 6638.
298. Nagelle, D. and Ringsdorf, H. (1977) *J. Polym. Sci., Polym. Chem. Ed.*, **15**, 2821.
299. Ackerman, R., Inacker, O. and Ringsdorf, H. (1971) *Kolloid Z. Z. Polym.*, **249**, 1118.
300. Hamid, S.M. and Sherrington, D.C. (1984) *Br. Polym. J.*, **16**, 39.
301. Wessling, R.A. and Pickelman, D.A. (1984) US Patent 4 426 489; (1984) *Chem. Abstr.*, **100**, 176904-v.
302. Reed, W., Guterman, L., Tundo, P. and Fendler, J.H. (1984) *J. Am. Chem. Soc.*, **106**, 1897.
303. Nome, F., Reed, W., Politi, M., Tundo, P. and Fendler, J.H. (1984) *J. Am. Chem. Soc.*, **106**, 8086.
304. Serrano, J., Mucino, S., Millen, S., Reynose, R., Fucugauchi, L.A., Reed, W., Nome, F., Tundo, P. and Fendler, J.H. (1985) *Macromolecules*, **18**, 1999.
305. Nieh, E.C.Y. (1983) US Patent 4 415 717; (1984) *Chem. Abstr.*, **100**, 70359-f.
306. Dorn, K., Patton, E.V., Klingbiel, R.T., O'Brien, D.F. and Ringsdorf, H. (1983) *Makromol. Chem. Rapid Commun.*, **4**, 513.
307. Kunitake, T., Nagai, M., Yanagi, H., Takarabe, K. and Naiashima, N. (1984) *J. Macromol. Sci. Chem.*, **A-21**, 1237.
308. Fukuda, H., Dien, T., Stefely, J., Kezdy, F.J. and Regen, S.L. (1986) *J. Am. Chem. Soc.*, **108**, 2321.

309. Regen, S.L., Shin, J.S. and Yamaguchi, K. (1984) *J. Am. Chem. Soc.*, **106**, 2446.
310. Bolikal, D. and Regen, S.L. (1984) *Macromolecules*, **17**, 1287.
311. Dorn, K., Klingbiel, R.T., Specht, D.P., Tyminsky, P.U., Ringsdorf, H. and O'Brien, D.F. (1984) *J. Am. Chem. Soc.*, **106**, 1627.
312. Cho, I. and Chung, K.C. (1984) *Macromolecules*, **17**, 2935.
313. Hamid, S. and Sherrington, D. (1986) *J. Chem. Soc. Chem. Commun.*, 936.
314. Akimoto, A., Dorn, K., Gros, L., Ringsdorf, H. and Schupp, H. (1981) *Angew. Chem., Int. Ed. Engl.*, **20**, 90.
315. Kusumi, A., Singh, M., Tirrell, D.A., Oehme, G., Singh, A., Samnel, N.K.P., Hyde, J.S. and Regen, S.L. (1983) *J. Am. Chem. Soc.*, **105**, 2975.
316. Umeda, T., Nakaya, T. and Imoto, M. (1982) *Makromol. Chem. Rapid Commun.*, **3**, 457.
317. Regen, S.L., Singh, A., Oehme, G. and Singh, M. (1981) *Biochem. Biophys. Res. Commun.*, **101**, 131; (1982) *J. Am. Chem. Soc.*, **104**, 791.
318. Paleos, C.M., Christias, C., Evangelatos, G.P. and Dais, P. (1982) *J. Polym. Sci., Polym. Chem. Ed.*, **20**, 2565.
319. Puterman, M., Fort, T. and Lando, J.B. (1974) *J. Colloid Interface Sci.*, **47**, 705.
320. Barraud, A., Raudel-Teixier, A. and Rosilio, C. (1975) *Semin. Chim. Etat Solide*, **9**, 21.
321. Day, D., Hub, H.H. and Ringsdorf, H. (1979) *Isr. J. Chem.*, **18**, 325.
322. Atik, S.S. and Thomas, J.K. (1981) *J. Am. Chem. Soc.*, **103**, 4279.
323. Lianas, P. (1982) *J. Phys. Chem.*, **86**, 1935.
324. Benz, R., Prass, W. and Ringsdorf, H. (1982) *Angew. Chem., Int. Ed. Engl.*, **21**, 368.
325. Kammer, V. and Elias, H.G. (1972) *Kolloid Z. Z. Polym.*, **250**, 344.
326. Mielke, I. and Ringsdorf, H. (1972) *Makromol. Chem.*, **153**, 307.
327. Ringsdorf, H. and Thuning, D. (1977) *Makromol. Chem.*, **178**, 2205.
328. Paleos, C.M. and Dias, P. (1978) *J. Polym. Sci., Polym. Chem. Ed.*, **16**, 1495.
329. Hub, H., Hupfer, B., Kock, H. and Ringsdorf, H. (1980) *Angew. Chem., Int. Ed. Engl.*, **19**, 938.
330. Johnson, D.S., Sanghera, J., Pons, M. and Chapman, D. (1980) *Biochim. Biophys. Acta*, **602**, 57; (1982) *Biochim. Biophys. Acta*, **604**, 461.
331. Lopez, E., O'Brien, D.F. and Whitesides, T.W. (1982) *J. Am. Chem. Soc.*, **104**, 305; (1982) *Biochem. Biophys. Acta*, **693**, 437.
332. Gros, L., Ringsdorf, H. and Schupp, H. (1981) *Angew. Chem., Int. Ed. Engl.*, **20**, 305.
333. Fendler, J.H., and Romero, A. (1977) *Life Sci.*, **20**, 1109.
334. Pudov, V.S. and Neiman, M.B. (1971) In *The Ageing and Stabilization of Polymers* (ed. A.S. Kuzminskii), Applied Science, Barking.
335. Lappin, G.R. (1971) In *Encyclopedia of Polymer Science and Technology*, vol. 14 (eds H.F. Mark, N.G. Gaylord and N.B. Bikalas), Wiley Interscience, New York, p. 125.
336. Reich, L. and Stival, S.S. (eds) (1972) *Elements of Polymer Degradation*, McGraw-Hill, New York.
337. Hawkins, W.L. (ed.) (1972) *Polymer Stabilization*, Wiley Interscience, New York.
338. Randy, B. and Rabek, F. (eds) (1975) *Photodegradation, Photooxidation and Photostabilization of Polymers*, Wiley, London.
339. Grassie, N. (ed.) (1977, 1980, 1981, 1982) *Developments in Polymer Degradation*, vols 1–4, Applied Science, London.
340. Jellinek, H.H.G. (ed.) (1978) *Aspects of Degradation and Stabilization of Polymers*, Elsevier, Amsterdam.
341. Allara, D.L. and Hawkins, W.L. (eds) (1978) *Stabilization and Degradation of Polymers*, Advances in Chemistry Series 169.
342. Scott, G. (ed.) (1979, 1980, 1981) *Developments in Polymer Stabilization*, vols 1–4, Applied Science, London.
343. Pappas, S.P. and Winslow, F.H. (eds) (1981) *Photodegradation and Photostabilization of Coatings*, Advances in Chemistry Series 151.

344.   Wilson, F.H. and Slam, T.J. (1974) US Patent 3 852 350.
345.   Parker, D.K. (1975) US Patent 3 907 893.
346.   Spoerke, R.W. (1977) US Patent 4 022 831.
347.   Shah, V.V. (1982) US Patent 4 322 551.
348.   Meyer, G.E., Kavchok, R.W. and Naples, F.J. (1973) *Rubber Chem. Technol.*, **46**, 106.
349.   Maxey, F. (1975) US Patent 3 867 334.
350.   Cox, W.L. (1977) US Patent 4 021 404.
351.   Horvath, J.W., Grimm, D.C. and Stevick, J.A. (1975) *Rubber Chem. Technol.*, **48**, 337.
352.   Kline, R.H. (1976) US Patent 3 953 411; (1977) US Patent 4 147 880.
353.   Tamura, M., Ohishi, T. and Sakurai, H. (1980) Br. Patent 2 053 911; (1981) US Patent 4 298 522.
354.   Parks, C.R. (1974) US Patent 3 817 916; (1975) US Patent 3 886 116; (1976) US Patent 3 979 436; (1978) US Patent 4 087 619.
355.   Andrieva, A.I. (1980) USSR Patent 926 660.
356.   Kline, R.H. (1978) US Patent 4 097 464.
357.   Steihl, R.T. (1971) US Patent 3 629 197.
358.   Kline, R.H. (1976) US Patent 3 953 402.
359.   Ueno, H., Ishikawa, H., Hamada, H. and Watanabe, T. (1973) US Patent 3753943.
360.   Scott, G. (1984) *Polym. Prepr., Am. Chem. Soc., Div. Polym. Chem.*, **25**(1), 62.
361.   Parker, D.K. (1978) US Patent 4 091 225; (1981) US Patent 4 260 832.
362.   Kline, R.H. (1976) US Patent 3 962 187; (1978) US Patent 4 117 240; (1981) US Patent 4 284 823.
363.   Kleiner, E.K. (1972) US Patent 3 645 970; (1972) Br. Patent 1 283 103; (1975) Can. Patent 969 192.
364.   Kline, R.H. (1979) US Patent 4 165 333, 4 152 319.
365.   Dexter, M., Spivack, J.D. and Steinberg, D.H. (1973) US Patent 3 708 520.
366.   Kleiner, E.K. and Dexter, M. (1976) US Patent 3 957 920.
367.   Kato, M. and Nakano, Y. (1972) *J. Polym. Sci., Polym. Lett. Ed.*, **10**, 157.
368.   Kato, M. and Takemoto, Y. (1975) *J. Polym. Sci., Polym. Chem. Ed.*, **13**, 1901.
369.   Shakh-Paron'Yants, A.M., Epstein, V.G. and Rumyantseva, Z.M. (1968) *Vysokomol. Soedin., Ser. B*, **10**, 805.
370.   Stuckey, J.C. and Tahan, M. (1976) Br. Patent 51 252.
371.   Hanauer, R.H. and Willette, G.L. (1978) US Patents 4 098 709, 4 096 319.
372.   Manecke, G. and Bourweig, G. (1966) *Makromol. Chem.*, **99**, 175.
373.   Jacquet, B., Mahieu, C. and Panatoniou, C. (1980) Br. Patent 2 036 012; (1981) US Patent 4 281 192; (1982) US Patent 4 339 561.
374.   Weinshenker, N.M. and Dale, J.A. (1977) US Patent 4 054 676; (1978) US Patent 4 098 829.
375.   Buysch, H.J., Krimm, H. and Margotte, D. (1970) US Patent 3 546 173.
376.   Cottman, K.S. (1976) US Patents 3 984 372, 4 168 387.
377.   Parker, R.G. (1981) US Patent 4 247 664.
378.   Tocker, S. (1962) Br. Patent 885 986; (1962) *Chem. Abstr.*, **57**, 1086-i; (1963) Belg. Patent 629 109; (1964) *Chem. Abstr.*, **60**, 13453-g; (1963) Belg. Patent 629 480; (1964) *Chem. Abstr.*, **61**, 8484-f; (1963) US Patent 3 107 199; (1964) *Chem. Abstr.*, **61**, 3267-x; (1964) US Patent 3 133 042; (1967) *Makromol. Chem.*, 101, 23.
379.   Oo, K.M. and Tahan, M. (1977) *Eur. Polym. J.*, **13**, 915.
380.   Osawa, Z., Matsui, K. and Ogiwara, Y. (1967) *J. Macromol. Sci. Chem.*, **A-1**, 581; (1971) *J. Macromol. Sci. Chem.*, **A-5**, 275.
381.   Goldberg, A.I., Skoultchi, M. and Fertig, J. (1964) US Patent 3 162 676; (1965) *Chem. Abstr.*, **62**, 7693-e.

382. Fertig, J., Skoultchi, M. and Goldberg, A.I. (1967) US Patent 3 340 231; (1967) Chem. Abstr., **67**, 100657.
383. Fertig, J., Goldberg, A.I. and Skoultchi, M. (1966) *J. Appl. Polym. Sci.*, **10**, 663.
384. Goldberg, A.I., Skoultchi, M. and Fertig, J. (1965) US Patent 3 202 716; (1966) *Chem. Abstr.*, **64**, 3425-e.
385. Fertig, J., Skoultchi, M. and Goldberg, A.I. (1965) US Patent 3 173 893; (1965) *Chem. Abstr.*, **63**, 1951-g.
386. Horton, R.L. and Brooks, H.G. (1967) US Patent 3 313 866; (1967) *Chem. Abstr.*, **67**, 121434.
387. Milionis, J.P. and Arthen, F.J. (1962) US Patent 3 049 503; (1963) *Chem. Abstr.*, **58**, 478-g.
388. Milionis, J.P. and Hardy, W.B. (1963) US Patent 3 072 585; (1963) *Chem. Abstr.*, **59**, 11508.
389. Tirrell, D., Bailey, D., Pinazzi, C. and Vogl, O. (1978) *Macromolecules*, **11**, 312.
390. Pinazzi, P. and Fernandez, A. (1973) *Makromol. Chem.*, **167**, 147.
391. Pinazzi, P. and Fernandez, A. (1973) *Makromol. Chem.*, **168**, 19.
392. Tocker, S. (1962) Br. Patent 893 507; (1962) *Chem. Abstr.*, **57**, 13955-e.
393. Dickey, J.B. and Stanin, T.E. (1950) US Patent 2 520 917.
394. Milionis, J.P., Hardy, W.B. and Baitinger, W.F. (1964) US Patent 3 159 646; (1965) *Chem. Abstr.*, **62**, 7951-f.
395. Heller, H., Rody, J. and Keller, E. (1968) US Patent 3 399 173; (1968) *Chem. Abstr.*, **69**, 87597.
396. Fertig, J., Goldberg, A.I. and Skoultchi, M. (1965) *J. Appl. Polym. Sci.*, **9**, 903.
397. Goldberg, A.I., Fertig, J. and Skoultchi, M. (1964) US Patent 3 141 903; (1964) *Chem. Abstr.*, **61**, 9440-a.
398. Goldberg, A.I. (1965) US Patent 3 186 968; (1965) Br. Patent 980 888; (1965) *Chem. Abstr.*, **62**, 11977-f.
399. Tocker, S. (1963) US Patent 3 113 907; (1964) *Chem. Abstr.*, **60**, 6948-f.
400 Hopff, H. and Lussi, H. (1956) *Makromol. Chem.*, **18–19**, 227.
401. Lussi, H. (1956) *Kunstst. Plast.*, **3**, 156.
402. Hopff, H. (1958) *Bull. Soc. Chim. Fr.*, 1283.
403. Skoultchi, M. and Meier, E. (1972) US Patent 3 666 732; (1972) *Chem. Abstr.*, **77**, 75749.
404. Bailey, D., Tirrell, D. and Vogl, O. (1976) *J. Polym. Sci., Polym. Chem. Ed.*, **14**, 2725.
405. Iwasaki, M., Tirrell, D. and Vogl, O. (1980) *J. Polym. Sci., Polym. Chem. Ed.*, **18**, 2755.
406. Tirrell, D. and Vogl, O. (1980) *Makromol. Chem.*, **181**, 2097.
407. Sumida, Y., Yoshida, S. and Vogl, O. (1980) *Polym. Prepr., Am. Chem. Soc., Div. Polym. Chem.*, **21** (1), 201.
408. Yoshida, S. and Vogl, O. (1980) *Polym. Prepr., Am. Chem. Soc., Div. Polym. Chem.*, **21** (1), 203; (1982) *Makromol. Chem.*, **183**, 259.
409. Pradellok, W., Gupta, A. and Vogl, O. (1981) *J. Polym. Sci., Polym. Chem. Ed.*, **19**, 3307.
410. Gupta, A., Liang, R., Coulter, D., Vogl, O. and Scott, G.W. (1984) *Polym. Prepr., Am. Chem. Soc., Div. Polym. Chem.*, **25** (1), 54.
411. Garner, A.Y., Abramo, J.G. and Chapin, E.C. (1962) US Patent 3 065 272.
412. Furukawa, J., Kobayashi, E. and Wakui, T. (1979) *Polym. Prepr., Am. Chem. Soc., Div. Polym. Chem.*, **20** (1), 551.
413. Maruzen Oil Co. (1980) Jpn. Patent Tokkyo Koho 8 051 094; (1980) *Chem. Abstr.*, **93**, 133082-v.
414. Robinowitz, R., Marcus, R. and Pellon, J. (1961) *J. Polym. Sci.*, **A-2**, 1241.
415. Garner, A.Y., Chapin, E.C. and Abramo, J.G. (1963) US Patent 3 075 011; (1963) *Chem. Abstr.*, **59**, 663.

416. Master, E.L. (1961) US Patent 2 980 721.
417. Dreisbach, R.R. and Lang, J.L. (1961) US Patent 2 992 942; (1961) *Chem. Abstr.*, **55**, 22857-e.
418. Razinskaya, I.N., Iskhakov, O.A., Sumenkov, K.F., Eliseava, L.A., Frumina, L.L. and Gudkov, V.I. (1984) *Plast. Massy*, (2), 39; (1984) *Chem. Abstr.*, **100**, 210909-n.
419. Younes, U.E. (1984) US Patent 4 444 969; (1984) *Chem. Abstr.*, **101**, 39378-h.
420. Kelen, T. (ed.) (1983) *Polymer Degradation*, Van Nostrand Reinhold, New York.
421. Kuczkowski, J.A. and Gillick, J.G. (1984) *Rubber Chem. Technol.*, **57**, 621.
422. Heller, H.J. (1969) *Eur. Poly. J., Suppl.*, 105.
423. Fitton, S.L., Howard, R.H. and Williamson, G.R. (1970) *Br. Polym. J.*, **2**, 217.
424. Cicchetti, O. (1971) *Adv. Polym. Sci.*, **7**, 70.
425. Trozzolo, A.M. (1972) In *Polymer Stabilization* (ed. W.L. Hawkins), Wiley Interscience, New York, p. 159.
426. Heller, H.J. and Blattmann, H.R. (1972) *Pure Appl. Chem.*, **30**, 145; (1974) *Pure Appl. Chem.*, **36**, 141.
427. Otterstedt, J.A. (1973) *J. Chem. Phys.*, **58**, 5716.
428. Allen, N.S. and McKellar, J.F. (1975) *Chem. Soc. Rev.*, **4**, 533; (1977) *Br. Polym. J.*, **9**, 302.
429. Carlsson, D.J. and Wiles, D.M. (1976) *J. Macromol. Sci., Rev. Macromol. Chem.*, **C-14**, 65, 155.
430. Bailey, D. and Vogl, O. (1976) *J. Macromol. Sci., Rev. Macromol. Chem.*, **C-14**, 267.
431. Scott, G. (1976) In *Ultraviolet Light Induced Reactions in Polymers* (ed. S.S. Labana), Advances in Chemistry Series, Washington, DC, p. 340.
432. Kloepffer, W. (1976) *J. Polym. Sci. Symp.*, **57**, 205; (1977) *Adv. Photochem.*, **10**, 311.
433. McKeller, J.F. and Allen, A.S. (eds) (1979) *Photochemistry of Man Made Polymers*, Applied Science, London.
434. Chakraborty, K.B. and Scott, G. (1979) *Polym. Degrad. Stab.*, **1**, 37.
435. Dewinter, W. (1969) *J. Appl. Polym. Sci., Appl. Polym. Symp.*, **9**, 195.
436. May, C.A. (ed.) (1980) *Resins for Aerospace*, Am. Chem. Soc. Symp. Ser. 132, Washington, DC.
437. Kieryla, W.C. and Papa, A.J. (eds) (1975) *Flame Retardancy of Polymeric Materials*, Marcel Dekker, New York.
438. Levin, M., Atlas, S.M. and Peerce, E.M. (eds) (1975) *Flame Retardant Polymeric Materials*, Plenum, New York.
439. Tesoro, G.C. (1978) *J. Polym. Sci., Macromol. Rev.*, **13**, 283.
440. Banfield, T.A. (1979) *Prog. Org. Coatings*, **7**, 253.
441. Muzyczko, T.M. (1970) US Patent 3 505 235; (1970) *Chem. Abstr.*, **72**, 122536-m.
442. Annand, R.R. (1969) US Patent 3 466 188; (1969) *Chem. Abstr.*, **71**, 102845-n.
443. Zavrazhina, V.I., Mikhallovskii, Y.N. and Zubov, P.I. (1968) *Zashch. Metal.*, **4**, 92; (1968) *Chem. Abstr.*, **69**, 11471-e.
444. McDonald, L.B. (1959) US Patent 2 885 312.
445. Eng, F.P. and Ishia, H. (1984) *Polym. Prepr., Am. Chem. Soc., Div. Polym. Chem.*, **25** (2) 156.
446. Wiley, R.M. (1962) US Patent 3 055 827.
447. Wheaton, R.M. (1958) US Patent 2 823 201.
448. Jones, G.D. (1954) US Patent 2 694 702.
449. Guebert, K.W. and Laman, J.D. (1966) US Patent 3 242 073.
450. Clark, J.T. (1957) US Patent 2 780 604.
451. Kelman, S. and Preising, C.P. (1966) US Patent 3 252 900.
452. Lang, J.L. (1966) US Patent 3 272 782.
453. Hatch, J. and McMaster, E.L. (1963) US Patent 3 078 259.
454. Jones, G.D. (1959) US Patent 2 909 508.
455. Marshall, C.A. and Mock, R.A. (1955) *J. Polym. Sci.*, **17**, 591.

456. Pugh, T.L. and Heller, W. (1960) *J. Polym. Sci.*, **47**, 219.
457. Heller, W. and Pugh, T.L. (1956) *J. Chem. Phys.*, **24**, 1107.
458. Mock, R.A., Marshall, C.A. and Slykhouse, T.E. (1954) *J. Phys. Chem.*, **58**, 498.
459. Cerny, L.C. (1957) *Bull. Soc. Chim. Belgs.*, **66**, 102.
460. Sullivan, E.J. (1964) Ger. Patent 1 177 626.
461. Smith, W.E. and Volk, H. (1967) US Patent 3 340 238.
462. Steigman, J. and Lando, J.L. (1965) *J. Phys. Chem.*, **69**, 2895.
463. Baumgartner, E., Liberman, S. and Lagos, A. (1968) *Z. Phys. Chem.*, **61**, 211.
464. Watanabe, M., Kiuchi, H. and Izumi, Z. (1963) Jpn. Patent 6392.
465. Tamaki, K. and Tsuchiga, H. (1964) *Kolloid Z. Z. Polym.*, **200**, 34.
466. Kimura, C., Serita, H., Murai, K. and Takahashi, Y. (1976) *Yukagaku*, **25**, 424; (1973) *Chem. Abstr.*, **85**, 148553-h.
467. Howard, G.J. and Leung, W.M. (1981) *Colloid Polym. Sci.*, **259**, 1031.
468. Ishikawa, M. (1976) *J. Colloid Interface Sci.*, **56**, 596.
469. Shariff, H., Sontheimer, H. and Vollmert, B. (1975) *Angew. Makromol. Chem.*, **42**, 167.
470. Eisenberg, A. (1967) *Adv. Polym. Sci.*, **5**, 59.
471. Otocka, E.P. (1971) *J. Macromol. Sci., Macromol. Rev.*, **C-5**, 275.
472. Holliday, L. (ed.) (1975) *Ionic Polymers*, Halstead, Wiley, New York.
473. Eisenberg, A. and King, M. (eds) (1977) *Ion Containing Polymers*, Academic Press, New York.
474. Bazuin, C.G. and Eisenberg, A. (1981) *Ind. Eng. Chem., Prod. Res. Devel.*, **20**, 271.
475. MacKnight, W.J. and Earnest, T.R. (1981) *J. Macromol. Rev.*, **16**, 41.
476. Longworth, R. (1983) In *Developments in Ionic Polymers-1*, Applied Science, London, Ch. 2.
477. Gierke, T.D. (ed.) (1982) *Perfluorinated Ionomer Membranes*, Am. Chem. Soc. Symp. Ser. 180, Washington, DC.
478. Sonnessa, A.J., Cullen, W. and Ander, P. (1980) *Macromolecules*, **13**, 195.
479. Noguchi, H. and Rembaum, A. (1972) *Macromolecules*, **5**, 253, 261.
480. Rembaum, A., Baumgartner, W. and Eisenberg, A. (1968) *J. Polym. Sci., Polym. Lett. Ed.*, **6**, 159.
481. Rembaum, A. (1969) *J. Macromol. Sci. Chem.*, **A-3**, 87.
482. Ciphjauskas, L., Piggott, M.R. and Woodhams, R.T. (1980) *Adv. Chem. Ser.*, **187**, 171.
483. Price, C.A. (1975) *Chem. Br.*, **11**, 350.
484. Lewin, S.Z. (1974) *Stud. Conserv.*, **19**, 24.
485. Sayre, E.V. (1971) *Proc. New York Conf. on the Conservation of Stone and Wood Objects*, vol. 1, p. 115, American Conservation Institute, Detroit.
486. Domaslowski, W. and Lehmann, J. (1972) *Proc. Meeting of the Joint Committee on the Conservation of Stone, Bologna*, p. 255.
487. Fowler, D.W., Houston, J.T. and Paul, D.R. (1973) In *Polymers in Concrete*, Publication SP-40, American Concrete Institute, Detroit, MI, pp. 93–117.
488. Munnikendam, R.A. (1967) *Stud. Conserv.*, **12**, 158.
489. Gauri, K.L. (1974) *Stud. Conserv.*, **19**, 100.
490. Munnikendam, R.A. (1973) *Stud. Conserv.*, **18**, 95.
491. Biczok, I. (ed.) (1967) *Concrete Corrosion and Concrete Protection*, Chem. Publ. Co., New York.
492. Orchard, D.F. (ed.) (1973) *Concrete Technology*, 3rd edn, vol. 1, Wiley, New York.
493. Taylor, W.H. (ed.) (1977) *Concrete Technology and Practice*, McGraw-Hill, New York.
494. Hewlett, P.C. (1979) *Mater. Eng. Appl.*, **1** (6), 335.
495. Clifton, J. and Fronhnsdorff, G. (1975) *Chem. Res. Prog.*, 173.

496. Swamy, R.N. (1979) *J. Mater. Sci.*, **14**, 1521.
497. Birchall, J.D. (1983) *Philos. Trans. R. Soc. London, Ser. A*, **310**, 31.
498. Froning, H.R., Fussell, D.D. and Heffern, E.W. (1982) *Encyclopedia Chem. Technol.*, **17**, 3rd edn, pp. 168–82.
499. Matheny, S.L. (1980) *Oil Gas. J.*, **79** (March 31).
500. Klein, J. (1984) *Angew. Makromol. Chem.*, **123–124**, 381.
501. Petterson, H.T. and Webb, I.D. (1954) US Patent 2 691 640.
502. Stowe, S.C. (1965) US Patent 3 190 925.
503. Murdock, S.A. (1963) US Patent 3 075 947.
504. Chen, C.S.H., Hosterman, E.F. and Stamm, R.F. (1965) US Patent 3 218 117.

Subject index

354   Subject index